WITHDRAWN

Systems Reliability and Risk Analysis

ENGINEERING APPLICATIONS OF SYSTEMS RELIABILITY
AND RISK ANALYSIS
Editor: E.G. Frankel, MIT, MA, USA

E.G. Frankel: Systems Reliability and Risk Analysis, second, revised edition. 1988.
ISBN 90-247-3665-X

Systems Reliability and Risk Analysis

Second, Revised Edition

By

Ernst G. Frankel

*Department of Ocean Engineering,
Massachusetts Institute of Technology,
Cambridge, MA, USA*

KLUWER ACADEMIC PUBLISHERS

DORDRECHT / BOSTON / LONDON

Library of Congress Cataloging in Publication Data

Frankel, Ernst G.
    Systems reliability and risk analysis.

    (Engineering applications of systems reliability
and risk analysis)
    Includes bibliographies and index.
    1. Reliability (Engineering) 2. Risk assessment.
I. Title. II. Series.
TA169.F73 1988   624   87-36155

ISBN 90-247-3665-X

Published by Kluwer Academic Publishers,
P.O. Box 17, 3300 AA Dordrecht, The Netherlands.

Kluwer Academic Publishers incorporates
the publishing programmes of
D. Reidel, Martinus Nijhoff, Dr W. Junk and MTP Press.

Sold and distributed in the U.S.A. and Canada
by Kluwer Academic Publishers,
101 Philip Drive, Norwell, MA 02061, U.S.A.

In all other countries, sold and distributed
by Kluwer Academic Publishers Group,
P.O. Box 322, 3300 AH Dordrecht, The Netherlands.

ISBN first edition 90-247-2895-9
Series EASR 1

All Rights Reserved
© 1988 by Kluwer Academic Publishers
No part of the material protected by this copyright notice may be reproduced or
utilized in any form or by any means, electronic or mechanical
including photocopying, recording or by any information storage and
retrieval system, without written permission from the copyright owner

Printed in The Netherlands

PREFACE

Ernst G. Frankel

This book has its origin in lecture notes developed over several years for use in a course in Systems Reliability for engineers concerned with the design of physical systems such as civil structures, power plants, and transport systems of all types. Increasing public concern with the reliability of systems for reasons of human safety, environmental protection, and acceptable investment risk limitations has resulted in an increasing interest by engineers in the formal application of reliability theory to engineering design. At the same time there is a demand for more effective approaches to the design of procedures for the operation and use of man-made systems, more meaningful assessment of the risks introduced, and use such a system poses both when operating as designed and when operating at below design performance.

The purpose of the book is to provide a sound, yet practical, introduction to reliability analysis and risk assessment which can be used by professionals in engineering, planning, management, and economics to improve the design, operation, and risk assessment of systems of interest. The text should be useful for students in many disciplines and is designed for fourth-year undergraduates or first-year graduate students.

I would like to acknowledge the help of many of my graduate students who contributed to the development of this book by offering comments and criticism. Similarly, I would like to thank Mrs. Sheila McNary who typed untold drafts of the manuscript, and Mr. Tasos Aslidis who proofread the manuscript. Finally, my appreciation goes to my wife, Tamar, and my son, Michael, for their patience and forebearance during the many months of preparation.

TABLE OF CONTENTS

PREFACE		v
LIST OF FIGURES		xi
GLOSSARY		xv
1.0	INTRODUCTION	1
1.1	Suggested Readings	10
2.0	FUNDAMENTAL CONCEPTS	11
2.1	Basic Concepts of Reliability	11
2.2	References	21
3.0	ASSESSMENT OF RELIABILITY FUNCTION	23
3.1	Non-parametric Reliability Function Assessment	23
3.2	Parametric Reliability Function Assessment	27
3.3	Exercises	30
3.4	References	31
4.0	RELIABILITY OF SERIES AND PARALLEL SYSTEMS	33
4.1	Simple Series Systems	33
4.1.1	Application to Components with Exponential Failure Densities	35
4.2	Simple Parallel Systems	36
4.2.1	Relation Between MTBF and $R(t)$ in Simple Exponential Redundant System	38
4.3	Variations of Simple Redundant Systems	42
4.4	Analysis of Complex Series Redundant System	48
4.5	Off-Line Redundant Systems	58
4.6	Exercises	61
4.7	References	63
5.0	FAILURE MODE AND EFFECTS ANALYSIS - FAULT TREE ANALYSIS	64
5.1	Common Cause Failure	65
5.2	Complex System Reliability Networks	66
5.3	Fault Tree Analysis	69
5.3.1	Min Cut Sets of Fault Trees	74
5.4	Exercises	75
5.5	References	76
Appendix 5A	Performance of a Failure Mode and Effects Analysis	78
Appendix 5B	Performance of a Maintainability Engineering Analysis (MEA)	81
6.0	MULTIVARIABLE PROBABILITY DISTRIBUTIONS AND STOCHASTIC PROCESSES	83
6.1	Multivariable Probability Distributions	83
6.2	Stochastic Processes	88
6.3	Markov Processes	94
6.4	Exercises	111
6.5	References	113

7.0	THE GENERALIZED FAILURE PROCESS FOR NON-MAINTAINED SYSTEMS	114
7.1	Solution Using Laplace Transforms	120
7.2	Stand-by (Off Line) Redundant System	122
7.3	Series Systems	127
7.4	Redundant (On-Line) Parallel Systems	129
7.5	State-Dependent Reliability Models	131
7.6	Linear Stress Model	132
7.7	The Effect of Switching	134
7.8	Exercises	140
7.9	References	140
8.0	ANALYSIS OF MAINTAINED SYSTEMS	141
8.1	Systems Availability	142
8.2	Markov Models for Maintained Systems	148
8.2	Maintained Series Systems	151
8.2	Maintained Parallel Systems	153
8.3	Development of the General Expression for the Mean Time to Failure of a Markov Chain	167
8.3.1	Mean Time to Failure and Variance of Time to Failure of Non-Maintained and Maintained Systems	170
8.4	Models of Maintained Systems with Redundant Off-Line Components	179
8.5	Exercises	197
8.6	References	200
9.0	STRATEGIES FOR REPAIR POLICIES	202
9.0.1	General Repair Strategy Determination	204
9.0.2	Cost of Scheduled Overhauls and Inspections	205
9.0.3	Spare Pare Inventory Provisioning	207
9.1	Use of Dynamic Programming in Systems Reliability	210
9.1.1	Complex system Reliability Analysis Under Constraints	214
9.1.2	Optimization of Multistage Decision Processes	215
9.1.3	Complex System with Component Stand-by	218
9.1.4	Complex System with Switching	218
9.1.5	Reliability of Complex System with Component Maintenance	221
9.1.6	Analysis of Component Failure	223
9.1.7	Conclusions	226
9.2	The Use of the Lagrange Multiplier Method	227
9.2.1	System Involving Two Types of Constraints	229
9.3	Optimum Maintenance Policies by Dynamic Programming	232
9.4	Spare Part Provisioning Models	236
9.5	System Performance Evaluation	243
9.6	Exercises	249
9.7	References	251
10.0	EFFECTS OF COMPONENT INTERACTION	253

10.1	Effect of Interaction of Component Reliability	254
10.2	Analysis of "Wear" Rates	256
10.3	Component Reliability	263
10.4	System Reliability	265
10.5	Use of Networks in the analysis of Interactive Systems Reliability, Maintainability, and Availability	266
10.6	Exercises	269
10.7	References	269
11.0	APPLICATION OF FAULT TREE AND OTHER NETWORK TECHNIQUES	271
11.1	Implementation of Fault Tree Analysis	274
11.1.1	Representing Fault Trees by Networks	275
11.2	Uncertainty in Reliability Analysis	277
11.2.1	GERT Reliability Networks with Uncertainty	278
11.3	References	282
12.0	RELIABILITY AND RISK IN PERSPECTIVE	284
12.1	General Considerations	284
12.1	Risk Attitudes	285
12.2	Analysis of Risk	286
12.2.1	Reliability and Risk Assessment	287
12.3	Issues and Concerns	288
Appendix A	BASIC CONCEPTS OF PROBABILITY AND STATISTICS	291
Appendix B	MATRIX ALGEBRA AND TRANSFORMATIONS	327
Appendix C	TESTING FOR MARKOV PROPERTIES	346
Appendix D	NON-MARKOVIAN SYSTEMS	372
Appendix E	INTRODUCTION TO FLOW GRAPHS AND GERT	376
Appendix F	STATISTICAL TABLES	390
INDEX		423

LIST OF FIGURES

1.1	The Concept of Integrated System Design	6
1.2	The Concept of Operability Analysis	7
1.3	The Concept of Maintainability Analysis	7
2.1	Typical Probability Density of Failure Rate (t)	13
2.2	Various Age Dependent Failure Rate Distribution	17
2.3	Life Characteristic of Bathtub Curve	18
2.4	Probability of Surviving Wear Failure	19
3.1	Transducer Test Results	25
3.2	Bounding Regions	29
4.1	Series System	33
4.2	Two-Component Series System	33
4.3	Two Event Venn Diagram	34
4.4	N-Component Parallel System with Different Components	37
4.5	Venn Diagram of Parallel Two-Component System	37
4.6	Variation of Reliability with Number of Components in a Simple Redundant System	42
4.7	N-Component Series Systems with Identical Components	45
4.8A	Two Component Parallel System	46
4.8B	N-Component Parallel System with Identical Components	47
4.9	Series System in Parallel	48
4.10	Parallel System in Series	49
4.11	Two Component Redundant System	58
5.1	Complex System	66
5.2	Network Representation	66
5.3	Elementary Fault Tree	71
5.4	Elementary Fault Tree	72

5.5	Example	73
5.6	Hierarchical Fault Min Cut Set Analysis	75
6.1	Probability of Intersection of Events	84
6.2	Number of Failures Subject to Wear x at Time t	90
6.3	Two State Markov Processes	102
6.4	Three State Markov Process	109
7.1	States of a Two-Component Standby (Off-Line) Redundant System	123
7.2	Comparison of the Reliability of Two-Component Off-Line and On-Line Systems	131
7.3	Reliability of n-Component On-Line Standby Parallel System	133
7.4	Two-Component System with Switching	134
8.1	Maintained Series System	151
8.2	2-Component System	180
8.3	3-Component System	181
8.4	N Parallel Redundant Off-Line System	182
8.5	2-Component System	183
8.6	3-Component System	184
8.7	N-Component System	186
8.8	State Number/State of the System	187
8.9	2-Components System	188
8.10	3-Components System	189
8.11	Two-Component Parallel System with Two Stand-by Components	191
8.12	States of Two-Component System with Two Stand-by Components (Model 3)	192
8.13	State Transition Diagram of One-Component System with One Stand-by Component (Model 4)	193
8.14	State Transition Diagram of One-Component System with Two Stand-by Components (Model 4)	194

		XIII
9.1	Series System with Redundant Components	219
9.2	System with Redundant Non-identical Stand-by and Main Components	220
9.3	Preventative Maintenance Policy	235
10.1	Instantaneous Failure Rate as Function of Derating Factor	255
10.2	Effect of Interaction of MTBF $(t) = 0.01 + Ct$	259
10.3	Effects of Wear Interaction of Diesel Cylinders, Liner, and Piston Rings	260
10.4	Factors Affecting Mechanical Component Failure Rate	268
11.1	Simple On-Line Redundant System	271
11.2	Fault Tree of Simple On-Line Redundant System	271
11.3	System Distribution	272
11.4	S-T Graph of Simple Fault Tree	276
11.5	Reliability Block Diagram for Parallel System with Parallel Subsystem	279
11.6	GERT Network for Parallel System with Parallel Subsystem	279
12.1	Operational Analysis of Systems Subject to Risk	288

LIST OF TABLES

3.1	Test Results	23
9.4.1	Diesel Engine Part Performance	238
9.4.2	Procedure for Spare Part Provisioning	242

GLOSSARY

FAILURE — The inability of a system, subsystem, or component to perform its required function.

RELIABILITY — The probability that a system or component will perform its intended function for a specified period of time, under required conditions. It can also be defined as the probability that a system, subsystem, or component will give specified performance for the duration of a mission when used in the manner and for the purpose intended, given that the system, subsystem, or component is functioning properly at the start of the mission.

AVAILABILITY — The probability or degree to which an equipment will be ready to start a mission when needed. Availability is divided into up-time availability, steady state availability, and instant availability.

DEPENDABILITY — The probability or degree to which an equipment will continue to work until a mission is completed.

MAINTAINABILITY — A characteristic of design and installation which is expressed as the probability that an item will conform to specified conditions within a given period of time when maintenance action is performed in accordance with prescribed procedures and resources.

MEAN-TIME-TO-REPAIR (MTTR) — The statistical mean of the distribution of times-to-repair. In other words, the summation of active repair times during a given period of time divided by the total number of malfunctions during the same time interval.

MEAN-TIME-BETWEEN-FAILURE (MTBF) OR MEAN TIME TO FAILURE — The average time between successive failures, estimated by the total measured operating time of a population of items divided by the total number of failures within the population during the measured time period. Alternatively, MTBF of a repairable item is estimated as the ratio of the total operating time to the total number of failures. Measured operating time of the items of the population which did not fail must be included.

FAILURE RATE — A value expressing the frequency of failure occurrence over any specified time interval or cycles of operation, i.e., the average frequency of failure occurrence per unit operating time.

FAILURE MODES — The various manner or ways in which failures occur and the resulting operating condition of the item at the time of failure.

COMMON CAUSE — A cause resulting in failure of all affected systems.

MODEL (MATHEMATICAL) — A mathematical relationship used as a means of evaluating reliability, performance, or system effectiveness.

OPERATING TIME — The time during which an item is performing a function. The time period between turn-on and turn-off of a system, subsystem, component or part during this operating is as specified.

RELIABILITY GOAL OR REQUIREMENT — The reliability desired or required of the system, subsystem, component, or part design.

REPAIR TIME — Time measured from the beginning of correction of a malfunction to the completion of such correction. That time during which one or more technicians are actually working to repair a failure. This time includes preparation time, fault location time, correction time, and checkout time.

FUNCTIONAL FLOW DIAGRAMS — A functional flow diagram is a schematic representation of the energy flow and physical interrelationships between subassemblies in an equipment or equipments in a system.

RELIABILITY BLOCK DIAGRAMS — Reliability diagrams do not attempt to depict the functional interconnections between equipments; they depict the contribution of each element in the system to the ultimate system effectiveness. Since they show the functional and mathematical relationship between the various parts, they can be used to predict the system reliability and availability of a system for different modes of operation.

1.0 INTRODUCTION

Ernst G. Frankel

The world we live in is imperfect and we increasingly have to live with the potential of failure of many of our traditional as well as advanced mechanical, electrical, structural, as well as economic and social systems. The reason is often not insufficiency of concept, design, or operational standards used in the system, but the fact that little or no consideration has been given to the desired reliability, availability, and maintainability of the system in its conception, design, and operation. It is not only that we often expect too much, but that we are ignorant of the actual operating environment and required performance imposed on a system. Usually little or no consideration is given to the basic fact that "nothing is perfect", and the system's design as well as operating conditions are subject to many deviations and uncertainties. It is therefore necessary to assign inherent or potential imperfections to systems so as to achieve desired performance.

The objective of this book is to provide an introduction to the effective design of required reliability, maintainability, and availability into physical, operational, social, economic, or other systems.

Reliability can be defined as the freedom from failure of a component or system equipment while maintaining a specific performance. It may also be defined as a measure of dependability or trustworthiness of a system in accomplishing a certain mission for a period of time. Reliability, therefore, differs from quality control or assurance in that it is concerned with the performance over time and not only performance at a time or at the time of test. We are usually interested in expressing this elusive concept of "reliability" in a quantitative manner so as to permit performance of effective fundamental design and the making of realistic operating decisions. Statistical techniques are usually employed to estimate the reliability or probability of success (or non-failure) of a particular component or system. Expressing reliability in a quantitative manner enables us to make fundamental design and operating decisions.

Engineers are renowned for their conservatism based upon

the overriding importance of reliability. In few professions does reliability play as great a role in defining and restricting design and operating criteria. It is true that the consequences of unreliability in engineering can be very costly and often tragic, but much too often the need for reliability is used as a reason for dismissal of novel concepts and designs whose capabilities have never been tested or properly evaluated in a particular environment or configuration and for which, as a result, empirical data is not available.

In many systems designs, conservative approaches are justified by the supposed overriding importance of reliability, which then plays an unduly important role in affecting the definition of and resulting restrictions on design and operating criteria. While it is true that the consequences of unreliability in physical, economic, and social systems can be very costly and often tragic, we often find that reliability is unnecessarily given as the justification for the use of outdated and often inefficient or obsolete systems when, with little additional effort, the increased risk, if any, in the use of more advanced systems could have been determined.

Reliability affects systems specification, design, operation, maintenance, spare part stocking, and, in fact, all aspects of a system. It is the factor that ultimately defines the tolerances of the performance and mission of a system. For this reason, it is unfortunate that more consideration is not given to formal analysis of the reliability of systems to assure better systems design and performance.

Most physical systems, in particular, mechanical equipments, consist of assemblies of components made by different manufacturers. Each maker again often has his own design criteria, philosophy, manufacturing tolerances, and testing procedures and evaluations. Components are normally fitted with safety or excursion devices designed to protect the particular component or subassembly without reference to the effect such devices have on the operational performance, fluctuations, and reliability of the system as a whole.

Safety margins should generally be based on risk factors but are more often based on guesswork. Safety factors are often factors of ignorance and in many cases have remained unchanged for long periods of time, even though performance data which could prove the feasibility of reducing or modifying these factors are often unavailable.

Reliability affects the probability that failure or excursion of a component results in failure or excessive deviation of the performance of the system. It should, therefore, be a major factor in the establishment of safety margins. Yet, this is seldomly done due to the difficulty in analyzing effects of component performance deviation on the system's reliability without using formal statistical methods.

Since we increasingly use automated or adaptive systems with short response times, this problem has assumed great urgency. On the one hand we attempt wholesale use of controls and attempt to integrate the diverse response functions of complex systems, knowing that the reliability of the controls is often far below that of the controlled system; while, on the other hand, we hesitate to accept, test, and evaluate radically new systems which could simplify the control problem to such an extent as to permit a comparable reliability of control and controlled component to be attained. As a result, we often find ourselves in a vicious circle in which a controlled system has a much lower performance than could be achieved by an integrated approach.

The adoption of formal reliability analysis requires replacement of bug-hunting with analytical methods, proper data logging, and failure analysis. The rate of feedback of information from all stages of system development, design, manufacture, installation, and operation though must be increased so as to improve the rate of attainment of maturity of a system. More often than not maintenance schedules today are based on educated guesswork, temperament, or prejudice. Spare part stocks are usually kept according to some old hearsay formula which seldomly has any relation to the reliability or failure rate of components of the system. As a result, we often scrap complete systems with an inventory full of original spares. On the other hand, other spares will often be unavailable when urgently needed.

Useful system life varies greatly, from highly inadequate, to vastly too great, and it is important to remember that overdesign with lifetimes and reliability greatly in excess of requirements can be very wasteful as well.

Another important reason for greater emphasis on reliability analysis and its use in systems design is the ever increasing rate of systems obsolescence. In fact the rate of obsolescence has increased to such an extent that we seldomly manage to develop a system to full maturity before new developments make it obsolete. Consequently, classical methods in which we use field results as feedback to improve systems design and dogmatic rule-of-thumb approaches in many of the phases of system developments will not serve any more. We require much more integrated feedback at all stages of development through which the system is continuously improved. In the long run it will always pay to design the right order of magnitude of reliability into the equipment. Test failures should not be rectified without a complete analysis of all causes. If a carefully designed component or system fails, it is of little use to change the parameters without fully investigating the causes of failure, which are often completely unrelated to the failure event. Although in many systems we may be unable to test to life as is done with most electronic components, proper sampling of failure data is usually possible.

"Inherent Reliability" is the probability that the system delivers specified performance, under test or design conditions for a specific time. Actual working conditions are seldomly comparable to test or design conditions. We, therefore, often correct the "Inherent Reliability" by environmental and operating factors which generally include allowances for improper handling, shipping defects, goodness of auxiliary equipment, ability of operators, deterioration with time, variation of inputs from designed values, etc. Reliability is not a static property of the system, but is affected by the dynamics of its environment.

Most systems are complicated series of large numbers of expensive components. As the reliability of a series system of independent components is equal to the product of the component reliabilities it is obvious that component reliability is very critical even if only moderate system reliability is to be achieved. Similarly, redundancy usually improves the reliability of unreliable systems. The main question is how to get the important parameters on which to base reliability predictions.

Where components or systems have short expected life, failure test results are often used to formulate future behavior. On the other hand, components with long expected lifetimes require organized data sampling. The data is derived from real life performance and not simulated tests to failure, and therefore fewer correction factors are required.

In any reliability analysis we infer failure to mean operation outside assigned tolerances within a given time period. Tolerances must be defined for each characteristic of the system and must include wear consideration and other factors as functions of time. We are normally given mean time before failure of a small sample which has to be amplified by analytical data. Obviously in this case much more stringent tests for confidence are required as in the case where any required number of components can be tested to failure.

Defining system failure as "Operation Outside Assigned Tolerances", as mentioned before, quantitative values of reliability can usually be determined. Tolerance limits are usually defined which include allowances for deterioration as a function of time. Data for future failure predictions is generally obtained from frequency rates of failure under operation. Statistically, the reliability index is computed from the mean time between failure obtained from normal failure for operating periods.

Although not all system or component failures are random, test data will usually be found to provide at least a start to meaningful reliability estimates. Unreliability is often the result of system complexity, incompatibility, faulty design, and ignorance of factors, apart from inherent component inadequacies. In many instances human errors are cause for unreliability. Therefore, indoctrination of system requirements for needed reliability, strict adherence to standards, maximum amount of feedback information, utilization of feedback information to improve performance, and effective simulation testing are prerequisites towards greater reliability. Only this will permit full exploitation of technological, economic, and social advances.

Although System Reliability as a subject is primarily concerned with physical systems and was initially developed to improve the performance of electrical, electronic, and mechanical operating systems, it is now increasingly used as a design input to the development of fixed structures as well as agricultural, biological, economic, social, political, and other types of systems. As we learn more about the interrelationship and response of components of a system, it becomes plausible to model total system performance. Few systems are of one kind or independent of other systems. Social systems interact with economic systems, physical systems interact with social and economic systems. In fact it is usually unrealistic to model a system solely in its primary domain such as physical, social, or economic, without reference to or consideration of its wider environment.

While we usually assume that physical performance is an effective measure of systems reliability and maintainability, it is found that a similar approach is also applicable to non-physical or mixed systems.

This book evolved from lecture notes used in an introductory course on 'Systems Reliability' taught over a number of years at the Massachusetts Institute of Technology. The book reviews first the "Basic Concepts of Reliability". Thereafter the "Assessment of the Reliability Function" is reviewed including the design of tests and experiments to determine meaningful failure, maintainability, and operability data. The approach used in this book is based on the Concept of Integrated System Design as shown in Figure 1.1. Systems design is not only based on systems performance and reliability requirements, but should include systems performance and reliability requirements. It is obvious that both the performance and reliability of a system design depend on how it can be and is operated. Similarly, system reliability and availability depend on how a system can be and is maintained. As a result, maintainability and operability analysis form an integral part of systems design and development.

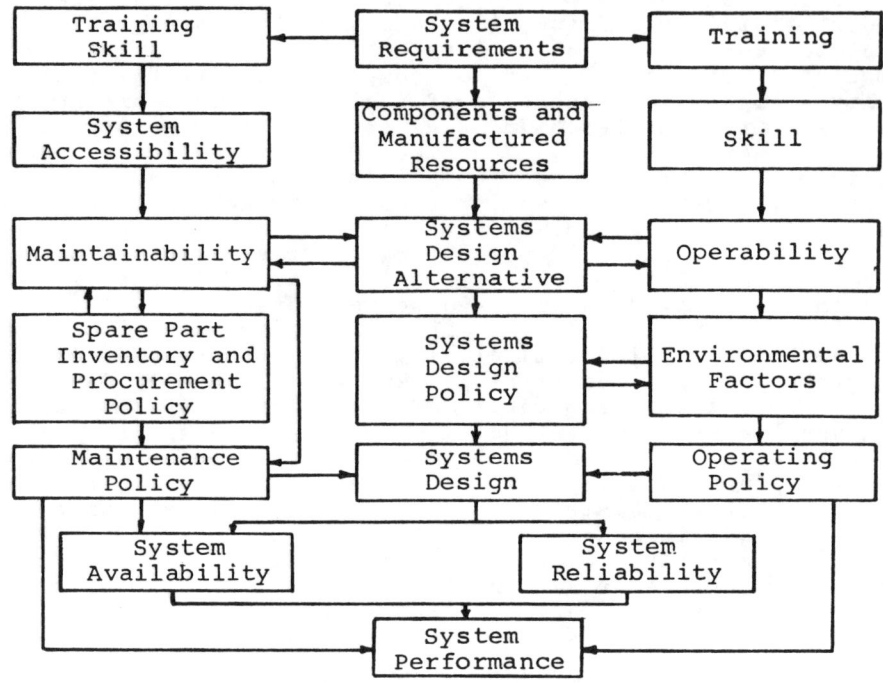

FIGURE 1.1: THE CONCEPT OF INTEGRATED SYSTEM DESIGN

Operability analysis includes consideration of labor, skill, environmental, and operating policy issues as shown in Figure 1.2. Conversely, maintainability analysis includes accessibility, spare part and tool availability, and other factors influencing maintainability of a system as shown in Figure 1.3.

Many otherwise excellent systems with superior performance and reliability, for example, have failed because they were not operable as a result of lack of skill or other requirements, or because they were not effectively maintainable due to inaccessibility.

The approach used in this book is to relate all the issues bearing on the performance of a system to the achievement of acceptable levels of reliability, availability, and operating performances as a function of cost which includes the cost of both maintenance and operations in terms of maintainability and operability.

Static and dynamic models are constructed for policies with or without maintenance of a system. The approach

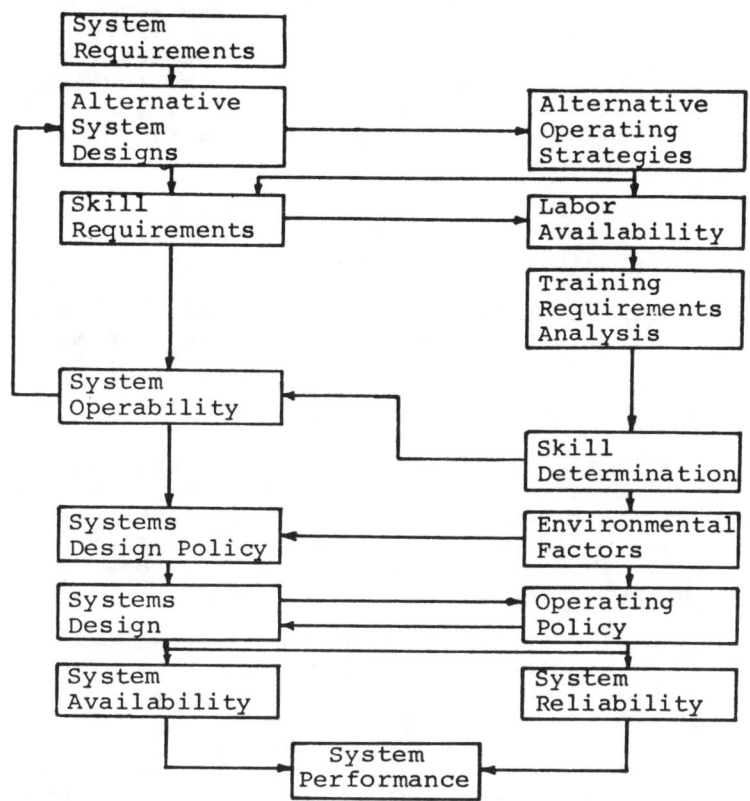

FIGURE 1.2: THE CONCEPT OF OPERABILITY ANALYSIS

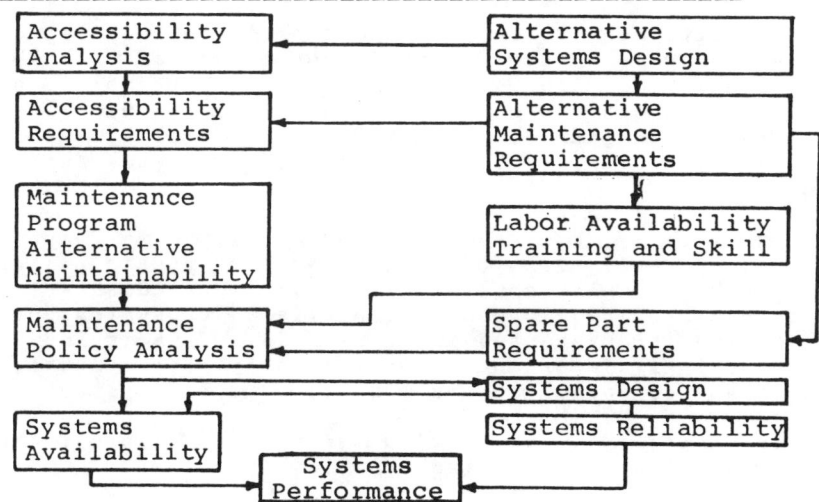

FIGURE 1.3: THE CONCEPT OF MAINTAINABILITY ANALYSIS

advocated and used in this book is to test the implications of various maintenance and maintainability policies as well as the effects of operability and consequent feasible operating policies. Conversely, both alternative maintenance and operating policies may be given inputs, and the system can be designed for maximum reliability at a given performance level and/or cost.

Various new system modelling techniques such as conditional stochastic networks for use in systems reliability analysis are presented.

Spare parts inventory, parts procurement, renewal and other supply problems are similarly modeled and methods for deriving a best spare-part stocking and renewal policy for a given system or as part of an integrated systems design are discussed.

After enlarging the problem to encompass multicriteria systems, we next introduce the analysis of risk. Decision Theory is used to represent the concept of risk acceptability and as a tool for modeling risk and performance in an integral model which inludes all the systems design and operating characteristics.

The book is designed as a fundamental text for both students and practitioners in engineering, systems analysis, operations management, and policy analysis. It treats systems reliability and risk analysis from the various viewpoints of the system's conception, design, manufacture, test, and operation. It is felt that at a time of increasing concern with the quality as well as the safety of our environment, systems reliability and risk must be analyzed in the wider non-parochial tests as presented in this book.

The approach taken in this book is to build up a solid understanding of the basic concepts of reliability, maintainability, and operability in Chapters 1 and 2, followed by a discussion of the assessment of the reliability function in Chapter 3. We next develop facility with static reliability models. Failure mode and effects analysis is discussed in Chapter 5.

Dynamic systems models are developed in Chapters 6-8 followed by development of strategies for repair policies in Chapter 9. This chapter also includes discussions of maintenance policy issues. The effect of system component interaction is discussed in Chapter 10. This is found to be an increasingly important issue not only for mechanical but also economic, social, operational, and other systems.

In Chapter 11, we review the use of network techniques in the solution of fault tree problems.

The problems inherent in the definition of risk, which

is usually defined as the product of the uncertainty of a consequence or damage occurring and the consequence of damage, are reviewed with particular attention to alternative methods for the quantitative determination of risk resulting from various identified scenarios. Methods for the analysis of common mode failures and causes leading to several consequences or damages are established and techniques for fault tree and probabilistic risk analysis are discussed.

It is shown how reliability analysis provides the tools for the improvements in the design, construction, and operation of systems by effective reliability and availability allocation which achieves the desired systems performance with an associated acceptable and balanced level of reliability and availability. Risk assessment, on the other hand, deals with the determination of the risk posed by the system itself when operating as designed as well as the risk posed when systems performance is degraded or the system fails to operate as designed.

Reliability analysis and risk assessment are therefore complementary activities designed to permit reasonable quantification of the expected performance of a system and the potential risk the system may pose to its operators, users, and its environment.

Nothing in the design, operation, use, and impact of man-made systems is certain. Instead of looking for certainty, we here develop methods and frameworks which should allow determination of better, more reliable probabilities.

In writing this book, I have attempted to provide a basic knowledge of reliability analysis techniques and risk assessment methods, which can be used to improve the design, manufacture, and operation on one hand, and the effective determination of the risk introduced by the system on the other hand. The emphasis is on quantitative analysis of random occurrences of undesirable outcomes, failures, or consequences during the life of a system. The system can be physical, economic, or social, and will generally be assumed to be man-made.

Man-made systems often suffer under faulty conceptualization, design, manufacture, test, operation, use, and placement. In fact, the bulk of failures of man-made systems can usually be traced to human error, compromise, lack of understanding, or other man-made causes. Similarly, the effects or consequences often affect humans significantly. It is the role of man in systems development and use which - probably more than nature - introduces uncertainty into the performance of man-made systems. Our job here will be to study, develop, and use approaches which permit their effective quantification which in turn can help to make more effective conceptual, design, manufacture, operating, and use decisions.

1.1 Suggested Readings

1. Bourne, A.J. and Green, A.E., "Reliability Technology", Wiley Interscience, New York, 1972.

2. Kaplan, S. and Garrick, B.J., "On the Quantitative Definition of Risk", Volume 1, Society for Risk Analysis, 1981.

3. Rowe, W.D., "An Anatomy of Risk", John Wiley & Sons Inc., New York, 1977.

4. Apostolagis, A., "Probability and Risk Assessment: The Subjectivistic Viewpoint and Some Suggestions", Nuclear Safety 19 (May-June) 1978.

5. Shooman, M.L., "Probabilistic Reliability: An Engineering Approach", McGraw-Hill, 1968.

6. Goldman, A.S. and Slattery, T.B., "Maintainability - A Major Element of System Effectiveness", John Wiley & Sons Inc., New York, 1969.

7. Barlow, R.E. and Hunter, L.C., "Mathematical Models for Systems Reliability", The Sylvania Technologist, Volume XIII, No. 1 and 2, 1960.

8. Ireson, W.G., "Reliability Handbook", McGraw-Hill, New York, 1966.

9. Jardine, A.K.S., "Maintenance, Replacement, and Reliability", John Wiley & Sons Inc., New York, 1973.

10. Smith, C.O., "Introduction to Reliability in Design", McGraw-Hill, New York, 1976.

11. Sandler, G.H., "System Reliability Engineering", Prentice-Hall, Englewood Cliffs, New Jersey, 1963.

12. Fussell, J.B. and Arendt, J.S., "System Reliability Engineering Methodology: A Discussion of the State of the Art", Nuclear Safety 20, 1979.

13. Myers, R.H., Wong, K.L., and Gordy, H.M. (editors), "Reliability Engineering for Electronic Systems", John Wiley & Sons Inc., New York, 1964.

2.0 FUNDAMENTAL CONCEPTS

Ernst G. Frankel

The reliability and the risk of failure of a system are complementary concepts. The reliability of a system is the probability that the system will not fail during a specified time period under given operation conditions, while the risk of failure is the probability that the system will fail during that period and operating conditions. Failure is a probabilistic event, and may occur as a result of inherent defects in the system, wear and tear, or imposition of unexpected internal or external factors. It may be the result of faulty design, insufficient maintenance, faulty operation, natural catastrophies, or other factors. Most systems interact with and are affected by other systems which may induce conditions or factors which increase the risk of or actually cause failure of the system.

Reliability theory encompasses the analysis of the structure of a system comprised of components, each of which is subject to failure in some way. It similarly includes the study of systems maintainability and the effect systems reliability and maintenance have on its availability. Availability is usually defined as the percentage of time the system is able to perform to its specification. Risk analysis and assessment is closely tied to reliability theory and reliability engineering. It involves estimating of the risk involved in the use of a system under given environmental conditions, under specific operating conditions, over a period of time. Reliability theory and risk analysis are based on the laws of probability. Probability can be interpreted in terms of the relative frequency connected with the occurrence of an event or it can be interpreted in a subjective sense.

The basic concepts of probability theory are reviewed in Appendix 1, which should be studied by readers without recent experience in the use of probability theory before proceeding with this chapter.

2.1 Basic Concepts of Reliability

The reliability of a component or system is its ability to perform within specified tolerances for a specified period of time. This definition, although containing the essence of the reliability concept, must be formulated in mathematical and probabilistic terms before it can be used successfully in systems and engineering applications. To do this, the following symbolic definitions are required:

$P(\tau=t) \equiv \phi(t) \equiv$ probability that the system will "fail" (perform outside specified tolerances), at time t, that is the probability density function (p.d.f.) for failure

$\phi(t)dt \equiv P(t \leq \tau \leq t+dt) \equiv$ the probability of system failure between time t and t+dt

$F(t) \equiv P(\tau \leq t) \equiv$ the probability that the system will fail at some time prior to or at time t

$R(t) \equiv P(\tau > t) \equiv$ the reliability of the system, e.g., the probability that the system will survive (perform within specified tolerances) to time t

$f(t)dt \equiv P(t \leq \tau \leq t+dt | \tau > t) \equiv$ age dependent failure rate or the probability that system will fail between time t and t+dt given that it has survived to time t. Also called the hazard rate.

From the basic principles of probability theory, the following relations are derived from the above definitions:

$R(t) + F(t) = 1$ $(P(\tau > t) + P(\tau \leq t) = 1$ since system must either survive to time t or fail previously   (2.1)

$F(t) = \int_0^t \phi(\tau)d\tau$ $(P(\tau \leq t) = \int_0^t P(\tau=x)dx)$ (2.2)

or

$F'(t) = \phi(t)$ $(\phi(t))$ is the failure density associated with failure distribution $F(t)$ (2.3)

If we define t as the time at which a failure occurs and denote $\phi(t)dt$ as the probability that the time of failure will be between t and t+dt, as before, we can also write:

$$\phi(t) = \lim_{\Delta t \to 0+} \frac{P(t \leq \tau \leq t + \Delta t)}{\Delta t}$$

and the distribution function F(t) of this failure time

density function can then be expressed as:

$$F(t) = P[\tau \leq t] = \int_0^t \phi(\tau)d\tau \text{ and } \phi(t) = F'(t)$$

which represents the cumulative probability that failure will occur at or before time t.

A typical failure density, $\phi(t)$, is shown in Figure 2.1.

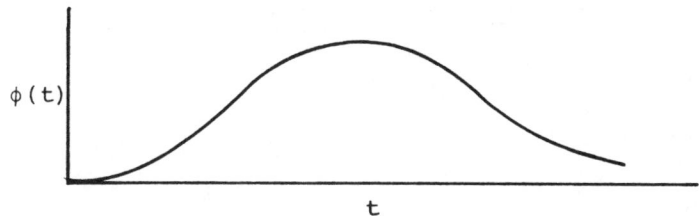

FIGURE 2.1: TYPICAL PROBABILITY DENSITY OF FAILURE RATE $\phi(t)$

Conversely, the reliability is the probability that failure will not occur before time t when:

$$R(t) = 1-F(t) = \int_t^\infty \phi(\tau)d\tau \text{ and } [-\phi(t)] = R'(t)$$

For $t \to 0$, F(t) will approach zero, while for $t \to \infty$, F(t) will approach unity. Similarly, $R(t) \to 1$ for $t \to 0$ and $R(t) \to 0$ for $t \to \infty$.

Another important general function is the age dependent failure rate or hazard function f(t)dt, which as stated before, is the probability that failure will occur between t and t+dt, given failure did not occur prior to time t. This can also be expressed as:

$$f(t) = \lim_{\Delta t \to 0+} \frac{P(t \leq \tau \leq t+\Delta t \mid \tau > t)}{\Delta t}$$

$$= \lim_{\Delta t \to 0+} \frac{P(t \leq \tau \leq t+\Delta t)}{\Delta t} \cdot \frac{1}{P(\tau > t)}$$

$$= \frac{\phi(t)}{R(t)} = \frac{R'(t)}{R(t)} = -\frac{d}{dt} \log R(t)$$

Now using Bayes' Theorem (P(B/A) = P(B) x P(A/B)/P(A)) with

$$P(B) = \phi(t)dt = P(t \leq \tau \leq t+dt)$$

$$P(A) = R(t) = P(\tau>t)$$

$$P(B|A) = P(t\leq\tau\leq t+dt|\tau>t) = f(t)dt$$

$$P(A|B) = P(\tau>t|t<\tau<t+dt) = 1$$

we obtain:

$$f(t)dt = \frac{\phi(t)dt \cdot 1}{R(t)} \tag{2.4}$$

Substituting (2.1) and (2.3) into (2.4):

$$f(t)dt = \frac{F'(t)dt}{1-F(t)} \tag{2.5}$$

Integrating (2.5) from t=0 to t=T:

$$\int_0^T f(t)dt = \int_0^T \frac{F'(t)dt}{1-F(t)} = \int_{F(0)}^{F(T)} \frac{dF(t)}{1-F(t)}$$

$$\int_0^T f(t)dt = \left[-\log(1-F(t))\right]\bigg|_{F(0)}^{F(T)}$$

$$\int_0^T f(t)dt = \left[-\log\frac{1-F(T)}{1-F(0)}\right]$$

Exponentiating and solving for F(T):

$$F(T) = 1 - (1-F(0))\exp\left[-\int_0^T f(t)dt\right] \tag{2.6}$$

Rearranging:

$$R(T) = 1 - F(T) = (1-F(0))\exp\left[-\int_0^T f(t)dt\right] \tag{2.7}$$

F(0) represents the probability of failure at time t=0. Of course, it could be any value between zero and one. However, for the discussion that follows, it is assumed that F(0)=0 so (2.7) becomes:

$$R(t) = \exp\left[-\int_0^t f(\tau)d\tau\right] \tag{2.8}$$

Differentiating (2.8) and with use of (2.1) and (2.3):

$$\frac{dR(t)}{dt} = \frac{d(1-F(t))}{dt} = \left[-d/dt \int_0^t f(\tau)d\tau\right] \cdot \exp\left[-\int_0^t f(\tau)d\tau\right]$$

and
$$\frac{dF(t)}{dt} = -f(t) \exp\left[-\int_0^t f(\tau)d\tau\right]$$
from which we obtain
$$\phi(t) = f(t) \exp\left[\int_0^t f(\tau)d\tau\right] \quad (2.9)$$

Substituting (2.8) into (2.9), we obtain the equivalent to (2.4):
$$\phi(t) = f(t)R(t) \quad (2.10)$$

In summary, we have defined the following relationships among our basic definitions:

$$R(t) = 1 - F(t) = \exp\left[-\int_0^t f(\tau)d\tau\right]$$

$$\phi(t) = F'(t) = f(t)R(t) = f(t)\exp\left[-\int_0^t f(\tau)d\tau\right]$$

Another important concept in reliability analysis is the expected lifetime or mean-time before failure, MTBF or $L(t)$. If the expected lifetime is taken from the origin of time, $t=0$, then from the definition of the expectation operator, the expected time to failure is given to be:

$$\text{MTBF} = L(0) = \int_0^\infty t\,\phi(t)dt \quad (2.11)$$

Combining (2.3)
$$\phi(t) = \frac{dF(t)}{dt}$$
and (2.1)
$$F(t) = 1 - R(t)$$
it can be derived that
$$\phi(t) = -\frac{dR(t)}{dt}$$

Substituting this result into (2.11) yields:
$$L(0) = -\int_0^\infty t\,\frac{dR(t)}{dt}\,dt \quad (2.12)$$

Integrating (2.12) by parts and using the fact that $R(\infty) = 1 - F(\infty) = 0$ since $F(\infty)$ must be one because $F(t)$ is a distribution function,

$$L(0) = \left[-tR(t) \Big|_0^\infty - \int_0^\infty R(t)\,dt \right] = \int_0^\infty R(t)\,dt \qquad (2.13)$$

Thus, from (2.13) it is seen that the MTBF taken from t=0 is equal to the area under the total reliability function curve.

If it is assumed that a system has survived a period t=T from the origin t=0, then the expected time to failure given the system has survived to time T, L(T), is a quantity of interest. To determine this quantity, R(t,T), the probability of nonfailure to time t, given the system survived to time T, must be used. In a derivation analogous to that for R(t), it can be found that

$$R(t,T) = \exp\left[-\int_T^t f(\tau)\,d\tau \right]$$

which may be re-expressed as:

$$R(t,T) = \frac{\exp\left[-\int_0^t f(\tau)\,d\tau \right]}{\exp\left[-\int_0^T f(\tau)\,d\tau \right]} = R(t)/R(T)$$

Substituting this expression into the form of L(T) analogous to (2.12)

$$L(T) = -\int_T^\infty (\tau-T) \frac{dR(\tau,T)}{d\tau}\,d\tau$$

yields:

$$L(T) = -\int_T^\infty \frac{(\tau-T)}{R(T)} \cdot \frac{dR(\tau)}{d\tau}\,d\tau = -\int_T^\infty \frac{(\tau-T)}{R(T)}\,dR(\tau)$$

Integrating by parts:

$$L(T) = -1/R(T) \left[(\tau-T)R(\tau) \Big|_T^\infty - \int_T^\infty R(\tau)\,d\tau \right]$$

$$L(T) = \int_T^\infty \frac{R(\tau)}{R(T)}\,d\tau \qquad (2.14)$$

As an example of application of these formulae, consider the case of a component whose failure density can be modeled adequately by an exponential distribution so that

$$\phi(t) = \rho e^{-\rho t}$$

In this case

$$R(t) = 1 - F(t) = 1 - \int_0^t \phi(\tau)d\tau = \int_t^\infty \phi(\tau)d\tau$$

$$= \int_t^\infty \rho e^{-\rho\tau}d\tau = -e^{-\rho\tau}\Big|_t^\infty$$

$$R(t) = e^{-\rho t} \qquad (2.15)$$

and also:

$$f(t) = \phi(t)\, R(t) = \rho e^{-\rho t}/e^{-\rho t} = \rho \text{ (constant)} \qquad (2.16)$$

$$L(0) = \int_0^\infty R(\tau)d\tau = \int_0^\infty e^{-\rho\tau}d\tau = 1/\rho \qquad (2.17)$$

$$L(T) = \left[\int_T^\infty R(\tau)/R(T)\right] d\tau = e^{\rho T}\int_T^\infty e^{-\rho\tau}d\tau = 1/\rho \qquad (2.18)$$

From (2.16), (2.17), and (2.18), it is seen that for exponentially distributed failure densities that f(t), the age dependent failure rate, is constant and that the MTBF remains constant no matter how old the system or component. Therefore, the probability of failure of a component or system is invariable with age. This ageless condition actually applies to a large number of electric and electronic components (light bulbs) after the initial "wear-in" failure period.

For other cases when it is found that f(t) increases with time, the component is said to undergo positive aging. Correspondingly, if f(t) decreases in time the component undergoes negative aging, as shown in Figure 2.2.

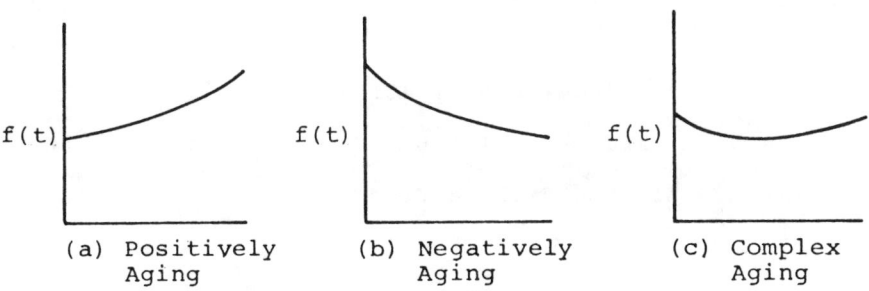

(a) Positively Aging   (b) Negatively Aging   (c) Complex Aging

FIGURE 2.2: VARIOUS AGE DEPENDENT FAILURE RATE DISTRIBUTION

Generally it is found that the age dependent rate over the life of a component or system can often be represented by a combination of negative aging, then constant or increasing failure rate followed by positive aging. The result is a bathtub-type curve as shown in Figure 2.3 in which the three regions are defined by initial or wear-in failure period, random failure period, and wear-out failure period.

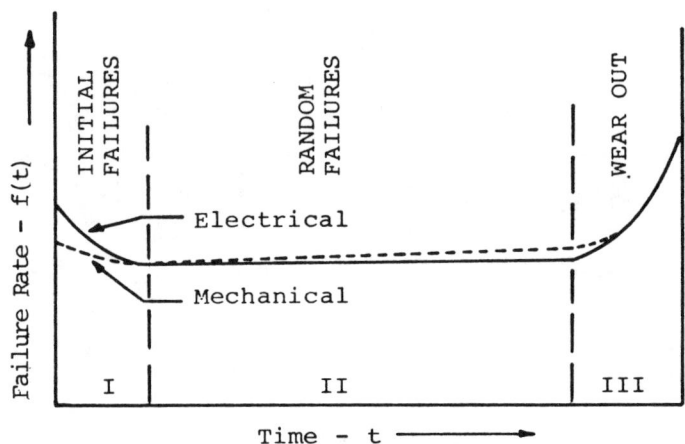

FIGURE 2.3: LIFE CHARACTERISTIC OF BATHTUB CURVE

Electrical and electronic components will normally have near instantaneous wear-in and wear-out failure periods. Mechanical components, however, generally exhibit a prolonged negatively aging wear-in period followed by slight positive aging during their random failure period followed by a fairly short and pronounced positive aging wear-out period. It has been found that the wear-out failure rate of mechanical components can usually be described by a normal density, so:

$$\phi_w(t) = \text{wear-our failure density} \simeq \frac{1}{\sqrt{2\pi}\,\sigma} \exp\left[-\frac{(t-\bar{t}_w)^2}{2\sigma^2}\right]$$

where

$\bar{t}_w$ = mean wear-out lifetime

σ = standard deviation of the time to wear-out

t = time to failure from t=0

and

$$\bar{t}_w = \sum_{i=1}^{N} t_i / N \qquad \sigma^2 = \sum_{i=1}^{N} (t_i - \bar{t}_w)^2 / (N-1)$$

when

t = time to failure from t=0 of the i sample

$\bar{t}_w$ = number of samples in the test

Given $\phi(t)$, it then follows that for the mechanical system, the probability of non-failure due to wear before time t, $R_w(t)_2$ will be:

$$R_w(t) = \int_t^\infty \phi_w(\tau) d\tau = \frac{1}{\sqrt{2\pi}\,\sigma} \int_t^\infty \exp\left[-\frac{(\tau-\bar{t}_w)^2}{2\sigma^2}\right] d\tau$$

The probability of surviving wear failure from t=0 to an operating time t_i is therefore,

$$R_w(t_i) = \int_{t_i}^\infty \phi_w(t) dt$$

and is shown by the shaded area of Figure 2.4.

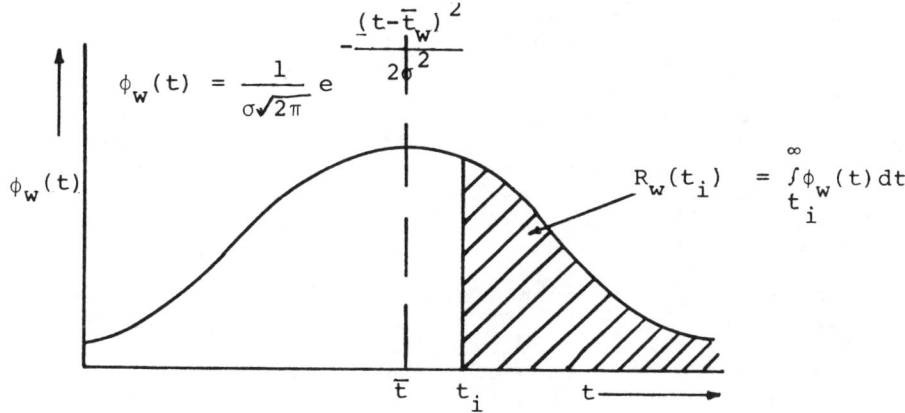

FIGURE 2.4: PROBABILITY OF SURVIVING WEAR FAILURE

If the effects of wear-in failure are ignored, since wear-in often takes place prior to delivery, then the total age dependent failure rate at any time t during the life of a mechanical component ill be just the sum of the random failure rate and wear-out failure rate:

$$f_T(t) = f_R(t) + f_w(t)$$

or, equivalently,

$$R_T(t) = R_R(t) \cdot R_w(t)$$

If we assume that $f_R(t) = \rho$ = constant, then

$$R_T(t) = e^{-\rho t} \cdot R_w(t)$$

$$R_T(t) = \frac{1}{\sqrt{2\pi}\,\sigma} e^{-\rho t} \int_t^\infty \exp\left[-\frac{(\tau - \bar{t}_w)^2}{2\sigma^2}\right] d\tau$$

Exercises

1. A component has a constant age dependent failure rate $f(t) = \delta = \lambda = 0.001/hr$. What is its reliability during its planned life of 1000 hours? What is its mean time before failure? If the component survives 500 operating hours, what is its mean time before failure then? Also compute its failure rate $\phi(t)$ at $t=0$ and $t=500$. Is the failure rate time varying when the age dependent failure rate is constant over time?

2. In the above exercise, what is the reliability of the component at $t = 0, 100, 200, 300, 400$, and 500? Similarly compute $\phi(t)$ and $F(t)$ for these values of time and plot the results.

3. A component with instantaneous wear-in has a random failure period of 1000 hours during which the age dependent failure rate is constant and equal to $f(t) = \delta = \lambda = 0.0005$ per hour. After 1000 hours a failure is due to wear out failure superimposed on random failure as shown in Figure 2.3. If the mean wear-out lifetime is 1100 hours with a standard deviation of the time to wear-out failure of 20 hours, what is the probability of non-failure (reliability) for $t = 0, 500, 1000, 1050$, and 1100 hours? Plot $R_T(t)$ for this wearing component.

4. The age dependent failure rate of a system is a linear positively aging function $f(t) = 2(1+t)$ where t is in hours. Determine the reliability and failure rate $\phi(t)$ of this system for the range of $t = 0$ to 1000.

5. Support that a component has the following failure rate distribution for the wear-in, random failure, and wear-out periods of its life:

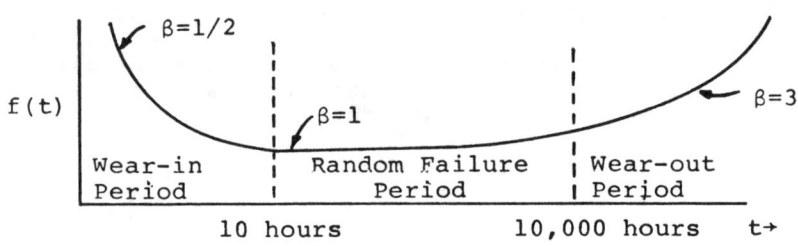

where

$$f(t) = \lambda^\beta \beta t^{\beta-1}, \qquad t > 0$$

Suppose that the system contains 100,000 of the components mentioned above.

    a. What is the expected number of components that will fail during time t = 10, 100, and 10,000 hours if λ = 0.00004/hr?

    b. Repeat (a) above for the period t = 10,000 hours to t = 20,000.

    c. What is the probability that the system will not malfunction due to a component failure during:

        (i)   its first hour of operation?
        (ii)   its first 10-hour period after an initial 100-hour "burn in" period?
        (iii) its first 10-hour period of wear-out?

For the conditions stated above:

        (i)   What is the MTBF of the system, if we ignore the early and wear-out failures?
        (ii)   What is the mean wear-out life of the components?
        (iii) What is the probability that during the wear-out period of the capacitors no more than five shutdowns of the system will be caused by a random (i.e., not early mortality or wear-out) failure of the components?

    6. A device fails with a hazard rate given by

$$\lambda(t) = at, \qquad t \leq T,$$

$$ = aT^2/t, \qquad t > T,$$

where a is a constant and $aT^2 > 1$. If $a = 10^{-2}\,\text{hr}^{-2}$ and T = 15 hr, (a) calculate the time after the device is placed in service before the probability of its failure is 0.95; (b) derive the equation for MTBF.

2.2 References

1. Abramowitz, M. and Stegun, I.A. (editors), "Handbook of Mathematical Functions", National Bureau of Standards, Washington, D.C., 1969, reprinted by Dover Book, New York, 1968.

2. Feller W., "An Introduction to Probability Theory and Its Applications", Volume I, 3rd edition, John Wiley & Sons Inc., New York, 1968.

3. Von Mises, R., "Probability, Statistics, and Truth", MacMillan, New York, 1957.

4. Lloyd, D.K. and Lapow, M., "Reliability: Management, Methods, and Mathematics", Prentice-Hall, Englewood Cliffs, New Jersey, 1962.

5. Bazovsky, I., "Reliability Theory and Practices", Prentice-Hall, Englewood Cliffs, New Jersey, 1961.

6. Von Alven, W.H. (editor), "Reliability Engineering", Prentice-Hall, Englewood Cliffs, New Jersey, 1964.

7. Parzen, E., "Modern Probability Theory and Its Applications", John Wiley & Sons Inc., New York, 1960.

8. Haugen, E.B., "Probabilistic Approaches to Design", John Wiley & Sons Inc., New York, 1961.

9. Cramer, H., "Mathematical Methods of Statistics", Princeton University Press, Princeton, New Jersey, 1946.

10. Roberts, N.H., "Mathematical Methods in Reliability Engineering", McGraw-Hill, New York, 1964.

3.0 ASSESSMENT OF RELIABILITY FUNCTION

Ernst G. Frankel

In the previous section, the reliability of a system was expressed in precise probabilistic terms. However, no indication was given on how to find the appropriate reliability function for a real component or system. There are two different approaches to this problem. Either the reliability function can be estimated from curve-fitting the failure data obtained from extensive life testing or it may be hypothesized to be a certain parameterized function (as was done for the mechanical system discussed in the previous section) with the parameters estimated via statistical sampling techniques. Below is an example of each of these approaches. However, it must be realized that these examples just give a taste of the methodology available and that reference to one of the standard statistics tests should be made prior to any testing for reliability assessment.

3.1 Non-parametric Reliability Function Assessment

Consider a continuous life test of 100 transducers. The transducers are tested to failure and counts of the failed transducers are made at discrete intervals. The results of this test are given in Table 3.1.

TABLE 3.1: TEST RESULTS

Period	1	2	3	4	5	6	7	8
n(t)	30	20	12	10	8	7	7	6

From this data compute the values of $\phi(t)$, $f(t)$, $F(t)$, and $R(t)$. Draw graphs of $\phi(t)$, $f(t)$, $F(t)$, and $R(t)$ against time. The following terms are defined:

$n(t)$ = number of transducers failed in time period t

$$N = \sum_{t=1}^{8} n(t) = 100$$

$r(t)$ = number of transducers remaining at the end of period t

f(t) = average of failures in period t, divided by the average number of transducers operating within the period

$$f(t) = n(t) \div \frac{r(t) + r(t-1)}{2}$$

f(t) is not constant and should be calculated from $R(t) = e^{-\int_0^t f(t)dt}$, but the intervals of time are not known. We will assume f(t) is constant over each period and calculate as above.

φ(t) = average failure density for period
= number of failures in period t, divided by N
= n(t)/100

$$R(t) = \phi(t)/f(t) = n(t)/100 \div \frac{n(t)}{\frac{r(t)+r(t-1)}{2}}$$

$$= \frac{r(t)+r(t-1)}{200}$$

$$F(t) = \text{cumulative function of } \phi(t) = \sum_{i=1}^{t-1} \phi(i) + \phi(t)/2$$

These values are tabulated as follows:

t	n(t)	r(t)	f(t)	(t)	R(t)	F(t)
1	30	70	0.353	0.3	0.85	0.15
2	20	50	0.333	0.2	0.60	0.40
3	12	38	0.273	0.12	0.44	0.56
4	10	28	0.303	0.10	0.33	0.67
5	8	20	0.333	0.08	0.24	0.76
6	7	13	0.424	0.07	0.165	0.835
7	7	6	0.632	0.07	0.095	0.905
8	6	0	2.000	0.06	0.03	0.97

and plotted in Figure 3.1.

A curve can be fitted to the graph of R(t) and this may be used as an estimate of the reliability function of the transducers. Alternatively, the graphs of R(t) and f(t) may be used to indicate which mathematical models might be appropriate to account for the reliability behavior of the components. For example, the life test data for lightbulbs should approximate that for an exponential failure density. Given that it appears that a certain distribution will account for the reliability behavior, then it is necessary to use the techniques of statistics to determine from test data

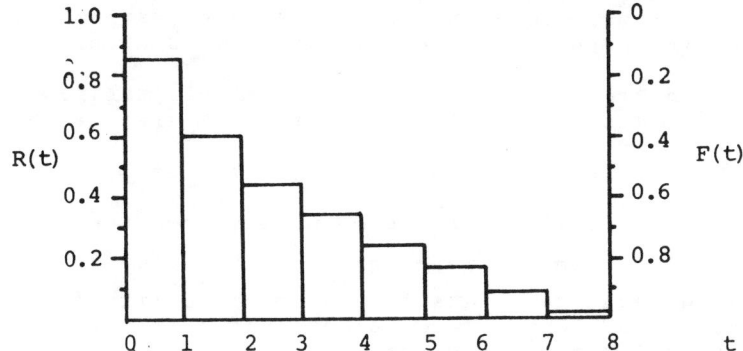

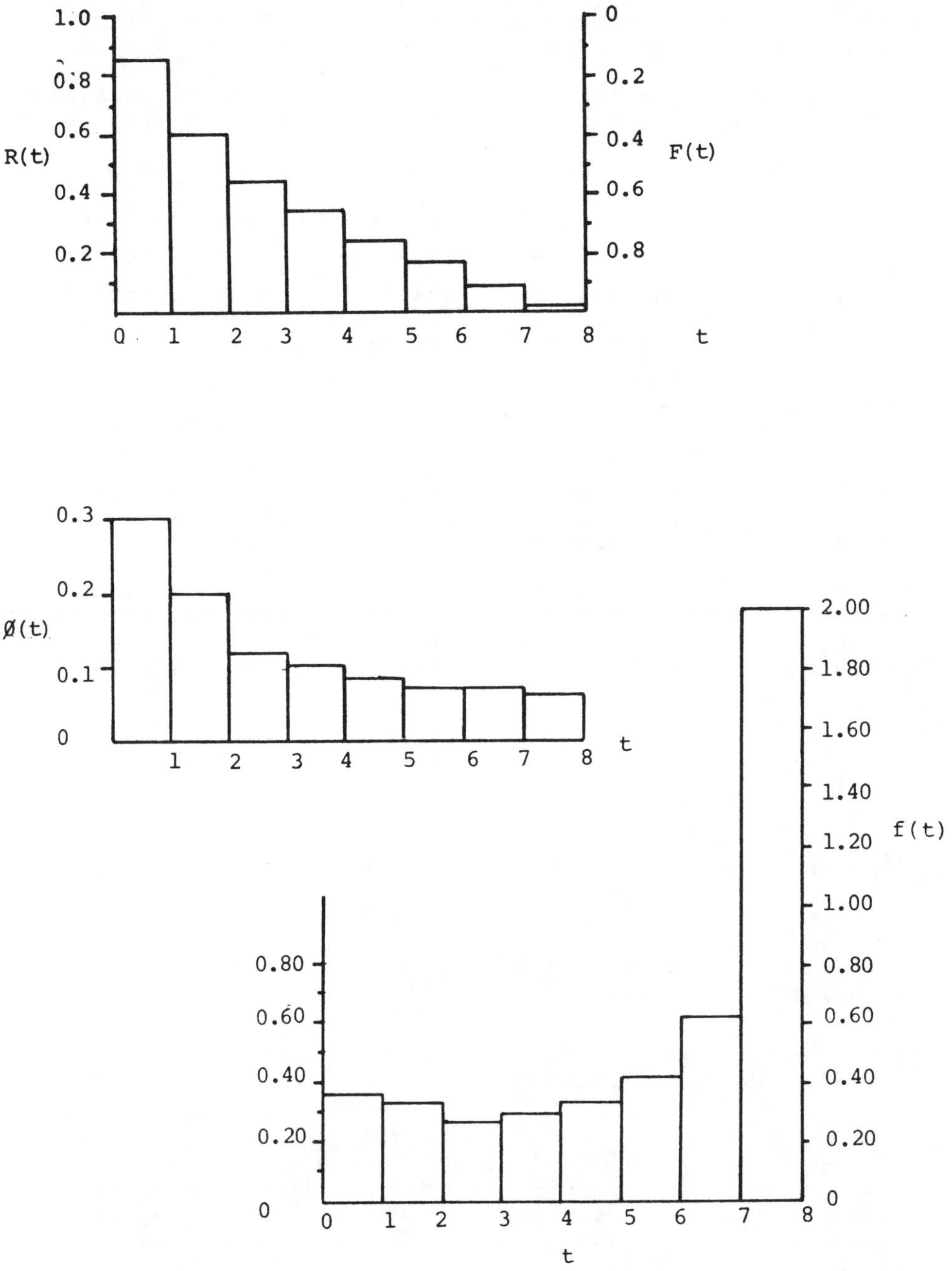

FIGURE 3.1: TRANSDUCER TEST RESULTS

the appropriate parameters of the distribution as well as the confidence that one may have in these parameter estimates. If, for example, the exponential failure density is assumed, the reliability function $R(t) = e^{-\rho t}$ has the single parameter ρ which must be estimated. If $t_1, t_2, \ldots t_i, \ldots t_r$ represent the recorded times of failure from the origin of time, $t=0$, of the first, second, $\ldots i^{th}, \ldots r^{th}$ components, then the recorded times occur in an ordered sequence $t_1 \leq t_2 \leq \ldots t_i \leq \ldots < t_r$. From this ordered sequence of data, the maximum likelihood estimate (unbiased, efficient, and consistent statistic) of $1/\rho$, $1/\hat{\rho}$ can be calculated from the formula:

$$1/\hat{\rho} = \sum_{i=1}^{r} \frac{t_i + (n-r)t_r}{r} \qquad (3.1)$$

where n is the number of samples in test, randomly selected from lot, and, r is the number of failures observed before terminating test (failed units are not replaced).

To give an indication of the methods of parameter estimation, (3.1) will be derived as follows. First, it must be remembered that the likelihood function for a sample of size n drawn from a population with a density $f(\chi; \theta)$ is

$$f(x_1, x_2, \ldots x_n; \theta) = \prod_{i=1}^{n} f(x_i; \theta)$$

where

$f(x_i; \theta)$ is the probability of sample i with the value of x_i

The maximum likelihood estimate of θ, $\hat{\theta}$, is the value of satisfying the equation:

$$\partial/\partial\theta \ \ln \prod_{i=1}^{n} f(x_i; \theta) = 0$$

In the case of the exponential function with an n sample test which is truncated after the first r (r n) failures, and with the data $t_1, t_2, t_3 \ldots t_r$, the likelihood function will be

$$f(t_1, t_2, \ldots t_r) = (\rho e^{-\rho t_1})(\rho e^{-\rho t_2}) \ldots (\rho e^{-\rho t_r}) e^{-(n-r)\rho t_r}$$

since the first r terms represent the probabilities of failure at times t_1 ... through t_r and the last term represents the reliability of the (n-r) survivors - that is the probability that at time t_r, (n-r) units are still operating satisfactorily. Taking logarithms and differentiating the above expression:

$$\partial/\partial\rho \ln\left[(\rho e^{-\rho t_1})(\rho e^{-\rho t_2})...(\rho e^{-\rho t_r})(e^{-(n-r)\rho t_r})\right] = 0$$

$$\partial/\partial\rho \ln\left[\rho^r e^{-\rho\left[\sum_{i=1}^{r} t_i + (n-r)t_r\right]}\right] = 0$$

$$\partial/\partial\rho \left[r\ln\rho - \rho\left[\sum_{i=1}^{r} t_i + (n-r)t_r\right]\right] = 0$$

$$r/\rho - \left[\sum_{i=1}^{r} t_i + (n-r)t_r\right] = 0$$

$$1/\hat{\rho} = \frac{\sum_{i=1}^{r} t_i + (n-r)t_r}{r}$$

which is the same as equation (3.1) presented previously.

For the case that the failed units are replaced by new units so that the sample size remains constant at the value n, it can be shown by a similar analysis that

$$1/\hat{\rho} = \frac{nt_r}{r}$$

3.2 Parametric Reliability Function Assessment

As an example of this parameter estimation technique, consider the life-test of a group of 480 fuses that were run at 90% of rated current. After about 1700 hours of operation 26 fuses had "blown". At this point the test was terminated. The data collection for fuse burnout times were as follows:

119	838	1300	1427	1504	1588	1665
121	896	1377	1429	1552	1592	1667
202	1154	1404	1479	1565	1648	
833	1183	1410	1483	1583	1660	

These entities represent the values of $t_1 \ldots t_r$, which substituted with the values of $r=26$ and $n=480$ into (3.1) yield:

$$1/\hat{\rho} = \frac{32,679 + (454)(1667)}{26} = 30,365 \text{ hours}$$

If the data were examined carefully, it would seem reasonable to assume that the first three failures should be considered initial failures. If this is true and these values are ignored, then $1/\hat{\rho} = 34,524$, a gain of over 4,000 in the mean life.

Instead of a point estimate for $\hat{\rho}$, statistical analysis may be used to obtain an interval estimate. That is, if $1/\rho_0$ is some acceptable (high) mean life, and $1/\rho_1$ is some unacceptable (low) mean life, then the hypothesis

$$H_0: 1/\rho = 1/\rho_0$$

is tested against the alternative

$$H_1: 1/\rho = 1/\rho_1 < 1/\rho_0$$

subject to the conditions that the probability of accepting H_0 when H_0 is true us $(1-\alpha) \cdot 100\%$ and the probability of accepting H_0 when H_1 is true is $\leq \beta \cdot 100\%$. It can be shown that given a sample of size n that H_0 should be accepted if

$$1/\hat{\rho} > 1/\rho_0 \, \chi^2_{1-\alpha}(2r)/2r$$

where r is the smallest integer such that

$$\chi^2_{1-\alpha}(2r)/\chi^2_{\beta}(2r) \geq \rho_0/\rho_1$$

Other interval tests have been devised for time truncated life tests when the length of the test is specified and the number of units tested will depend on the values of and desired.

Finally, sequential testing procedures have been developed which allow the testing of the hypothesis continually while the testing occurs in the bounding regions as shown in Figure 3.2.

Details concerning this and other parameter point and interval estimates can be found in References (1) and (2).

It should be noted that point and interval estimates for the various testing procedures have been developed for most of the failure densities that may be used in models of real systems.

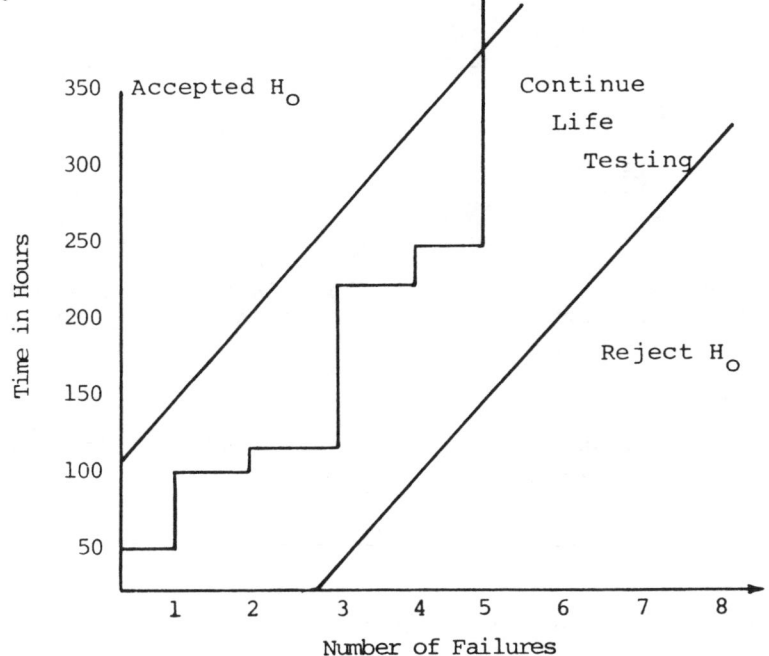

FIGURE 3.2: BOUNDING REGIONS

Generally non-parametric estimates are not quite as efficient as parametric estimates. Non-parametric estimates require larger samples of data to achieve as reliable results as parametric estimates.

Estimates for purposes of reliability analysis can be:

    a. single value estimates of a particular parameter of interest, using a maximum likelihood estimating approach,

    b. interval estimates based on confidence-interval estimates,

or

    c. probability distribution estimates based on curve fitting as discussed before.

Let us review these statistical testing procedures next. Starting with non-parametric point and interval estimating methods, we may want to consider a very simple point estimate of a systems reliability over a time interval t, say R(t). In this case we calculate the ratio of parts that survive the interval t out of a total of N parts tested. Assuming n points fail before t then

$$\bar{R}(t) = \frac{N-n}{N}$$

Obviously the number of parts under test remains N during the period of the test. The confidence interval about $\bar{R}(t)$ is computed assuming the point estimates correspond to a binomial parameter

$$\bar{R}_{L,a} = (1 + \frac{(n+1)}{(N-n+1)} F_a(2n+2, 2N-2n))^{-1}$$

where $F_a(2n+2, 2N-2n)$ is the upper a% point of the F distribution with (2n+2) and (2N-2n) degrees of freedom. (Tables of the F distribution are presented at the end of the book.)

Numerical Example

Assume N=40 parts are put on a test and n=10 failures are observed before t=50 hours.

1. Point estimate of $\bar{R}(50) = \frac{40-10}{40} = 0.75$

2. Lower 90% Confidence Limit = $\bar{R}_{L,0.1} = (1+\frac{11}{30}F_{0.1}(22,60))$

and

From Tables $F_{0.1}(22,60) = 1.6389$

$$\bar{R}_{L,0.1} = \frac{1}{1 + \frac{11}{30} \times 1.6389} = \frac{1}{1.6} = 0.625$$

3.3 Exercises

1. It is known that tubes exhibit wear-out characteristics. Seven tubes are tested to destruction with the following elapsed times to failure:

	Elapsed Time
1st failure	9,300 hours
2nd failure	9,500 hours
3rd failure	9,885 hours
4th failure	10,000 hours
5th failure	10,115 hours
6th failure	10,500 hours
7th failure	10,700 hours

a. What is the "best estimate" of the mean life of tubes?
b. With 90% confidence, what is mean life of tubes?
c. With 95% confidence?
d. What confidence do you have that a shipment of tubes from the same lot would have a mean life of 9,000 hours?
e. Suppose that the 7 tubes above were 7 failures out of a total of 20 tubes on test. Using a non-parametric approach, with 95% confidence, what is the MTBF of tubes? (Assume the test was truncated at 11,000 hours.)

2. Results of tests of a component indicate a failure rate of $\phi(t)$/hr as follows:

t(hrs)	$\phi(t)$
100	0.25
200	0.20
300	0.17
400	0.13
500	0.12
600	0.10
700	0.06
800	0.04
900	0.03
1000	0.01

Determine the reliability and age dependent failure rate of the component.

3. In a test of 50 items, failures occurred at the following elapsed hours: 7, 18, 25, 27, 35, 41, 47, 50, 54, 60. Obtain the observed reliability function at:

a. every 5 hour period up to 60 hours,
b. every 10 hour period up to 60 hours, and
c. each failure time.

Plot results and comment on differences.

4. Five electric motors were tested to failure and failed at times 8, 11, 16, 22, and 34 ($\times 10^3$) hours respectively. Determine the parameters a and b and fit a log-normal distribution.

5. Using the data of problem 4, determine the Weibull parameters a and b using curve fitting methods.

3.4 References

1. Benjamin, J.R. and Cornell, C.A., "Probability Statistics and Decision for Civil Engineers", McGraw-Hill Book Company, New York, 1970.

2. Mosteller, F., Rourke, R.E.G., and Thomas, G.B., Jr., "Probability and Statistics", Addison-Wesley Publishing Company, Reading, Mass., 1961.

3. Weatherburn, C.E., "Mathematical Statistics", University Press, Cambridge, Mass., 1961.

4. Parzen, E., "Modern Probability Theory and Its Applications", John Wiley and Sons Publishers, New York, 1963.

5. Bury, K.V., "Statistical Models in Applied Science", John Wiley and Sons Publishers, New York, 1963.

6. Feller, W., "An Introduction to Probability Theory and Its Applications", John Wiley and Sons Publishers, New York, 1960.

7. Ang, A.H., Tang, S., and Tang, W.H., "Probability Concepts in Engineering Planning and Design", John Wiley and Sons Publishers, New York, 1975.

8. Hahn, G.J. and Shapiro, S.S., "Statistical Models in Engineering", John Wiley and Sons, New York, 1967.

9. Tribus, M., "Rational Descriptions, Decisions, and Designs", Pergamon Press, Oxford, 1959.

10. Mann, N.R., Schafer, R.E., and Singpurwalla, N.D., "Methods for Statistical Analysis for Reliability and Life Data", John Wiley and Sons Publishers, New York, 1974.

4.0 RELIABILITY OF SERIES AND PARALLEL SYSTEMS

Ernst G. Frankel

Now that techniques for determining the reliability of a component or system have been discussed, the effect of combining components in series or in parallel- redundant groups should be considered. First, series systems will be discussed.

4.1 Simple Series Systems

A series system, as shown in Figure 4.1, generally consists of a group of independent components (the failure of one component is independent of the failure of another component) in which the failure of one component causes system failure. If $R_i(t)$ is the reliability of the ith component, it will be shown that the series system reliability, $R_T(t)$, is the product of the component reliabilities.

$$R_T(t) = \prod_{i=1}^{n} R_i(t)$$

The argument goes as follows. Consider first a two-component series system shown in Figure 4.2.

FIGURE 4.1: SERIES SYSTEM

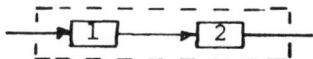

FIGURE 4.2: TWO-COMPONENT SERIES SYSTEM

It is apparent that the system fails if either component one (1) fails (event F_1), component two (2) fails (event F_2) or both fail (event $F_1 \cap F_2$). This statement may be expressed pictorially in a logic or Venn diagram, as shown in Figure 4.3.

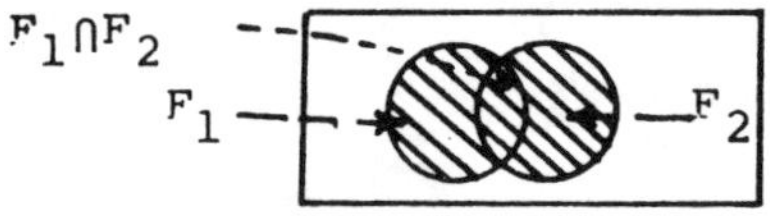

FIGURE 4.3: TWO EVENT VENN DIAGRAM

U   = failure space

F   = event that component one (1) fails

F   = event that component two (2) fails

$F_1 \cap F_2$ = event that component one and component two fail simultaneously

The shaded area of the diagram is the event of total system failure, F_T, i.e., the union of the individual component failures. It follows that:

$$F_T = F_1 + F_2 - F_1 \cap F_2 \qquad (4.1)$$

Now the probability of component one (1) failing through time t is $F_1(t)$, or more specifically:

$$P_r(F_1) = F_1(t)$$

Similarly:

$$P_r(F_2) = F_2(t)$$

$$P_r(F_1 \cap F_2) = F_1(t)F_2(t) \quad \text{(from product law of probability function)}$$

$$P_r(F_T) = F_T(t)$$

Thus, by the fundamental definition of probability functions, (4.1) is equivalent to:

$$F_T(t) = F_1(t) + F_2(t) - F_1(t)F_2(t) \qquad (4.2)$$

If the reliability function definition is substituted into (4.2):

$$(1-R_T(t)) = (1-R_1(t)) + (1-R_2(t)) - (1-R_1(t))(1-R_2(t))$$

$$1-R_T(t) = 2-R_1(t) - R_2(t) - 1 + R_1(t) + R_2(t) - R_1(t)R_2(t)$$

$$R_T(t) = R_1(t)R_2(t) \qquad (4.3)$$

Through inductive reasoning, the above argument can be extended to show that:

$$R_T(t) = \prod_{i=1}^{n} R_i(t) \qquad (4.4)$$

Thus, the reliability of a series system is equal to the intersection of the reliabilities of all the series system components.

The MTBF of a series system easily follows from (2.13)

$$L(0) = \int_0^\infty R_T(\tau)d\tau = \int_0^\infty \prod_{i=1}^{n} R_i(\tau)d\tau \qquad (4.5)$$

4.1.1 Application to Components with Exponential Failure Densities

For a system of components with exponential failure densities, the reliabilities of the components are expressed as:

$$R_i(t) = e^{-\rho_i t}$$

The effect on the system reliability and MTBF, of varying the number of components and their reliabilities is seen in the three following systems.

Two Component Series System with Different Components

$$R_T(t) = R_1(t)R_2(t) = e^{-\rho_1 t} e^{-\rho_2 t} = e^{-(\rho_1+\rho_2)t}$$

$$L(0) = \int_0^\infty R_T(\tau)d\tau = \int_0^\infty e^{-(\rho_1+\rho_2)\tau} d\tau = \frac{1}{\rho_1+\rho_2}$$

Two Component Series System with Identical Components

$$R_T(t) = R_1(t)R_1(t) = e^{-2\rho_1 t}$$

$$L(0) = 1/2\rho_1$$

n-Component Series System with Different Components

$$R_T(t) = \prod_{i=1}^{n} R_i(t) = \prod_{i=1}^{n} e^{-\rho_i t} = e^{-\sum_{i=1}^{n} \rho_i t}$$

$$L(0) = \int_0^\infty R_T(\tau) d\tau = \int_0^\infty e^{-\Sigma \rho_i \tau} d\tau = \frac{1}{\sum_{i=1}^{n} \rho_i}$$

In an n-component series system, a time independent system age dependent failure rate, ρ_T, can be defined from solving

$$R_T = e^{-\rho_T t} = e^{-\sum_{i=1}^{n} \rho_i t}$$

for ρ_T so that

$$\rho_T = \sum_{i=1}^{n} \rho_i$$

4.2 Simple Parallel Systems

Next consider a parallel arrangement of components in a system as shown in Figure 4.4. In this case, it is assumed that only one of the system's components is required for system operation and all other components are redundant and on-line standbys. No switching is needed to replace a failed component by an operable one. The failure rate of the components is assumed to be independent of the number of components still operating. In this case, the failure distribution will be the product of the components' failure distributions:

$$F_T(t) = \prod_{i=1}^{n} F_i(t)$$

The argument goes as follows. First consider a two-component parallel system. It is apparent that the system fails only if both component one and component two have failed (event

$F_1 \cap F_2$). This event is shown by the shaded region of the Venn diagram (Figure 4.5).

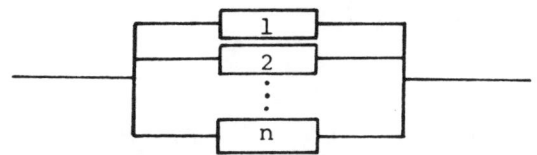

FIGURE 4.4: N-COMPONENT PARALLEL SYSTEM
WITH DIFFERENT COMPONENTS

$F_1 \cap F_2$

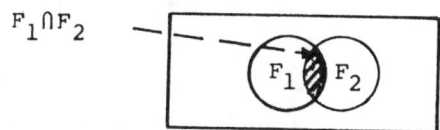

FIGURE 4.5: VENN DIAGRAM OF PARALLEL TWO-COMPONENT SYSTEM

Thus, the system failure event is the intersection of individual component failure events. Replacing events by their probabilities of occurrence:

$$F_T(t) = F_1(t) F_2(t) \qquad (4.6)$$

Again, through inductive reasoning, the above argument may be extended to prove that for an n-component parallel system:

$$F_T(t) = \prod_{i=1}^{n} F_i(t) \qquad (4.7)$$

If the reliability function definition is substituted into (4.7):

$$(1 - R_T(t)) = \prod_{i=1}^{n} (1 - R_i(t))$$

$$R_T(t) = 1 - \prod_{i=1}^{n} (1 - R_i(t)) \qquad (4.8)$$

Thus, the total system reliability is equal to the union of all component reliabilities.* For example, if n=2, then

$$R_T(t) = 1-(1-R_1(t))(1-R_2(t)) = R_1(t) + R_2(t) - R_1(t)R_2(t)$$

And, the MTBF of a parallel system follows from (4.8) and (2.3):

$$L(0) = \int_0^\infty R_T(\tau)d\tau = \int_0^\infty \left[1 - \prod_{i=1}^n (1-R_i(\tau))\right] d\tau \qquad (4.9)$$

4.2.1 Relation Between MTBF and R(t) in Simple Exponential Redundant System

In a simple exponential series system, we have seen that the equivalent system failure rate, ρ_T, was calculated by:

$$\rho_R = \sum_{i=1}^n \rho_i$$

The same does not apply to redundant systems, as can easily be seen in the following systems:

---

* To show this explicitly, remember that according to the laws of logic, if $\bar{A}$ is the complement of A, such that $A \cap \bar{A} = \phi$ and $A \cup \bar{A} = U$, then it follows that:

$$\bar{A}_1 \cap \bar{A}_2 \ldots \cap \bar{A}_n = \overline{A_1 \cup A_2 \cup \ldots \cup A_n}$$

Since by definition $F_i = \bar{R}_i$ then the above expression is equivalent to:

$$\bar{R}_T = F_T = F_1 \cap F_2 \cap \ldots \cap F_n = \overline{R_1 \cup R_2 \cup \ldots \cup R_n} \quad \text{(extending 4.5)}$$

or

$$R_T = R_1 \cup R_2 \cup \ldots \cup R_n$$

which in terms of probabilities because

$$P_r(R_T) = P_r(R_1 \cup R_2 \cup \ldots \cup R_n)$$

can be expressed as

$$R_T(t) = \sum_{i=1}^n R_i(t) - \prod_{i,j} R_i(t)R_j(t) + \ldots \prod_{i=1}^n R_i(t)$$

Two Component System with Different Components

$$R_T(t) = 1-(1-R_1(t))(1-R_2(t)) = R_1(t)+R_2(t)-R_1(t)R_2(t)$$

$$= e^{-\rho_1 t} + e^{-\rho_2 t} - e^{-(\rho_1+\rho_2)t}$$

$$R_T(t) = e^{-\rho_T t} = e^{-(\rho_1+\rho_2)t}[e^{\rho_1 t} + e^{\rho_1 t} - 1]$$

$$\rho_T = \rho_1 + \rho_2 - \ln[e^{\rho_2 t} + e^{\rho_1 t} - 1]$$

$$L(0) = \int_0^\infty R_T(\tau)d\tau = \int_0^\infty [e^{-\rho_1 \tau} + e^{-\rho_2 \tau} - e^{-(\rho_1+\rho_2)\tau}]d\tau$$

$$= \frac{1}{\rho_1} + \frac{1}{\rho_2} - \frac{1}{\rho_1+\rho_2}$$

Two Component System with Identical Components

In this case $R_1 = R_2$ and consequently $\rho_1 = \rho_2$. This gives us:

$$R_T = 2e^{-\rho\tau} - e^{-2\rho\tau}$$

$$\rho_T = 2\rho - \ln[2e^{\rho t}-1]$$

$$L(0) = \frac{2}{\rho} - \frac{1}{2\rho} = \frac{3}{3\rho}$$

If these results are compared to those from the series systems, we can see that the MTBF for two identical components with exponential failure densities is three times greater when arranged in parallel than when arranged in series.

n-Component Redundant System with Different Components

$$R_T(t) = 1 - \prod_{i=1}^{n}(1-R_i(t)) = 1 - \prod_{i=1}^{n}\left[1-e^{-\rho_i t}\right]$$

$$L(0) = \int_0^\infty R_T(\tau)d\tau = \int_0^\infty \left[1 - \prod_{i=1}^{n}\left[1-e^{-\rho_i \tau}\right]\right]d\tau$$

Example 4.1

A single component system requires a reliability of $R(t) = 0.999$. After 100 hours of operation

a. What is $\phi(t)$, $f(t)$, and $L(0)$ if it is subject to an exponential failure rate?

b. What would the required $\phi(t)$, $f(t)$ be if you were allowed to introduce two-component, on-line redundancy?

c. or three-component, on-line redundancy? What is the resulting MTBF = $L(0)$ again assuming exponential failure rates, and all identical components?

Given:

$$R(t) = 0.999 = e^{-\lambda t} \qquad t = 100$$

a. $R(t) = e^{-100\lambda} = .999$

$\lambda = 10^{-5}$

from equation 2.11d

$$\phi(t) = \frac{-dR(t)}{dt}$$

$$\phi(t) = \lambda e^{-\lambda t} = 10^{-5} e^{-10^{-5} t} = 10^{-5} e^{-10^{-3}}$$

$$f(t) = \phi(t)/R(t) = \lambda = 10^{-5}$$

$$L(0) = \int_0^\infty R(t) dt = \frac{1}{\lambda} = 10^5$$

b. $R_T(t) = .999 = 2R(t) - R(t)^2$

$R(t)^2 - 2R(t) + .999 = 0$

$R(t) = .968$

$\quad = e^{-10^2 \lambda}$

$10^2 \lambda = .03256$

$\lambda \quad = 0.0003256 \simeq 3 \times 10^{-4}$

$$\phi(t) = 3 \times 10^{-4} e^{-3 \times 10^{-4} t}$$

$$f(t) = 3 \times 10^{-4}$$

$$L(0) = \frac{3}{2\lambda} = \frac{3 \times 10^4}{2 \times 3.0} = 0.5 \times 10^4$$

c. $R_T(t) = 1 - (\prod_{i=1}^{3} (1-R_i(t)))$

$$= R^3 - 3R^2 + 3R$$

$$= .999$$

$$R(t) = .9$$

$$= e^{-10^2 \lambda}$$

$$10^2 \lambda = .10436$$

$$\lambda \simeq 10^{-3}$$

$$\phi(t) = 10^{-3} e^{-10^{-3} t}$$

$$f(t) = 10^{-3}$$

$$L(0) = \frac{11}{6\lambda} \simeq 2 \times 10^3$$

Note: In both redundant cases above the <u>overall system</u> reliability is kept at 0.999 and <u>component</u> reliabilities are changed appropriately.

Example 4.2

Consider systems of one, two, and three identical on-line components with simple redundancy with exponential failure distributions with MTBF, L(0)=50,000 hours. What will the system reliabilities be at t=10,000 hours?

For a one-component system:

$$R_T(t) = e^{-\rho t}$$

$$L(0) = \frac{1}{\rho} = 50,000 \text{ hours} \qquad \rho = 2 \times 10^{-5}/\text{hour}$$

$$R(10,000) = e^{-2 \times 10^{-5} \times 10^4} = 0.819$$

For a two-component parallel system:

$$R(t) = 2e^{-\rho t} - e^{-2\rho t}$$

$$L(0) = \frac{3}{2\rho} = 50,000 \text{ hours}; \quad \rho = 3 \times 10^{-5}/\text{hour}$$

$$R(10,000) = 2e^{-0.3} - e^{-0.6} = .933$$

For a three-component parallel system:

$$R(t) = 3e^{-\rho t} - 3e^{-2\rho t} + e^{-3\rho t}$$

$$L(0) = \frac{11}{6\rho} = 50,000 \text{ hours}; \quad \rho = 3.67 \times 10^{-5}/\text{hour}$$

$$R(10,000) = 3e^{-0.367} - 3e^{-0.733} + e^{-1.1} = 0.971$$

A graphical representation of the above is shown in Figure 4.6 which shows the variation of reliability of a parallel system with the number of redundant components.

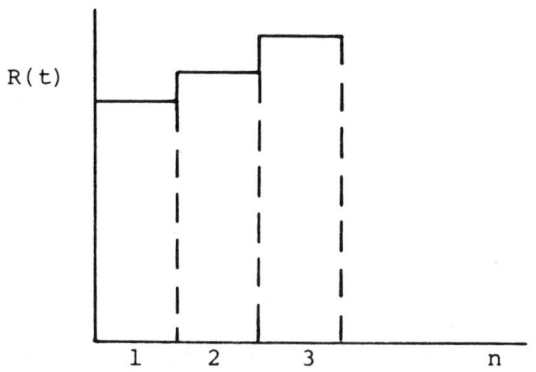

n = # of redundant components

FIGURE 4.6: VARIATION OF RELIABILITY WITH NUMBER OF COMPONENTS IN A SIMPLE REDUNDANT SYSTEM

4.3 Variations of Simple Redundant Systems

There are several important variations of simple, redundant systems.

One of these is the staggered system, where the redundant on-line standby components are put on-line at some time $t_i > 0$.

Consider a two-component parallel system.

Assuming an exponential failure density for the components, then the original component will have a reliability of:

$$R_1(t) = e^{-\rho_1 t}$$

and the redundant component a reliability of:

$$R_2(t) = e^{-\rho_2 (t-t_1)}$$

since it is put on-line at $t = t_1$.

Thus, the system reliability will be that of a single component for $t < t_1$ and a two-component parallel system for $t \geq t_1$ or

$$R_T(t) = \begin{cases} R_1(t) = e^{-\rho_1 t} & t < t_1 \\ R_T(t) + R_2(t) - R_1(t) R_2(t) \\ \quad = e^{-\rho_1 t} + e^{-\rho_2 (t-t_1)} - e^{\rho_2 t} e^{(\rho_1 + \rho_2) t} & t \geq t_1 \end{cases}$$

If $\rho_1 = \rho_2 = \rho$ then:

$$R_T(t) = \begin{cases} e^{-\rho t} & t < t_1 \\ e^{-\rho t} + e^{-\rho(t-t_1)} - e^{+\rho t_1} e^{-2\rho t} & t \geq t_1 \end{cases}$$

And the MTBF:

$$L(0) = \int_0^\infty R_T(\tau) d\tau = \int_0^{t_1} e^{-\rho\tau} d\tau + \int_{t_1}^\infty \left[e^{-\rho\tau} + e^{-\rho(\tau-t_1)} - e^{+\rho t_1} e^{-2\rho\tau} \right] d\tau$$

$$= -\frac{1}{\rho}(e^{-\rho t_1} - 1) - \frac{1}{\rho}(0 - e^{-\rho t_1}) - \frac{1}{\rho} e^{\rho t_1} \left[0 - e^{-\rho t_1} \right]$$

$$+ (1/2\rho)e^{+\rho t_1}\left[0 - e^{-2\rho t_1}\right]$$

$$L(0) = \frac{2}{\rho} - \frac{1}{2\rho} e^{-\rho t_1} = \frac{1}{2\rho}[4 - e^{-\rho t_1}]$$

If we let t_j, the time at which the second component is put on-line, equal zero, the above expression for MTBF reduces to that for a two-component on-line redundant system:

$$L(0) = \frac{3}{2\rho}$$

However, if the second component isn't turned on until

$$t_1 = \frac{1}{\rho}$$

we get

$$L(0) = \frac{1.8161}{\rho}$$

or an improvement of 21% in MTBF over the simple redundant system.

In a staggered system we still end up at $t \geq t_1$ with all the components on-line. If, however, a new component is switched on only when a previous component fails, we have an off-line redundant system. This is discussed in Section 4.2.

Another variation of the simple parallel system is a redundant standby on-line system for which at least r out of the k identical independent components must work for the system to be operational. The reliability of such a system can be reasoned from the following argument. The probability that the system will operate at time t will be the sum of the probabilities that k, k-1, k-2.. and r of the k components will be operating at time t. The probability that say, ℓ of the k components will operate at time t will be equal to the probability that there are ℓ components surviving times the probability that k-ℓ components fail times the number of combinations ℓ of k components will survive. More symbolically:

P (ℓ of k components survive to time t) =

$\binom{k}{\ell}$ P (ℓ components survive to time t) x P(k-ℓ fail prior to time t)

But

P (ℓ components survive to time t) =

[P (a single component survives to t)]$^\ell$ = $(R(t))^\ell$

and

P (k-ℓ components fail prior to time t) =
[P (a single component fails prior to t)]$^{k-\ell}$ =
$(1-R(t))^{k-\ell}$

So

P (ℓ of k components survive to time t) = $\binom{k}{\ell}(R(t))^\ell \cdot (1-R(t))^{k-\ell}$

Thus, the reliability of the system:

$$R_T(t) = \sum_{i=\ell}^{k} P \text{ (i out of k components survive to time t)}$$

$$= \sum_{i=\ell}^{k} \binom{k}{i} (R(t))^i (1-R(t))^{k-i} \quad \ldots \ldots \ldots \ldots \quad (4.10)$$

We now summarize the above results, for systems with exponential failure densities.

I. n-component series systems (Figure 4.7)

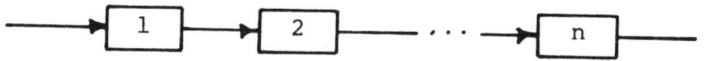

FIGURE 4.7: n-COMPONENT SERIES SYSTEMS WITH IDENTICAL COMPONENTS

1. with different components

$$R_T(t) = \prod_{i=1}^{n} R_i(t) = e^{-t\sum_{i=1}^{n} \rho_i}$$

$$L(0) = \left[\sum_{i=1}^{n} \rho_i\right]^{-1}$$

2. with identical components

$$R_T(t) = \prod_{i=1}^{n} R_i(t) = e^{-n\rho t}$$

$$L(0) = \frac{1}{n\rho}$$

II. Parallel Systems

1. n-component parallel system with different components

$$R_T(t) = 1 - \prod_{i=1}^{n} (1-R_i(t))$$

$$L(0) = \int_0^\infty [1 - \prod_{i=1}^{n} (1-R_i(\tau))] d\tau$$

2. Two different components (Figure 4.8a)

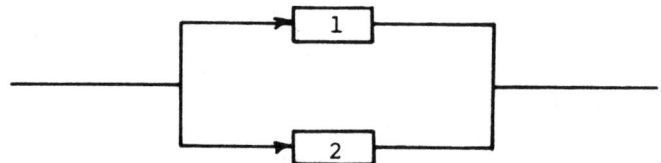

FIGURE 4.8A: TWO COMPONENT PARALLEL SYSTEM

$$R_T(t) = 1 - (1-R_1(t))(1-R_2(t)) = R_1(t) + R_2(t) - R_1(t)R_2(t)$$

$$= e^{-\rho_1 t} + e^{-\rho_2 t} - e^{-(\rho_1+\rho_2)t}$$

$$L(0) = \int_0^\infty R_T(\tau) d\tau = \int_0^\infty \left[e^{-\rho_1 \tau} + e^{-\rho_2 \tau} - e^{-(\rho_1+\rho_2)\tau} \right] d\tau$$

$$= 1/\rho_1 + 1/\rho_2 - \frac{1}{\rho_1+\rho_2}$$

3. Two Identical Components

$$R_T(t) = 1 - (1-R(t))^2 = 2R(t) - R^2(t) = 2e^{-\rho t} - e^{-2\rho t}$$

$$L(0) = 2/\rho - 1/2\rho = 3/2\rho$$

Note that the expression for L(0) can be logically derived by considering that the initial age dependent failure rate of the system will be $2\rho_1$ until the first component fails when it reduces to ρ_1. So

L(0) = time to first failure + time to second failure

$= 1/2\rho_1 + 1/\rho_1 = 3/2\rho_1$

4. n-identical components (Figure 4.8b)

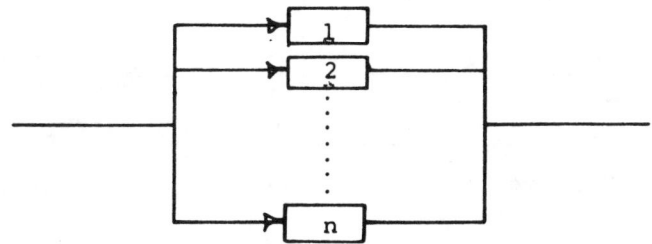

FIGURE 4.8B: N-COMPONENT PARALLEL SYSTEM WITH IDENTICAL COMPONENTS

In general for a redundant system with n-identical components:

$$R_T(t) = 1 - (1-R(t))^n = [1 - [e^{-\rho t}]^n$$

$$L(0) = \int_0^\infty R_T(\tau)d\tau = \int_0^\infty \left[1 - \left[1 - e^{-\rho\tau}\right]^n\right] d\tau$$

i.e. For n=3

$$R_T(t) = 1 - \left[1-e^{-\rho t}\right]^3 = 3e^{-\rho t} - 3e^{-2\rho t} + e^{-3\rho t}$$

$$L(0) = 3/\rho - 3/2\rho + 1/3\rho = 11/6\rho$$

For n=4

$$R_T(t) = 1 - \left[1-e^{-\rho t}\right]^4 = 4e^{-\rho\tau} - 6e^{-2\rho t} + 4e^{-3\rho t} - t^{-4\rho t}$$

$$L(0) = 4/\rho - 6/2\rho + 4/3\rho - 1/4\rho = 25/12\rho$$

etc......................

4.4 Analysis of Complex Series Redundant System

A complex system in which a series of one of each of n components is required for operation, and where there are K such identical series in parallel is shown in Figure 4.9.

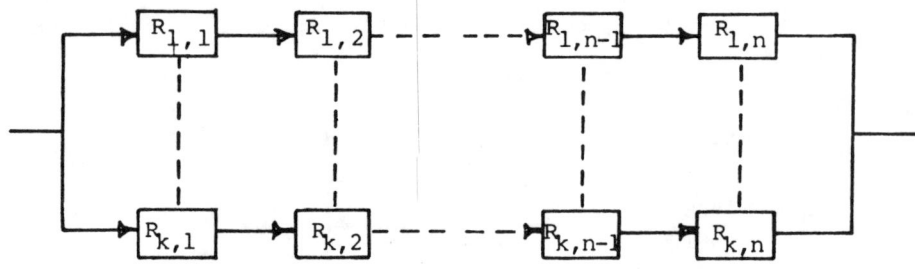

FIGURE 4.9: SERIES SYSTEM IN PARALLEL

Each of the K series will have a segment reliability of

$$R_i(t) = \prod_{j=1}^{n} R_{i,j}(t)$$

where

$R_{i,j}(t)$ is the reliability of the jth component of the ith series.

Thus, the system reliability will be:

$$R_T(t) = 1 - \prod_{i=1}^{k} (1-R_i(t)) \qquad (4.11)$$

Expanding, leaving out the time arguments:

$$R_T = 1 - \prod_{i=1}^{k} (1-R_i) = \left[1 - \prod_{i=1}^{k} \left(1 - \prod_{i=1}^{n} R_{i,j} \right) \right]$$

Since it is assumed that corresponding components in each series are identical, then the jth component of the series may be assigned the same symbol:

$$R_j = R_{ij} = R_{1j} = R_{2j} = R_{3j} = \ldots = R_{kj}$$

so (4.11) becomes

$$R_T = 1- \prod_{i=1}^{k}\left[1- \prod_{j=1}^{n} R_j\right] = 1-(1- \prod_{i=1}^{n} R_j)^k$$

If it is further assumed that the components of each series segment are identical, so that $R=R_1=R_2=R_3=\ldots R_{in}$, then (4.12) becomes:

$$R_T = 1- \left[1- \prod_{i=1}^{n} R\right]^k = [1-(1-R^n)^k]$$

If only r of the k n-component series links are necessary for the system to operate, then from (4.10) we get:

$$R_T(t) = \prod_{i=r}^{k} \binom{k}{i}(R_1 \cdot R_2 \ldots R_n)^i (1-R_1 R_2 \ldots R_n)^{k-i}$$

Next consider a system which has n subsystems of k parallel components in series. Such a system is shown in Figure 4.10.

The jth parallel subsystem will have reliability of

$$R_j = 1- \prod_{i=1}^{k} (1-R_{i,j}) \qquad (4.14)$$

and the total system made up of n of these subsystems in series will have reliability:

$$R_T = \prod_{j=1}^{n} R_j = \prod_{j=1}^{n}\left[1- \prod_{i=1}^{k} (1-R_{i,j})\right] \qquad (4.15)$$

If it is assumed that corresponding components in each parallel subsystem are identical, $R_i = R_{i1}=R_{i2}=R_{i3}=\ldots R_{in}$,

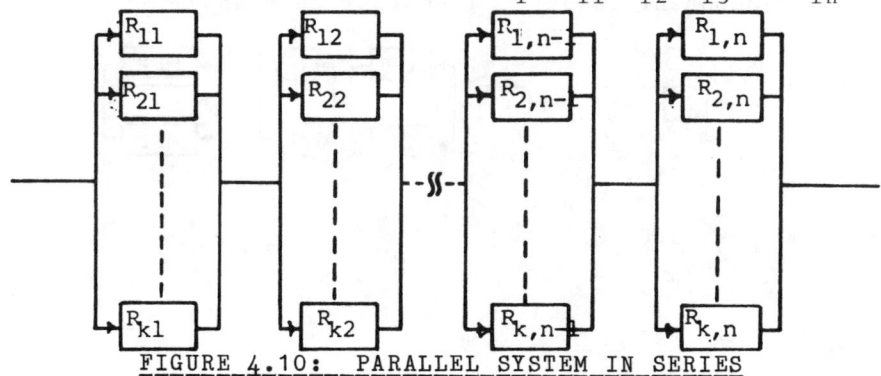

FIGURE 4.10: PARALLEL SYSTEM IN SERIES

then (4.15) becomes:

$$R_T = \prod_{j=1}^{n}\left[1 - \prod_{i=1}^{k}(1-R_i)\right] = \left[1 - \prod_{i=1}^{k}(1-R_i)\right]^n \quad (4.16)$$

and if it is further assumed that the components of each parallel subsystem are identical so that $R=R_1=R_2=\ldots=R_n$, then (4.16) becomes:

$$R_T = \left[1 - \prod_{i=1}^{k}(1-R)\right]^n = (1-(1-R)^k)^n \quad (4/17)$$

Example 4.3 - Subsystems of Parallel Components in Series

A steam propulsion system as shown has the following reliabilities

$R_B = 0.96$ $R_T = 0.99$ $R_C = 0.995$ $R_{CP} = 0.98$

$R_H = 0.999$ $R_{MFP} = 0.97$

Assuming one of the CP and one of the H and two of the MFP are required for full power operation:

a. What is the system reliability between scheduled overhauls?

b. If the system is designed for 730 days of operation between scheduled overhauls and has $f(t_s) = \lambda_s =$ constant, what is λ_s? What is the MTBF using λ_s?

Also what are the corresponding age-dependent component failure rates (all constant)?

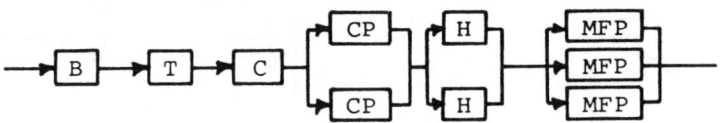

a. $R_B = .960$

$R_T = .990$

$R_C = .995$

From 4.10 $R_H(t) = \sum_{i=r}^{k} \binom{k}{i}(R(t))^i (1-R(t))^{k-i}$

R_{CP} (1 of 2) = $C(^2_2)(.98)^2 + C(^2_1)(.98)(.02) = .999$

R_H (1 of 2) = $.999$

R_{MFP} (2 of 3) = $C(^3_3)(.97)^3 + C(^3_2)(.97)^2(.03) = .997$

$R_{TO} = R_B R_T R_C R_{CP} R_{FH} R_{MFP} = .942$

b. $R_{TO} = e^{-\lambda_s t}$

$.942 = e^{-730\lambda_s}$

when t = 730 days = 2 years

$\lambda_s = 7.94 \times 10^{-5}/\text{day}$

L(0) over two year period = $\int_0^t R \, dt'$

$= \int_0^t e^{-\lambda t'} dt'$

$= \frac{1}{\lambda} - \frac{e^{-\lambda t}}{\lambda}$

$= 1.26 \times 10^4 - 1.26 \times 10^4 \times$

$(.942)$

= 730 days which is what the system was designed for

$R_B = .96 \times e^{-\lambda_B t}$

$\lambda_B = 5.5 \times 10^{-5}/\text{day}$

$R_T = e^{-\lambda_T t}$ $\qquad \lambda_T = 1.37 \times 10^{-5}/\text{day}$

$R_c = e^{-\lambda_c t}$ $\qquad \lambda_c = 1.1 \times 10^{-5}/\text{day}$

$$R_{cp} = R_{FH} = e^{-\lambda'_{cp}t} = e^{-2/3\lambda_{cp}t} = .999$$

$$\lambda_{cp} = \frac{3}{2}\frac{.0001}{730}$$

$$\lambda_{cp} \simeq 2 \times 10^{-6}/day$$

$$R_{MFP} = e^{-\lambda'_{mfp}t} = e^{-6/11\lambda_{mfp}t} = .997$$

$$\lambda_{mfp} \simeq 5 \times 10^{-6}/day$$

and

$$\lambda_s = 7.94 \times 10^{-5}/day = .029/yr$$

$$P_n(N) = \frac{(\lambda_s t)^n e^{-\lambda_s}}{N!} \qquad N = 1,2,3,\ldots$$

$$P_n(N) = \frac{(.029)^n e^{-.029}}{N!}$$

Example 4.4 Mixed System with Alternative Operating Conditions

Another application of the static type of model is the following problem. A redundant system consists of 6 parallel on-line components. Three of these components have a constant failure rate C_1, while the other three have a constant failure rate $C_2 = \frac{2}{3}C_1$.

Assuming the system requires either two of component 2 and one of component 1, or one of component 2 and 3 of component 1, find the total system's reliability and MTBF.

Here, we can proceed as follows. Assume that the six components are on-line until they fail and their failure rate is C_1 and $\frac{2}{3}C_1$, respectively, and that to operate no more than 2 of type 1 and 1 of type 2, may fail, or none of type 1 and 2 of type 2 may fail. Then given R_1 and R_2 are the component reliabilities of type 1 and type 2 components respectively, the system reliability is the sum of the probabilities that either of the two limiting conditions mentioned in the problem statement occurs.

Condition I

Probability of non-failure of at least two of component 2 and one of component 1.

$$R_I = [\binom{3}{2}R_2^2(1-R_2) + \binom{3}{3}R_2^3][\binom{3}{1}R_1(1-R_1)^2 +$$

$$\binom{3}{2}R_1^2(1-R_1) + \binom{3}{3}R_1^3]$$

Condition II

Probability of non-failure of at least one of component 2 and three of component 1.

$$R_{II} = [\binom{3}{1}R_2(1-R_2)^2 + \binom{3}{2}R_2^2(1-R_2) + \binom{3}{3}R_2^3]$$

$$[\binom{3}{3}R_1^3]$$

Total reliability of the system is

$$R_T = R_I + R_{II} - R(I\ II) = 9R_1R_2^2 - 9R_2^2R_1^2 + 6R_1^2R_2^3 - 6R_1R_2^3 -$$

$$R_1^3R_2^3 + 3R_1^3R_2 - 3R_1^3R_2^2$$

but

$$R_1 = e^{-C_1 t}$$

$$R_2 = e^{-2/3C_1 t}$$

and the total systems reliability is the sum of the reliability under these two possible conditions. Therefore

$$R_T = 9e^{-(C_1 + \frac{4}{3}C_1)t} - 9e^{-(2C_1 + \frac{4}{3}C_1)t} + 6e^{-(2C_1 + \frac{6}{3}C_1)t}$$

$$-6e^{-(C_1 + \frac{6}{3}C_1)t} - e^{-(3C_1 + \frac{6}{3}C_1)t} + 3e^{-(3C_1 + \frac{2}{3}C_1)t}$$

$$= 9e^{-\frac{7}{3}c_1 t} - 9e^{-\frac{10}{3}c_1 t} + 6e^{-\frac{12}{3}c_1 t} - 6e^{-\frac{9}{3}c_1 t} -$$

$$e^{-\frac{15}{3}c_1 t} + 3e^{-\frac{11}{3}c_1 t}$$

$$\text{MTBF} = \int_0^\infty R_T(t)\,dt = \frac{1}{c_1}[27/7 - 2.7 + 1.5 - 2 - 0.2 + 9/11]$$

$$= \frac{1.2753}{c_1}$$

Example 4.5: Series System Allocation

Taking a simple series system allocation problem, let us assume that three identical components are required in series, and that there are 10 of these components available. The problem is then how to arrange them to maximize the

a. system's reliability, and
b. system's MTBF.

Assuming any on-line arrangement of these 10 components is possible, we proceed as follows:

As low level redundancy has been shown to provide larger marginal increase in reliability:

$$R_s(t) = [1-(1-R_1(t))^{n_1}][1-(1-R_2(t))^{n_2}][1-(1-R_3(t))^{n_3}]$$

All components are equal,

$$R_s(t) = [1-(1-R(t))^{n_1}][1-(1-R(t))^{n_2}][1-(1-R(t))^{n_3}]$$

and $n_1 + n_2 + n_3 = 10$ let $(1-R(t)) = x$

a. $R_s(t) = [(1-x^{n_1})(1-x^{n_2})(1-x^{10-n_1-n_2})]$

For $R_s(t)$ max $\quad n_1 = n_2 = 3 \quad n_3 = 4$

or $\quad n_1 = n_3 = 3 \quad n_2 = 4$

or $\quad n_2 = n_3 = 3 \quad n_1 = 4$

b. $R_s(t)_{max} = [1-(1-R)^3]^2[1-(1-R)^4]$

$$= [1-x^3]^2[1-x^4] = [1-2x^3+x^6][1-x^4]$$

$$= [1-2x^3+x^6-x^4+2x^7-x^{10}]$$

Another component allocation problem could be as follows:

a. A system with 3 components in series is to attain a reliability of 90% in time t. If the third component has twice the failure rate of the second and the second twice the failure rate of the first, what must be their failure rates, if t=100 hours?

b. What is the mean time before failure of your system?

c. What is the probability of having 0, 1, and 2 failures in 100 hours then?

d. What failure rate of component 1, 2, and 3 would you require if you demanded 90% reliability for the duration of MTBF = L(0) (use same ratio of failure rates)?

e. If the redundancy is allowed for all 3 components and if each component cost is the same, how much and where would you impose redundancy if you required a reliability of 95% for a duration of 1000 hours?

Assuming constant failure rates of λ_1, λ_2, and λ_2 respectively, where $\lambda_3 = 2\lambda_2 = 2(2\lambda_1) = 4\lambda_1$,

a. Let $R_s = R_1 \cdot R_2 \cdot R_3$ = Reliability of the system

$$= e^{-\lambda_3 100} \, e^{-\lambda_2 100} \, e^{-\lambda_1 100} = e^{-7\lambda_1 100}$$

To obtain a 90% reliability

$$0.90 = e^{-700\lambda_1}$$

and ln 0.9 = $-700\lambda_1$ $\qquad \underline{\lambda_1} = 1.505 \times 10^{-4}$/hr

$$\lambda_2 = 3.01 \times 10^{-4}/hr$$

$$\lambda_3 = 6.02 \times 10^{-4}/hr$$

b. $\text{MTBF} = \int_0^\infty R_s(t)dt = \int_0^\infty e^{-7\lambda_1 t}dt = \frac{1}{7\lambda_1} = 949$ hours

c. Probability of 0 failures in 100 hours = 0.90 = $R^2(100)$.

Probabilty of 1 failure in 100 hours = P (1 fails and 2, 3 operate) + P (2 fails and 1, 3 operate) + P (3 fails and 1, 2 operate)

$= (1-R_1(100)R_2(100)R_3(100)) + (1-R_2(100))R_1(100)R_3(100) + (1-R_3(100))R_1(100)R_2(100)$

$= (.0149 + .9703 \times .9416) + (.0297 \times .9851 \times .9416)$
$+ (.0584 \times .9851 \times .9703)$
$= 0.0136 + 0.0275 + 0.0558 = .097$

Probability of 2 failures in 100 hours =

P (1 and 2 fails, 3 operates) +

P (2 and 3 fails, 1 operates) +

P (1 and 3 fails, 2 operates)

$= (0.0149 \times 0.0297 \times 0.9416) + (0.0149 \times 0.0584 \times 0.9703) + (0.0397 \times 0.0584 \times 0.9)$

$= 0.00297$

Therefore, probability of series system failing as a result of more than one component failing is negligible. This probability is a second order effect and therefore usually ignored.

d. If t-=949 hours, then for a reliability of 90% log $0.9 = -7\lambda_1 t = -7 \times 949 \times \lambda_1$. The required failure rates are:

$\lambda_1 = 1.585 \times 10^{-5}$ failures/hours

$\lambda_2 = 3.170 \times 10^{-5}$ failures/hours

$\lambda_3 = 6.340 \times 10^{-5}$ failures/hours

e. Considering that low level redundancy offers a higher marginal increase in reliability a redundancy of n_1, n_2, and n_3 is considered for the

3 series components. Assume component failure rates as computed in a.

$$R_2(1000) = 0.95 \leq [1-(1-R_1)^{n_1}][1-(1-R_2)^{n_2}]$$
$$(1-(1-R_3)^{n_3})$$

The problem is to satisfy the above weak inequality subject to minimizing the sum $n_1 + n_2 + n_3 = n$ or

$$0.95 \leq [1-(1-e^{-0.151})^{n_1}] \cdot (1-(1-r^{-0.302})^{n_2}$$
$$(1-(1-e^{-0.604})^{n_3}]$$

$$= (1-0.1401)^{n_1}(1-0.261)^{n_2}(1-0.452)^{n_3}$$

This problem can be solved by

1. use of Lagrangian Multipliers
2. Marginal Analysis
3. Trial and Error Analysis

The simplest approach in this discrete analysis is the use of marginal or trial and error analysis:

n_i	$1-(1-R_1)^{n_i}$	$1-(1-R_2)^{n_i}$	$1-(1-R_3)^{n_i}$
1	0.8599	0.7390	0.5480
2	0.9803	0.9319	0.7957
3	0.9972	0.9823	0.9076
4	-	0.9953	0.9582
5	-	-	0.9811
6	-	-	0.9914

These results indicate that we need at least

$n_1 \geq 2$
$n_2 \geq 3$
$n_3 \geq 4$

For the above combination $R_s = R_1(2) \, R_2(3) \, R_3(4)$
$$= 0.9226$$

As the marginal increase in systems reliability is largest for a unit increase in redundancy of component 3,

$$R_s = R_1(2)\ R_2(3)\ R_3(5) = .9446$$

and finally

$$R_s = R_1(2) R_2(3) R_3(6) = 0.9546$$

Increasing component 1 by one unit gives

$$R_s = R_1(2) R_2(3) R_3(6) = 0.9609$$

Therefore a total of 11 components are required and the best allocation is

$n_1 = 3$
$n_2 = 3$
$n_3 = 5$

4.5 Off-Line Redundant Systems

In on-line redundant systems, such as those discussed above, it has been assumed that all components are constantly on-line even when not needed for systems operation. Consequently, all non-failed components are assumed to be subject to failure independent of the fact they may or may not be required for service.

Another approach may be to assume that only the required number of parallel components are actuall on-line with all redundant components disconnected and therefore not subject to failure until switched on. Redundant off-line components will be brought on-line as on-line components fail. If we assume that this switching of an off-line component to on-line is instantaneous and occurs with perfect reliability, then system reliability for components with exponential failure density may be derived as follows.

Consider initially a two-component redundant system with only one component required for operation and the standby redundant component off-line, as shown in Figure 4.11.

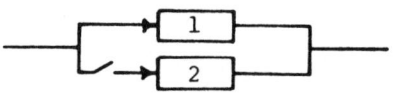

FIGURE 4.11: TWO COMPONENT REDUNDANT SYSTEM

Then, system failure may be expressed as:

$$P_r(\text{Failure of system}) = \int_0^t P_r(\text{1st component failing at time } \tau) \times P_r(\text{2nd component failing prior to time } t-\tau) \times d\tau$$

$$F_T(t) = \int_0^t \phi(\tau) F(t-\tau) d\tau$$

$$= \int_0^t \rho e^{-\rho\tau}(1-e^{-\rho(t-\tau)}) d\tau \quad \text{(Assuming exponential failure density and identical components)}$$

$$= \rho \int_0^t e^{-\rho\tau} d\tau - \rho \int_0^t e^{-\rho\tau} d\tau$$

$$= 1 - e^{-\rho t} - \rho t\, e^{-\rho t}$$

$$P_r(\text{System Failure}) = F_T(t) = 1 - e^{-\rho t}(1+\rho t)$$

$$R_T(t) = 1 - F_T(t) = 1 - (1 - e^{-\rho t}(1+\rho t)) = e^{-\rho t}(1+\rho t) \quad (4.18)$$

By a similar argument (4.18) can be extended for an n-component parallel system for which only one component is required on line. The appropriate expression for the reliability of the system made up of identical components with constant age-dependent failure rates (exponential failure densities) will be:

$$R_T(t) = e^{-\rho t} \sum_{i=0}^{n-1} (\rho t)^i / (i)! \quad (4.19)$$

Example 4.6

Let us assume that there are two identical components which can be put either into on-line or off-line standby redundancy with perfect switching.

If we consider the single on-line component only with the second component in off-line standby, then

$$R(t) = e^{-ct}$$

and

$$L(0) = \int_0^\infty R(t) dt = \int_0^\infty e^{-ct} dt = \frac{1}{c}$$

Together the on-line and standby components comprise the following event space

	System Works
On-line works $t_1 > t$	Standby off-line	Yes
On-line doesn't work $t_1 < t$	Standby works $t_2 > t$	Yes
–	Standby doesn't work	No

Therefore

$$R(t) = 1 - \int_0^t c e^{-ct}(1-e^{-c(t-z)})dz = e^{-ct}(1+ct)$$

and

$$L(o) = \int_0^\infty e^{-ct}(1+ct)dt = \frac{1}{c} + \frac{1}{c} = \frac{2}{c}$$

Considering next the two components with on-line standby we obtain

$$R(t) = 1-(1-e^{-ct})^2 = 2e^{-ct} - e^{2ct}$$

and

$$L(0) = \int_0^\infty (2e^{-ct} - e^{2ct})dt = \frac{2}{c} - \frac{s}{2c} = \frac{3}{2c}$$

Comparison of these two alternative arrangements of two components

$$\frac{R_T(\text{component redundancy})}{R_T(\text{standby})} = \frac{2e^{-ct} - e^{-2ct}}{e^{-ct}(s+ct)} = \frac{2-e^{-ct}}{1+ct}$$

for a time $t = \frac{1}{c}$ it becomes

$$\frac{R_T(s \cdot b)}{R(c \cdot r)} = \frac{2}{2-e^{-1}} = 1.23 \rightarrow \text{an improvement of } 23\%$$

$$\frac{L(0)(\text{standby})}{L_c(s \cdot \text{red} \cdot)} = \frac{2/c}{3/2c} = \frac{4}{3} = 1.33$$

Hence, standby redundancy has a 33% improvement in MTBF.

4.6 Exercises

1. Three identical components are required in series. Assuming you have 10 of these components available, how would you arrange them to maximize the system's reliability and the system's MTBF. Assume any on-line arrangement of these 10 components is possible.

2. a. A 3-component series system has complete on-line redundancy. If the age dependent failure rates are equal for all components ($\lambda = 0.01$/hour), what is the probability that the system is still operative after 10 hours? (λ is independent of the number of components operating and initially all components are operating.)

b. What is the probability that the system is still operative after 10 hours if the redundant series is off-line with instantaneous 100% reliable switching ($\lambda = 0$ while components off-line)?

c. What is the MTBF of the two systems?

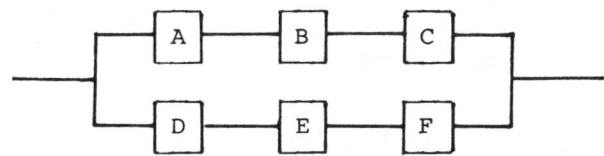

3. A non-maintained on-line series system with parallel path requires either path 1 or path 2 and path 3 or all paths to operate

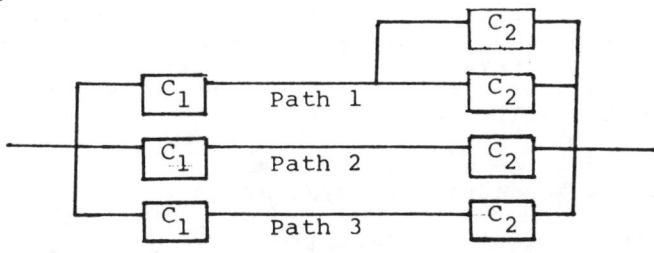

a. What is the reliability of the system?

b. If the age dependent failure rates of the components of the remaining path(s) doubles if any path becomes inoperative, find the reliability of the system now.

4. a. A system with 3 components in series is to attain a reliability of 90% in time t. If the 3rd component has twice the failure rate of the 2nd and the 2nd twice the failure rate of the first, what must be their failure rates, if t=100 hours?

b. What is the mean time before failure of your system?

c. What is the probability of having 0, 1, or 2 failures in 100 hours then?

d. What failure rate of component 1, 2, and 3 would you require if you demanded 90% reliability for duration of MTBF = L(0) (use same ratio of failure rates)?

5. A parallel system consists of two independent on-line components whose time to failure distributions is exponentially distributed age dependent failure rates λ_1 and λ_2 respectively where $\lambda_1 \neq \lambda_2$. Does the time to failure for the combined total system have positive aging (increasing failure rate)?

6. Five identical components, each with a constant age dependent failure rate λ, are available. For operation the design of the system requires three of the components to operate in series. What is the best arrangement of the remaining two components to provide the highest MTBF for the system in terms of maintaining three components working in series?

7. A power plant requires a pumping capacity of 1000 gallons/minute, with a reliability of 99.6% for the time between scheduled overhauls. The best 1000 gallon/minute pump available has a reliability of 99.4% and costs $10,000. Another make is available which offers a 1000 gallon/minute pump with a reliability of 99% at a cost of $6,000 and a pump with a 500 gallon/minute capacity with a reliability of 99.2% at a cost of $4,000. If your objective is to provide the lowest cost arrangement of pumps which will provide 1000 gallons/minute with a reliability of 99.6%, which pumps would you buy and how will you arrange them

    a. with an off-line system
    b. with an on-line system?

Assume perfect (100% reliability) switching from pump to pump. Does the off-line or on-line system provide the cheaper solution, and do the respective cheapest solutions have the same arrangement.

8. Three components in a parallel off-line system, each with its own switch and sensor, comprise a system. If the failure of the switches and components is completely independent, what is the reliability of the system and its MTBF if:

    a. one component only is required for the system to operate?

b. two components are required for the system to operate?

Assume that all components and their switches have the same constant age dependent failure rate λ and λ_s respectively. If you needed only a central switch for transfer, would the reliability and MTBF be improved?

9. Given you are to design a series system of three components for which you have a total amount of 16. You need at least one of each component. The component cost and reliability is as follows:

Component	Reliability	Cost
1	0.95	1
2	0.80	2
3	0.75	3

Determine the allocation of your investment which will give you the highest reliability of the three component series system for your investment of 16 units. What is the redundancy of components 1, 2, and 3?

4.7 References

1. Lloyd, D.K. and Lipow, M., "Reliability: Management, Methods, and Mathematics", Prentice Hall, Inc., Englewood Cliffs, New Jersey, 1962.

2. Tsokos, C.P. and Shimi, I.N. (editors), "The Theory and Application of Reliability", Academic Press Inc., New York, 1977.

3. Kapur, K.C. and Lamberson, L.R., "Reliability in Engineering Design", John Wiley and Sons, Inc., New York, 1977.

4. Goldman, A.S. and Slattery, T.B., "Maintainability", John Wiley and Sons, Inc., New York, 1964.

5. Zelen, M. (editor), "Statistical Theory of Reliability", The University of Wisconsin Press, Madison, Wisconsin, 1964.

6. Bourne, A.J. and Green, A.E., "Reliability Technology", Wiley Interscience, Newe York, 1972.

7. Smith, C.O., "Introduction to Reliability in Design", McGraw-Hill, New York, 1976.

8. Shooman, M.L., "Probabilistic Reliability: An Engineering Approach", McGraw-Hill, New York, 1968.

5.0 FAILURE MODE AND EFFECTS ANALYSIS - FAULT TREE ANALYSIS

Ernst G. Frankel

The purpose of failure mode and effects analysis is to identify the different failures and modes of failure that can occur at the component, subsystem, and system level and to evaluate the direct and consequential effects of these failures. It involves a formal analysis to determine the effect of subsystem, component, or part failure on system performance or the ability to meet performance requirements or objectives. Such an analysis is usually performed upstream during the conceptual or development phases of a system, to assure that all possible modes of failure have been considered and that the proper design and/or operating provisions have been incorporated to eliminate the potential or cause for the failure or that the magnitude and effect of the failure mode have been reduced to an acceptable level.

Logical tree-type networks which relate the various failure events, both causal and consequential, are called fault trees. Fault trees use simple relationships of AND and OR to allow effective representation of the actual fault relationships of the system under study.

To develop a fault tree of a system, we must understand in detail how the system works and what faults could possibly occur. For this purpose flow graphs and system logic diagrams are usually used to show the relationships of all the components and events of states of the system. An important tool used is a "Failure Mode and Effects Analysis" (FMEA) in which each component or subsystem is analyzed to determine all the failure modes that could occur to it and identify their effects on other components or the system as a whole. Expected failure frequencies, consequential failure of effects, methods of detecting the failure are also determined and listed. Finally we identify correcting, adjusting, or other factors which can be expected to be available or be introduced to correct the originating failure or the consequential events. All the above information is entered into a form, which must be continuously updated by induction and as the system develops as well as when more information on failure modes, effects, and corrective approaches become available.

5.1 Common Cause Failure

Events that cause failure in several components or a total system are called common cause, and the resulting failures are denoted common cause or common mode failures. There are a number of possible methods for analyzing common cause failures.

In the simplest case, components of a system may be subject to failure as a result of shock, overheating, abnormal stress, or other extreme event causing the components of the system to react to such common cause.

Often the occurrence of the common cause, if discrete, can be represented by a Poisson process. If n is the number of occurrences of the common cause event in a time period t of interest, then the probability of exactly n occurrences in time 0 to t is

$$P(n,t) = \frac{(\lambda_1 t)^n e^{-\lambda_1 t}}{n!}$$

where

$\lambda_1 dt_1$ = probability of common cause event occurring in the time interval t to t+dt given it has not occurred before t

Assuming a single occurrence of the common cause event results in the failure of the components and therefore system, the probability of non-failure of the system from the common cause event is then

$$R_c(t) = e^{-\lambda_1 t}$$

and the total system reliability would be the product of the reliability of the system and the common cause probability of non-failure.

Marshal and Olkin (Ref. 1) developed a Markov model which considers common cause failures. Failures are caused by independent or common causes but all are assumed to be represented by exponential distributions for the time to first failure. Failure events subject to common and independent failures are often described in terms of ratio of the age dependent common cause failure rate over the sum of the age dependent common cause and the independent failure rate. This ratio is often called the beta factor where

$$B = \frac{\lambda_c}{\lambda_c + \lambda}$$

λ_c = age dependent common cause failure rate = common cause hazard function

λ = age dependent random failure rate = hazard function

Common cause failure rates usually affect all states of a system and result in a degradation to non-operating state from any operating state.

5.2 Complex System Reliability Networks

Complex systems made up of interfacing components each with a reliability $R_i(t)$ can usually be represented by a network of links where each link represents a component. Consider a system as shown in Figure 5.1, composed of 8 components, and assume that the system will operate if any sequence or series of components from 1 to 8 is operable.

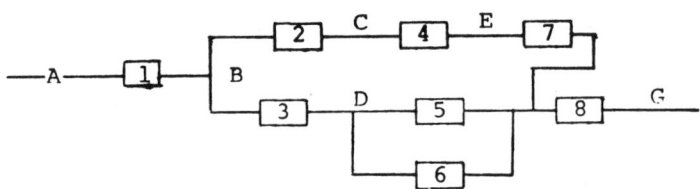

FIGURE 5.1: COMPLEX SYSTEM

We can then represent such a system by a network as shown in Figure 5.2, in which the system will operate now if

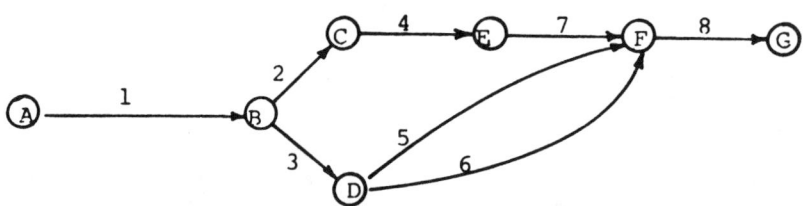

FIGURE 5.2: NETWORK REPRESENTATION

there is a flow from A, the source, to G, the sink of the system of directed graphs. We could say that a link has a capacity of 1 if it is operable and a capacity of 0 if it is not operable. Therefore if X_i designates the capacity of link i then $P(X_i=1) = p_i$ is the probability that $X_i=1$ and

$P(X_i=0) = 1-p_i=q_i$ is the probability that the network does not operate successfully. Each realization of a path from A to G is disjoint from any other realization of a path from A to G. For example, a path A to B to C to E to F to G with link BD inoperative has a probability of

$$P\{X_1=1, X_2=1, X_3=0, X_4=1, X_5=0, X_6=0, X_7=1, X_8=1\} \text{ and}$$
is equal to $P_1 P_2 (1-P_3) P_4 (1-P_5)(1-P_6) P_7 P_8$

If all components have the same probability of successful operations this probability will be $p^5(1-p)^3$. If p is the probability of success of any link p_i (i=1,2,3,...8) then the total probability of success of the network is simply the sum of all the path probabilities which contain at least one successful path from A to G. There are three possible paths from A to G (A,B,C,E,G,G,-A,B,D,F,G,-A,B,D,F,G, say Path I, Path II, and Path III). Breakthrough is obtained if any single or combination of path is operable. Both paths can be operable with some or all links not constructing the path not operating.

It is rather obvious that this implies the evaluation of many paths. Instead of this approach, it is advantageous to consider the problem as one of having at least one operating path which can be expressed as the probability of one or more paths operating (reliability of the system), ADP and in turn as the probability that the sum of the path probabilities is larger or equal to one. This can also be described as the probability of the minimum flow from source to sink which must be equal to or larger than one. If X_i is the flow in link i ($X_i = 0,1$) then we can define minimal path as a minimal set of links which, if operable, assures operation of the system. The minimal paths of our system are:

I   : X_1 X_2 X_4 X_7 X_8

II : X_1 X_3 X_5 X_8

III : X_1 X_3 X_6 X_8

Similarly a minimal cut is one or more cuts of links whose failure causes the system to fail. The minimal cuts are therefore:

1. X_1

2. X_2 X_3

3. X_2 X_5 X_6
4. X_3 X_4
5. X_3 X_7
6. X_4 X_5 X_6
7. X_5 X_6 X_7
8. X_8

Using the algebra of sets the system can be shown to operate if the union of the probabilities of the minimum paths is equal or larger than one.

System Reliability = $\underline{P}(X_1 X_2 X_4 X_7 X_8 = 1)$
$+ \underline{P}(X_1 X_3 X_5 X_8 = 1) + \underline{P}(X_1 X_3 X_6 X_8 = 1)$
$- \underline{P}(X_1 X_2 X_3 X_4 X_5 X_7 X_8 = 1) - \underline{P}(X_1 X_2 X_3 X_6 X_7 X_8 = 1)$
$- \underline{P}(X_1 X_3 X_5 X_6 X_8) + \underline{P}(X_1 X_2 X_3 X_4 X_5 X_6 X_7 X_8 = 1)$
$= p_1 p_2 p_4 p_7 p_8 + p_1 p_3 p_5 p_8 + p_1 p_3 p_6 p_8$
$- p_1 p_2 p_3 p_4 p_5 p_7 p_8 - p_1 p_2 p_3 p_4 p_6 p_7 p_8$
$- p_1 p_3 p_5 p_6 p_8 + p_1 p_2 p_3 p_4 p_5 p_6 p_7 p_8$

and if $p_i = p$ for all i then

System Reliability = $2p^4 - 2p^7 + p^8$
$= p^4 [2 - 2p^3 + p^4]$

The number of terms in the reliability equation is equal to $m = 2^n - 1$ where n is the number of distinct paths. In our case n=3 and m=7. Another approach which is sometimes more efficient is the use of minimal cuts. The system will operate if at least one of the links in each cut operates. Cuts operate a series system and the cuts as listed for a system can therefore be structured as a series (product) of series of cuts in which at least one component operates.

There are many methods for the computation of the exact reliability of a system. Most of the combinatorial or network methods discussed though become quite complex when a multi-component system of some size is considered. For this reason it is often useful to apply methods such as upper and lower bounds.

Using the well known result for the probabilities of functions of binary variables, and given $x_1, x_2, x_3, \ldots x_n$ are independent binary variables which assume values 1 or 0 and then
$$Z_i = \prod_{k \in n}$$

$$P(Z_1=0, Z_2=0 \ldots Z_{n}=0) \geq P(Z_1=0) P(Z_2=0) \ldots P(Z_n=0)$$

Using this result we could now develop an upper and lower bound for our reliability problem. Such bounding facilitates solution of complex reliability problems.

5.3 Fault Tree Analysis

Fault tree analysis is a method used to relate the occurrence and sequence of events that act together and/or in a chain to cause other events and finally faults or failures. While the purpose of failure mode and effects analysis is to identify the different failure modes and effects which could occur, fault tree analysis is designed to provide the structure whereby simple logical relationships are used to establish the probabilistic relationships among all the different other events and ultimately faults or failures. Fault tree analysis is usually preceded by a 'Failure Mode and Effects' analysis in which the design, method of operation, and environment in which the system works are evaluated and the cause and effect relationships leading to faults or failures are identified. This is an essential step in the understanding of the system without which a formal fault tree analysis and probabilistic risk assessment cannot be performed. To build a fault tree, we must know how the system functions. The functioning of a system is usually described diagramatically by a function or flow diagram in which the flow of information, signals, materials, services, or other transmissions is shown. Such a diagram may for example show the flow of all the various inputs of a power plant such as fuel from receipt to the emission from the smokestack, cooling water from the intake to the final discharge and similarly for all other flows in such a system. This is then amplified to include the various functional sequences from input to output. A logic diagram is the next step. Here we translate the functional relationships into logical relationships among the component parts of the system. After these steps have been undertaken and the systems operation in terms of all the cause and effect relationships is clear, the building of a fault tree can commence.

In fault tree analysis we usually start with the identification of the top events, usually the set of most severe, ultimate, or final failure events that could occur. Next, events directly contributing to the top events are identified and connected to the top events by logical links. The process continues until we reach the lowest or most basic event. The connection of events is performed using logical gates which indicate the relationships among events or their respective contributions. The principal gates used are AND and OR logical connectors or switches. Events, for the purpose of fault tree analysis, are usually divided into independent and dependent events. An example of dependent basic events are 'common cause failure events'. In the case where 'common cause failures' result in consequential failures at the next or subsequent event level, these consequential failures are no longer independent.

The purpose of fault tree analysis, therefore, is to construct a fault tree structure which relates top events to basic events, and which can be used to perform a quantitative analysis of the failure sequences. This is usually done by reducing a fault tree structure to a logically equivalent form which can be evaluated by the use of 'minimal cut set' theory. To obtain actual estimates of the probabilities of the occurrence of the top events from the probabilities of the basic events, equivalent conditional probabilities are associated with all the links of the fault tree structure. The basic logic of fault tree analysis is based, as mentioned before, on two types of gates which can be expressed in probability terms as:

$$\text{AND } P(A_1, A_2, \ldots A_n) = \prod_{i=1}^{n} P(A_i)$$

$$\text{OR } P(A_1 + A_2 + \ldots A_n) = \sum_{i=1}^{n} P(A_i)$$

The above formulations are used to determine the relations between different adjacent levels of events.

To construct a fault tree we have to develop an understanding of the logical structure of fault sequences. In other words we must identify interacting events which in turn produce other events by use of simple logical relationships. The first step is always the determination of the top failure event. All other events are assumed to contribute to the occurrence of the top event. The top event is usually a major system or component failure. Going to the next level of events we introduce gates as shown in Figure 5.3 which indicate if the next level of events cause jointly (AND) or individually (OR) the occurrence of the top event. Fault trees are used in other qualitative or quantitative analysis. Qualitative analysis would be used to help improve a systems performance in a relative sense while quantitative analysis would be designed to quantify system or component failure rates. Failures can usually be defined as:

1. primary failures
2. secondary failures
3. command failures
4. common cause failures

and contributing failure events causing top failure events can be the result of any of these possibilities. A primary failure is usually defined as the failure or non-operating state of a component, which occurs under normal operating conditions, including the effects of natural environmental conditions and aging. A secondary failure is similar to a primary failure except that the component is not directly responsible for the failure. It may be caused by past or present excessive stresses to which the component was subjected or by stresses caused by 'out-of-tolerance' conditions. Command faults are defined as failure events causing non-working states due to improper controls. Most command failures are temporary and do not require repairs to correct the condition.

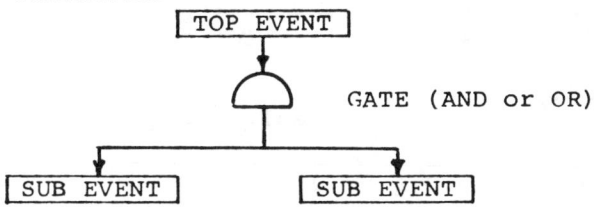

FIGURE 5.3: ELEMENTARY FAULT TREE

In summary the failure characteristics can be differentiated, as in the following table:

Failure Characteristics	Caused By	Sample Origins
Primary Failure	Random or Natural Causes	Natural environment and aging
Secondary Failure	Excess Stress	Interaction with other components
Command Failure	Incorrect Control	Inadvertant effects
Common Cause Failure	Common Events	Major effects

Once the structure of a fault tree is established, probabilities must be assigned to each fault. Combining these probabilities using the logic gates described before, we obtain the probabilities of higher, and finally the top, event. In addition to the gradation or character of the failures described above, failure events must also be defined by their occurrence and existence. This is quite important as the differentiation affects the quantification of the probability of the top event.

Results from fault tree analysis in qualitative form indicate the combinations of failure events of components or subsystems which cause system failure. They also permit a qualitative ranking of the contribution of component or subsystem failures to system failure. Finally, these results indicate the susceptibility of systems failure to single failure causes. When quantitative results are obtained, they usually provide the probabilities of system failure in a structured quantitative cause and effect relationship, as well as the quantitative ranking of the contributions of component or subsystems failures to the system failure.

Let us next evaluate a fault tree. We first structure the tree and then describe it by a set of Boolean or theoretic algebraic equations. A separate equation is required for each gate which, as described before, expresses the relations of an event to next lower or causal events. Consider the problem shown in Figure 5.4 in which the following standard symbols are used:

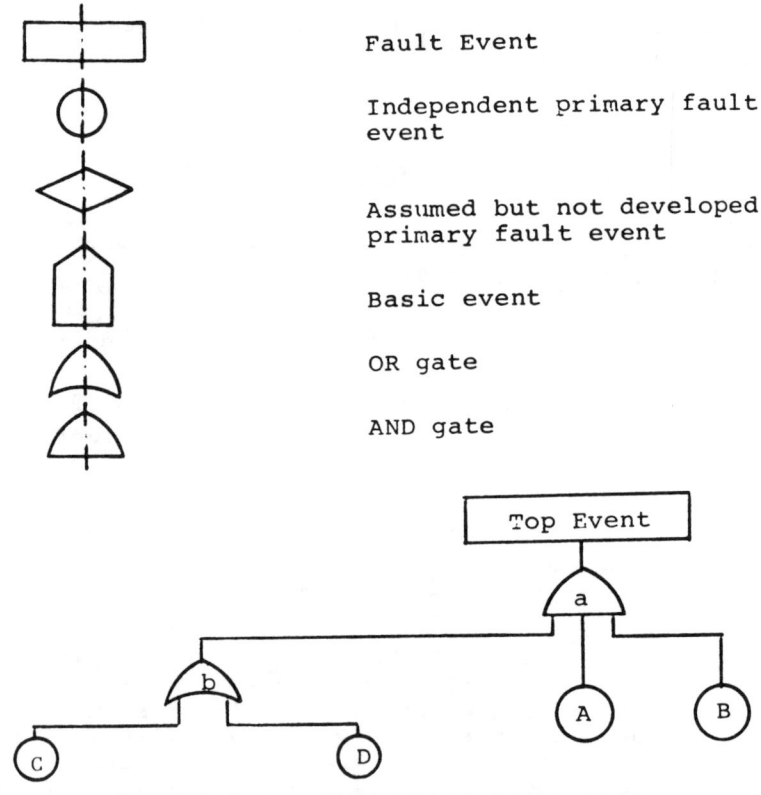

FIGURE 5.4: ELEMENTARY FAULT TREE

There are additional symbols to describe relations with secondary, command, and common cause events. Similarly some symbols (triangle) are used to indicate continuation of the tree. The algebraic equations for this tree can be written as

a = ABb   b = C+D

a = AB(C+D) = ABC+ABD

Let us next consider an example.

Example 5.1

A ship has a hydraulic steering engine with a hydraulic ram supplied by a pump driven by its own motor. This is obviously a grossly simplified example given to describe the process of developing a fault tree structure. Note that the structure is not carried down to the basic events.

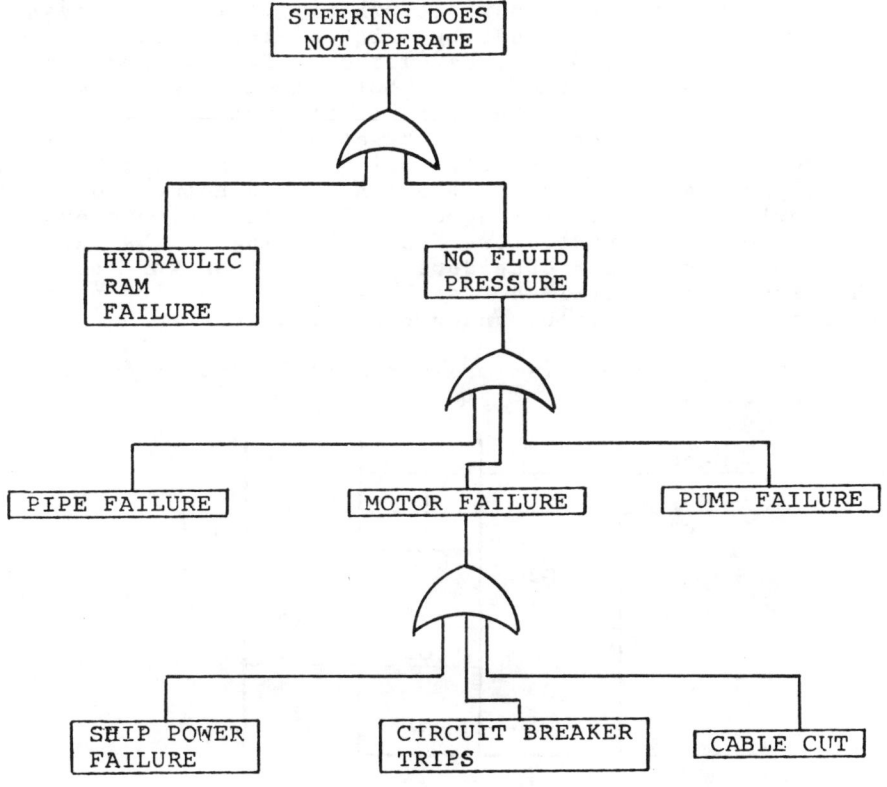

FIGURE 5.5: EXAMPLE

Here the steering will not operate if the hydraulic ram fails OR the hydraulic fluid pressure drops. The hydraulic pressure in turn would drop below acceptable working pressures as a result of pipe rupture, motor failure, or hydraulic pump failure. The motor in turn would fail to operate if there is a ship power failure, a circuit breaker trip, or a cut in the electric cable.

In most systems we have many levels of failure events, often designated as top event or primary failure, second level events or secondary failures, and so on. Secondary and lower level failures occur due to command failures, random failures, environmental effects, inadequate maintenance and repair, and more. Command failures are usually considered dependent fault events and are therefore represented by triangles, while basic failures caused by independent events such as environmental impacts are represented by circles.

5.3.1 Min Cut Sets of Fault Trees

A min cut set is the smallest unreducible collection of basic events required to insure occurrence of the top event. There are a number of methods by which a minimal cut set can be determined. The simplest, developed by Barlow and others (Ref. 9-11), can be used for both manual and computer analysis of fault trees. The basis of the method's rationale is that AND gates increase and OR gates decrease the number of cut sets, and that a min cut set can be developed efficiently by a hierarchial algorithm. The method consists of a stagewise vertical listing of all basic failure events and gates of the next lower state leading to the failure event at the stage under consideration. We then work our way down hierarchially until we reach all the basic or primary faulty events of the system under consideration.

Considering the problem presented in Figure 5.6, we can proceed as follows.

Step	1	2	3
	1	1	1
		2	2
	G2		
		G4	3,4
		G5	5,6
	G3		
		G6	7 8

The first step was to list vertically all the basic events and gates leading to the first level OR gate G1. In

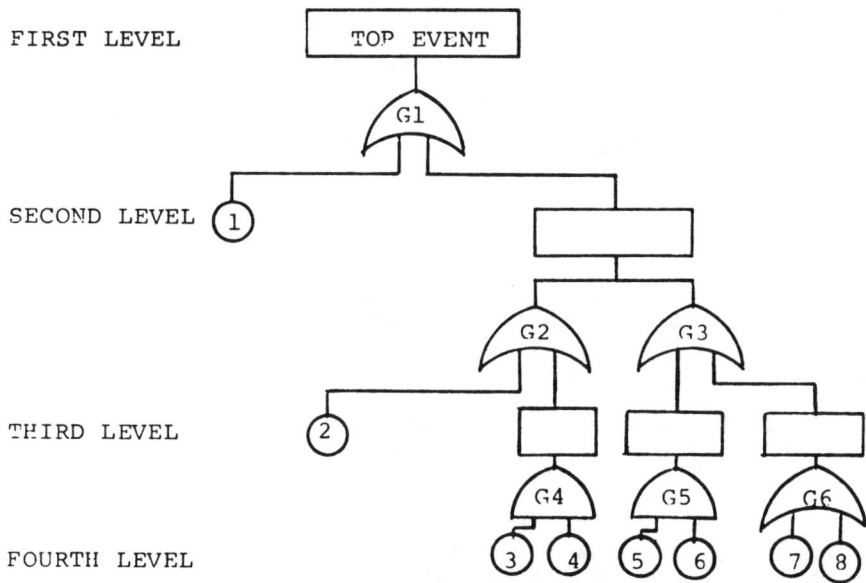

FIGURE 5.6: HIERARCHIAL FAULT MIN CUT SET ANALYSIS

the second step we expand second level gates G2 and G3 by vertically listing all basic events or third level gates leading to them. Whenever an AND gate is met, basic events and gates leading to it are listed horizontally, as shown in the above table. Essentially what we have done is to sequentially or hierarchially replace each gate by its input basic events or gates until all fault tree gates are replaced with the basic event entries.

The final min cut set listed under step 3 consists of events 1, 2, 7, and 8 and joint events 3, 4, and 5, 6. If in a cut set in which all gates have been eliminated a basic event is both a member of the cut set as a single event as well as part of joint events, then we would eliminate the joint event from the cut set to derive the min cut set. In other words, a single event is a "more" basic event than a joint event made up of such single events. The above hierarchial procedure to derive a min cut set can readily be programmed for computer solution.

5.4 Exercises

1. Assume a system can be represented by the following network and that each of the five components is independent and has a probability p_i that it will not fail before time t.

If $p_i = 0.95$ what are

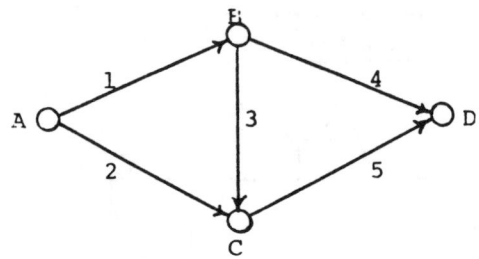

  a. the minimal paths and cuts (list all of them)?
  b. the reliability of the system?
  c. the upper and lower bounds of the reliability of the system?

2. If the component BC could flow in either direction (see figure above), what is the reliability of the system now?

3. Construct a fault tree for a student preparing for an examination, in which the top event is failure in the examination. There should be at least 3 levels of events. Try to identify the gates and give a reasonable estimate of the probabilities.

4. Develop a fault tree for an airline passenger trying to catch a plane where the top event is "missing the plane". Assume the passenger departs from home by private cate.

5. Given a fault tree has an OR gate, G1, at the first level fed by a basic event 1 and an AND gate, G2, which in turn is fed by two OR gates, G3 and G4, at the third level with inputs of basic events 2 and 3, as well as 4 and 5. What is the minimal cut of this fault tree?

5.5 References

1. Marshall, A.W. and Olkin, A. "A Multivariate Exponential Distribution", Journal of American Statistical Association 62, 1967.

2. Apostolakis, G.E. "The Effect of a Certain Class of Potential Common Mode Failures on the Reliability of Redundant Systems", Nuclear Engineering Design 36, 1976.

3. Vesely, W.E., Goldberg, F.F., Roberts, N.H., and Haasl, D.F. "Fault Tree Handbook", Nuclear Regulatory Commission Report, NUREG-0492, 1981.

4. Jordan, W.E. "Failure Modes, Effects, and Criticality Analysis", Proceedings - Annual Reliability and Maintainability Symposium, San Francisco, California, Institute of Electrical and Electronics Engineers, 1972.

5. Lambert, H.E., "Systems Safety Analysis and Fault Tree Analysis", Lawrence Livermore Laboratory Report, VCID-16238, 1973.

6. Lieberman, G.J. "The Status and Impact of Reliability Theory", Academic Press, New York, 1969.

7. Ross, S. "Introduction to Probability Models", Naval Research Logistics Quarterly, 16(1), 1969.

8. Gnedenko, B.V., Yu, B.V., Belyagev, K., and Solovyev, A.D. "Mathematical Methods of Reliability Theory", Academic Press, New York, 1969.

9. Barlow, R.E. and Proschan, F. "Statistical Theory of Reliability and Life Testing - Probability Models", Rinehart and Winston, New York, 1975.

10. Barlow, R.E. Fussell, J.B., and Singpurwalla, N.D., "Reliability and Fault Tree Analysis", SIAM, Philadelphia, 1975.

11. Barlow, R.E. and Lambert, H.E. "Introduction to Fault Tree Analysis", SIAM, Philadelphia, 1975.

12. Bazovsky, I. "Fault Tree, Block Diagrams, and Markov Graphs", Proceedings of the Annual Reliability, Maintainability Symposium, IEEE, New York, 1975.

13. Dhillon, B. and Singh, C., "Bibliography of Literature on Fault Trees", Microelectronic Reliability 17, 1978.

APPENDIX 5A PERFORMANCE OF A FAILURE MODE AND EFFECTS ANALYSIS

In order to maintain a uniform approach in the subsystem and component analyses, a formal approach is usually followed, and each component, its function, and potential failure mode is listed. We usually start with the identification of each component and its function followed by other relevant information such as:

1. Component Identification

This contains the identification of the component, major assembly, subassembly, or part being considered.

2. Function

This contains a concise statement of the component's intended purpose in the system or subsystem.

3. Failure Mode and Failure Cause

This contains the physical or operational description of the manner in which a failure of the component occurs, and the operating condition of the equipment or part at the time of failure. This should also contain a short statement as to the probable cause(s) of failures. These causes include chance, wear-out, maintenance induced, over-stressed, etc.

4. Failure Mode Frequency

Here we provide a quantitative estimate on the frequency of each failure mode as described. This estimate may be expressed in failures per 10^3, 10^4, or even 10^6 hours. In a preliminary analysis a qualitative estimate for the failure mode frequency may be used. This can be high, low, infrequent, etc.

5. Method of Detection

Here we describe the way(s) in which a failure would be discovered.

6. Corrective Measures

a. Here a short statement as to the corrective maintenance required to restore the system to an operative condition is presented. In some instances a design or a specification change may be desired to correct a potential problem area.

b. This is followed by an estimate of the total time required to effect the corrective maintenance and includes fault isolation time, restoration time, and system verification time. This elapsed time should exclude delay times.

7. Failure Mode Effects on System

a. Equipment and Subsystem Performance

Here we provide an explanation of the failure mode(s) effects on the equipment and/or subsystem performance. It contains an explanation of the alteration of the equipment and subsystem performance or reduction of performance.

b. Mission

An explanation of the failure mode(s) effects on the mission performance. This should be a statement of the impact on the ability to complete the mission or mission phase (delays, aborts, etc.).

c. Safety

This contains a statement as to whether the failure mode(s) causes an unsafe condition to exist from the point of view of the operators, passengers, people in the vicinity of the system or people affected by the system.

d. Mission Delay in Hours

This contains the amount of time the system will be behind schedule per failure incident or its consequence on percentage of available performance.

8. Failure Class

Here we state whether the failure is critical, major, or minor. To standardize on the Failure Class only three categories as defined are usually used. If the failure causes a severe safety problem, it automatically falls into the critical classification.

1. Critical Failure

A failure is considered critical if the failure results in an unsafe condition or safety hazard, causes the mission to be aborted, or causes excessive maintenance actions.

2. Major Failure

A failure is considered major if the failure materially degrades system performance or the accomplishment of the system's mission.

3. Minor Failure

A failure is considered minor if the failure has no significant effect on performance or accomplishment of the mission.

9. Preventative Measures Required in System Design and/or Maintenance

If possible a brief description of the preventative maintenance required to minimize the occurrence of the failure mode(s) should be provided. In some cases a description should be made as to the design, maintenance, or procedural change which is required to minimize the future occurrence of failure or to lessen the impact of failure.

The above categories are usually presented in tabular form and each component and subsystem evaluated in a hierarchial manner. We may, for example, consider failure modes and effects of each component of a pump, and after that is done, consider failure modes and effects of subassemblies which make up the pump. Finally, a failure mode and effects analysis of the complete pump is made considering that the pump is still a subsystem of power or other plant composed of many pieces of machinery.

APPENDIX 5B PERFORMANCE OF A MAINTAINABILITY ENGINEERING ANALYSIS (MEA)

The MEA should be performed as part of the design stage of a system and updated iteratively as the design proceeds to final construction or manufacturing plans. The MEA analysis progressively lists all the maintenance requirements for repair, replacement, adjustment and inspection, personnel and training, tools and support equipment, spare parts, technical publications, and maintenance facilities required for each part, subsystem, or the system as a whole. A MEA also serves as a framework for the determination of operating and maintenance costs. We usually start with the preparation of a listing of maintenance requirements for each system and component. This involves:

1. recording of relevant design data;

2. description of potential failure events of components, assemblies of components and systems, using a Failure Mode and Effects analysis framework;

3. examination of reliability and maintainability aspects such as possible changes in design or operations which may increase reliability and maintainability;

4. preliminary determination of essential and desirable maintenance tasks, their timing, special tool, test equipment, or spare part requirements;

5. maintenance task need analysis; and,

6. analysis of critical components and systems.

As part of the MEA, simple trade-off studies are usually performed for each of the alternate concepts, component selection under consideration, and alternate policies for maintenance such as: repair vs. throw-away and replacement; operator repair vs. contractor maintenance, preventive vs. corrective maintenance. Spare parts requirements, test and checkout methods and the standardization potentials are also being compared. Next the minimum maintenance requirements are listed. Here we consider such questions as:

A. What maintenance tasks are necessary, desirable etc.? Each listed maintenance task is described briefly including an explanation of why this task is necessary, and what it contributes to system reliability.

B. Next we develop a maintenance program in terms of the timing of maintenance tasks, considering

the requirements of both system reliability and
system operating requirements.

The effect of lack of performance of
maintenance tasks on the system, subsystem,
or components is also estimated both if the
task is delayed or not performed at all.

The MEA is essentially a framework for the effective integration of systems design and operations by developing a maintenance program which is best suited to the design and operating requirements of the system. Maintenance tasks are usually organized by their functional relations and frequency, and an effort is made to coordinate functionally related tasks to the maximum degree possible.

6.0 MULTIVARIABLE PROBABILITY DISTRIBUTIONS AND STOCHASTIC PROCESSES

Ernst G. Frankel

Up to this point, the reliability of a system has been determined by use of simple probability concepts and logical arguments. However, for more complex systems, especially those that are maintained, the use of the more powerful techniques of stochastic process theory will be necessary in order to obtain effective solutions. Before discussing stochastic processes though, it may be useful to review multivariable probability distributions.

6.1 Multivariable Probability Distributions

Probability distributions in several dimensions arise when it is desired to express the probability of several events simultaneously. We then obtain a joint probability density of these events. If the events are denoted by subscripts, then

$$f(x_1, x_2, \ldots x_n) = P(x_1 \leq \xi_1 \leq x + dx_1, \ldots x_n \leq \xi_n \leq x_n + dx_n)$$

Similarly, the joint distribution function is

$$F(x_1, x_2, \ldots x_n) = P(\xi_1 \leq x_1, \ldots \xi_n \leq x_n)$$

where $F(x_1, \ldots, x_n)$ is a nondecreasing and continuous function, which has the property that:

$$\lim_{x_k, \ldots x_n \to \infty} F(x_1, \ldots, x_n) = G(x_1, \ldots x_{k-1})$$

In other words, if any set of variables, say $x_k, \ldots, x_n$, approaches ∞, then the limiting value of the joint distribution is the distribution function of the remaining variables. If all x_i tend to ∞, then the joint distribution function tends to unity. Similarly, if <u>any</u> of the variables tends to $-\infty$, then the joint distribution function tends to zero.

To calculate the probability of the intersection of the

events

$(x_1 < \xi \leq x_2)$ and $(y_1 < \eta \leq y_2)$, or

$P(x_1 < \xi \leq x_2, y < \eta \leq y_2) = P(\xi \leq x_2, \eta \leq y_2)$

$-P(\xi \leq x_2, \eta \leq y_1) - P(\xi \leq x_1, \eta \leq y_2) + P(\xi \leq x_1, \eta \leq y_1)$

as shown in Figure 6.1.

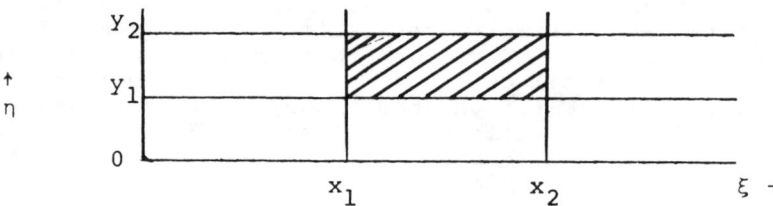

FIGURE 6.1: PROBABILITY OF INTERSECTION OF EVENTS

or

$P(x_1 < \xi \leq x_2, y < \eta \leq y_2) = F(x_2, y_2) - F(x_2, y_1) - F(x_1, y_2)$

$+ F(x_1, y_1)$

To find the joint probability density function of the two variables in terms of their joint distribution function we write

$x_2 = x_1 + dx_1$ and $y_2 = y_1 + dy_1$ when

$P(x_1 < \xi \leq x_1 + dx_1, y_1 < \eta \leq y_1 + dy_1) = F(x_1 + dx_1, y_1 + dy_1)$

$-F(x_1 + dx_1, y_1) - F(x_1, y_1 + dy_1) + F(x_1, y_1)$

Dividing this expression by dx_1, dy_1, if the partial derivative of $\partial^2 F / dx_1 dy_1$ exists at point (x_1, y_1), then as $dx_1\, dy_1 \to 0$

$$\frac{P(x_1<\xi\leq x_1+dx_1, y_1<\eta\leq y_1+dy_1)}{dx_1 dy_1} = \{F(x_1+dx_1,y_1+dy_1) - F(x_1+dx_1,y_1)$$
$$- [F(x_1,y_1+dy_1) - F(x_1,y_1)]\}/$$
$$dx_1 dy_1$$

can be written as:

$$f(x_1,y_1) = \frac{\partial^2 F}{\partial x_1 \partial y_1}$$

For the case of n dimensions,

$$f(x_1,\ldots,x_n) = \frac{\partial^n F}{\partial x_1 \partial x_2 \ldots \partial x_n}$$

If the random variables, x_i are all independent,

$$F(x_1 \ldots x_n) = \prod_{j=1}^{n} F_j(x_j)$$

where $F_j(x_j)$ = distribution function of x_j.

Returning to our two-dimensional example above, we find that if independence of x and y is assumed,

$$P(x_1<\xi\leq x_1+dx_1, y_1<\eta\leq y_1+dy_1) = [F_1(x_1+dx_1) - F_1(x_1)]$$
$$\cdot [F_2(y_1+dy_1) - F_2(y_1)]$$

where

$$F_1(x_1) = \int_{-\infty}^{x_1} dt [\int_{-\infty}^{x_1} f(t,n) \, dn] = \int_{-\infty}^{x_1} f_1(t) dt, \text{ etc.}$$

When the joint probability density function exists, then in the n dimensional case,

$$F(x_1 \ldots x_n) = \int_{-\infty}^{x_1} \int_{-\infty}^{x_2} \ldots \int_{-\infty}^{x_n} f(\xi_1, \xi_2, \ldots \xi_n) d\xi_1 \ldots d\xi_n$$

Conditional Multivariable Distributions

If ξ is smaller than x given η is larger than y and smaller than y+dy, then we can write

$$P(\xi \leq x | y < \eta \leq y+dy) = \frac{P(\xi < x, y < \eta \leq y+dy)}{P(y < \eta \leq y+dy)} = \frac{\int_{-\infty}^{x} \int_{y}^{y+dy} f(\xi, \eta) d\xi d\eta}{\int_{y}^{y+dy} \left[\int_{-\infty}^{+\infty} f(\xi, \eta) d\xi\right] d\eta}$$

In the limit, as $dy \to 0$, this can be written as follows:

$$P(\xi \leq x | \eta = y) = \frac{\int_{-\infty}^{x} f(\xi, y) d\xi}{f_2(y)} \quad \text{where } f_2(y) = \int_{-\infty}^{\infty} f(\xi, y) d\xi$$

If ξ and η are independent,

$$P(\xi \leq x | \eta = y) = f_2(y) \int_{-\infty}^{x} f_1(\xi) \frac{d\xi}{f_2(y)} = P(\xi \leq x) \quad (6.1)$$

By differentiating the conditional distribution function with respect to x, we obtain the conditional probability density function given y, or

$$f(x|y) = \frac{f(x, y)}{f_2(y)}$$

Considering next, the covariance of a two-dimensional distribution:

$$\text{Cov}(\xi, \eta) = \int_{-\infty}^{\infty} \int_{-\infty}^{\infty} (x-\bar{x})(y-\bar{y}) f(x, y) dx dy$$

$$\bar{x} = E[\xi] \qquad \bar{y} = E[\eta]$$

Similarly, for discrete distributions:

$$\text{Cov}(\xi, \eta) = \sum_{i} \sum_{j} (x_i - \bar{x})(y_j - \bar{y}) P(\xi = x_i, \eta = y_i)$$

In general, we may write:

$$\text{Cov}(\xi, \eta) = E[\xi\eta - \bar{x}\eta - \bar{y}\xi + \bar{x}\bar{y}] = E[\xi\eta] - \bar{x}\bar{y} \quad (6.2)$$

The correlation coefficient,

$$\rho(\xi, \eta) = \frac{\text{Cov}(\xi, \eta)}{\sqrt{\text{Var } \xi \cdot \text{Var } \eta}}$$

The Multinomial Distribution

If the outcome of an experiment may result in k mutually exclusive events, each of which has a probability of p_i $i=1,\ldots,k$ where

$$\sum_{i=1}^{k} p_i = 1$$

and if n_i is the number of times the ith event occurs in N trials where

$$\sum_{i=1}^{k} n_i = N$$

then

$$P(n_1 \ldots n_{k-1}) = \frac{N! \; p_1^{n_1} \ldots p_{k-1}^{n_{k-1}} (1-p_1 \ldots p_{k-1})^{N-n_1-\ldots n_{k-1}}}{n_1! \ldots n_{k-1}! \; (N-n_1-\ldots-n_{k-1})!}$$

where

$$0 \leq \sum_{j=1}^{k-1} n_j \leq N \; , \quad \sum_{j=1}^{k-1} p_j \leq 1 \tag{6.3}$$

Similarly,

$$P(n_1 \ldots n_{k-2}) = \frac{N! \; p_1^{n_1} \ldots p_{k-2}^{n_{k-2}} (1-p_1-\ldots-p_{k-2})^{N-n_1-\ldots n_{k-2}}}{n_1! \ldots n_{k-2}! \; (N-n_1-\ldots-n_{k-2})!}$$

For example for k=3 it is:

$$P(n_1, n_2) = \frac{N! \; p_1^{n_1} p_2^{n_2} (1-p_1-p_2)^{N-n_1-n_2}}{n_1! \; n_2! \; (N-n_1-n_2)!}$$

$$\text{Cov}(n_1, n_2) = E(n_1 n_2) - E(n_1) E(n_2)$$

but

$$E(n_1 n_2) = \sum_{n_1=0}^{N} \sum_{n_2=0}^{N-n_1} n_1 n_2 \; P[n_1 n_2]$$

Summation with respect to n_2

$$\frac{N! \, p_1^{n_1} \, n_1}{n_1!} \sum_{n_2=0}^{N-n_1} \frac{n_2 p_2^{n_2}(1-p_1-p_2)^{N-n_1-n_2}}{n_2!(N-n_1-n_2)!} = \left(\frac{N! p_1^{n_1} \, n_1}{n_1!}\right)$$

$$\times \left(\frac{p_2(1-p_1)^{N-n_1-1}}{(N-n_1-1)!}\right)$$

Summing now over n

$$E[n_1 n_2] = \frac{N!}{(N-2)!} \, p_2 \sum_{n_1=0}^{N-2} \frac{p_1^{n_1+1}(1-p_1)^{N-n_1-2}(N-2)!}{n_1!(N-n_1-2)!} = p_2 p_1 \, N(N-1)$$

$$\text{Cov}(n_1 n_2) = p_2 p_1 \, N(N-1) - N^2 p_1 p_2 = N \, p_1 p_2$$

and the correlation coefficient is

$$\rho(n_1 n_2) = \frac{-N p_1 p_2}{[N p_1 (1-p_1) N p_2 (1-p_2)]^{1/2}} = -\frac{p_1 p_2}{(1-p_1)(1-p_2)}$$

Because of the symmetry, similar expressions for the pairs (n_1, n_3) and (n_2, n_3) can be obtained by cyclical permutation of the indices.

6.2 Stochastic Processes

If a random variable and its associated probability distribution function are dependent on time (time varying), they are said to characterize a random or stochastic process. Stochastic processes may be continuous in time and discrete or continuous in the parameter of the variable; conversely, they may be discrete in time and discrete or continuous in the parameter of the variable. A stochastic process continuous in time and parameter may be described by a function $x(t)$ and its probability density function

$$w(x,t) dx dt = P[x < \xi \le x+dx, \, t < \tau \le t+dt]$$

which is the probability that our parameter assumes a value

between x and x+dx when time is between t and t+dt. Such a process is shown in Figure 6.2. Most processes investigated in reliability theory are random processes. A typical example might be the measurement of wear of a mechanical component. If a large number of continuous records for identical components are available, these can be analyzed to obtain the probability distribution function of the cumulative wear as a function of time:

$$F(x, t) = \int_0^x w(\xi, t) dt$$

where wear $x(t)$ is a function of time and $w(\xi, t)$ is the probability density function of wear, or the time dependent wear rate. Such information permits improvements in system design and maintenance replacement scheduling to be incorporated. In most practical examples, we will be content to define

$$w_1(x, t) dx = P[x<\xi<x+dx, \tau = t]$$

as the probability that $x<\xi<x+dx$ at time t.

In other cases, it might be advisable to consider a process with discrete parameters as, for instance, when the variable is the number of failures subject to wear at a certain time t. In this case we can define:

$$w(x, t) = P[\xi=x, \tau=t]$$

as shown in Figure 6.2.

A stochastic process $\{x(t), \tau \varepsilon\ T\}$ can therefore be described as a family of random variables in which for a given value of the time parameter t of the index set T, the stochastic process of the function $x(t)$ is a simple random variable. Thus, $x(t)$ for a given t will have a cumulative probability distribution function:

$$P(x(t);t) = F(x;t) = \int_0^x w(x;t) dx(t) \qquad (6.4)$$

and the probability density $w(x;t)$ is defined as:

$$w(x;t) = \frac{d}{dx} F(x;t) \qquad (6.5)$$

Frequently, in analyzing stochastic processes, we are interested in the joint probability density function:

$$w(x_1, t_1; x_2, t_2; \ldots x_n, t_n) dx_1 dx_2 \ldots dx_n$$

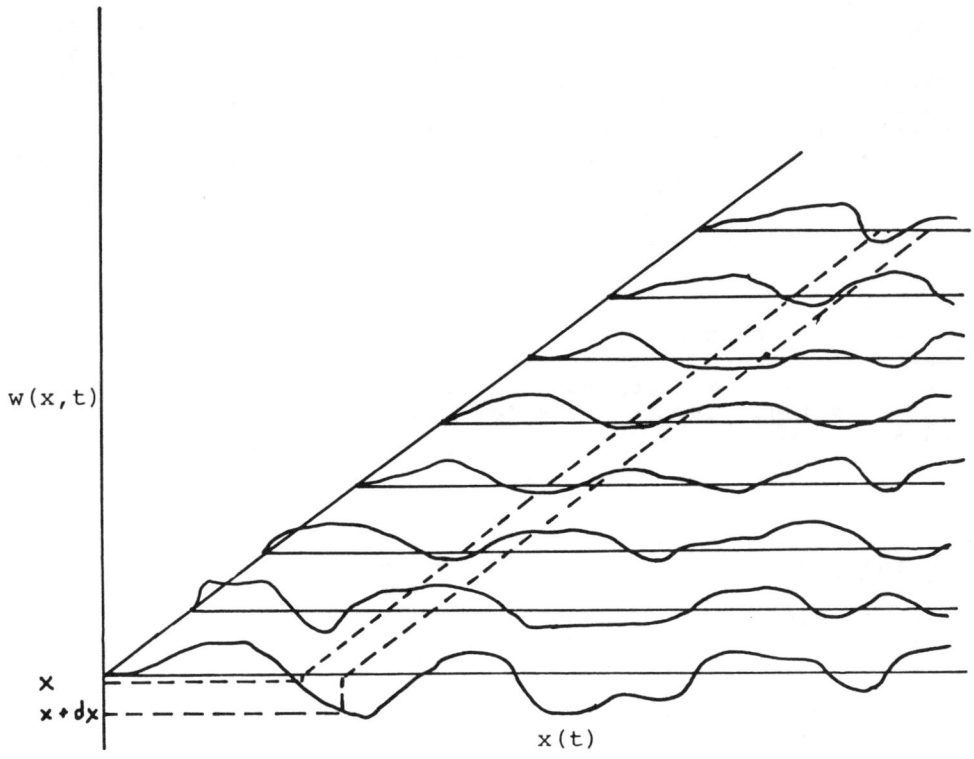

FIGURE 6.2:   NUMBER OF FAILURES SUBJECT TO WEAR x AT TIME t

Such a stochastic process may also be described in its joint probability distribution:

$$F(x_1, x_2 \ldots x_n, t_1 t_2 \ldots t_n) =$$

$$P[\xi \leq x_1, \ \tau = t_1; \ \xi \leq x_2, \tau = t_2; \ \ldots \xi \leq x_n, \ t = t_n]$$

which is the probability that our variable is smaller than x_1 at t_1, smaller than x_2 at t_2 and so forth. Obviously,

instead of being continuous in the parameter of the variable x and discrete in time the process could have been discrete in x and continuous in time, or continuous in both the variable x and time.

Even though joint distributions of the type defined above are required for the complete description of the process, most of the information required in practice can be derived through conditional probability distribution functions based on some information of the process available for specific values of the time parameter. For example, a conditional cumulative probability distribution might be:

$$F(x_n; t_n \mid x_{n-1}, x_{n-2} \ldots x_o; t_{n-1}, t_{n-2} \ldots t_o) =$$

$$P(\xi \leq x_n, \tau = t_n \mid \xi \leq x_{n-1}, \tau = t_{n-1}; \xi \leq x_{n-2}, \tau = t_{n-2} \ldots ;$$

$$\xi \leq x_o, \tau = t_o)$$

or the probability that the variable is smaller than x_n at t_n given it was smaller than x_{n-1} at t_{n-1}, smaller than x_{n-2} at t_{n-2} and so on, to smaller than x_o at t_o. If the limiting state of our variable x at t_n only depends on its limiting value at t_{n-1} then this relationship will later be seen to be useful in the discussion of Markov processes.

A stochastic process is said to be stationary if:

$$F(x_o, x_1 \ldots x_n; t_o, t_1 \ldots t_n) = F(x_o, x_1 \ldots x_n; t_o+h, t_1+h \ldots t_n+h)$$

A stochastic process of a variable x whose value is $x(t_i)$ at time $t=t_i$ is said to have independent increments if for all choices of $t_o t_1 \ldots t_n$, the n random variables:

$$x(t_1)-x(t_o), \quad x(t_2)-x(t_1), \ldots, x(t_n)-x(t_{n-1})$$

are independent. The process is said to have stationary independent increments if in addition $x(t_2+s) - x(t_1+s)$ has the same distribution as $x(t_2) - x(t_1)$ for all t_1, $t_2 \epsilon T$ and $s>0$.

Similarly to the definition of statistical measures in probability theory we define different moments to represent the statistical measures of a stochastic process. For instance, the **mean value** for a group of functions x(t) at time t is:

$$\overline{x(t)} = \int_{-\infty}^{\infty} x(t) \; w(x,t) dx \qquad (6.6)$$

The <u>time average</u> of a stochastic process is:

$$\bar{x} = \lim_{T \to \infty} \frac{1}{2T} \int_{-T}^{T} \int_{\infty}^{\infty} x(t) dt \qquad (6.7)$$

The expected value of a group of stochastic function is:

$$E[x(t)] = \lim_{T \to \infty} \frac{1}{2T} \int_{-T}^{T} \int_{\infty}^{\infty} x(t) w(x,t) dx dt$$

For a discrete stochastic process $x(t)$ $t=1,\ldots\ldots T$. The <u>time average</u>

$$\bar{x} = \frac{1}{T} \sum_{t=1}^{T} x(t) = \text{sample wear}$$

It can be shown that if a random process is stationary which implies that

$$w(x_1, t_1; \ldots x_n, t_n) = w(x_1, t_1+\tau, \ldots x_n, t_n+\tau)$$

or the probability distributions of two groups of values of a variable

$$[x(t_1)\ldots x(t_n)] \text{ and } [x(t_1+\tau),\ldots x(t_n+\tau)]$$

are identical, where the number n and the interval τ are selected arbitrarily, then the mean value of a random process gives the same result as the time average or sample wear of a process independent of the choice of functions making up this group. In other words, $x(t)=\bar{x}$. The same applies to higher order moments of $x(t)$

$$E\{x^n(t)\} = \int_{-\infty}^{\infty} x^n(t) w(x,t) dx$$

and

$$E\{x^n\} = \lim_{T \to \infty} \frac{1}{2T} \int_{-T}^{T} [x(t)]^n dt$$

where $\overline{x^n} = \bar{x}^n$.

If a stationary stochastic process has a sample wear whose variance is zero, then the process is said to be ergodic. Condition of ergodicity, $\lim_{T \to \infty} \text{Var } [\bar{x}] = 0$. As in the case of time invariant random variables, we are also concerned with conditional probabilities in the analysis of random processes. Considering a single process of the variable x(t),

$$P[x_n \leq \xi \leq x_n+dx_n, t_n/x_{n-1} \leq \xi \leq x_{n-1}+dx_{n-1}, t_{n-1}; \ldots x_1 \leq \xi \leq x_1+dx_1, t_1]$$

$$= \frac{P[x_n \leq \xi \leq x_n+dx_n, t_n; \ldots x_1 \leq \xi \leq x_1+dx_1, t_1]}{P[x_{n-1} \leq \xi \leq x_{n-1}+dx_{n-1}, t_{n-1}; \ldots x_1 \leq \xi \leq x_1+dx_1, t_1]}$$

$$= P[x_{n-1} \leq \xi \leq x_{n-1}+dx_{n-1}, t_{n-1}, \ldots x_1 \leq \xi \leq x_1+dx_1, t_1/x_n \leq \xi \leq x_n+dx_n, tn]$$

$$= \frac{P[x_n \leq \xi \leq x_n+dx_n, t_n]}{P[x_{n-1} \leq \xi \leq x_{n-1}+dx_{n-1}, t_{n-1}; \ldots x_1 \leq \xi \leq x_1-dx_1, t_1]}$$

= probability that the variable assumes a value between x_n and x_n+dx_n at t_n, given it has assumed values between x_{n-1} and $x_{n-1}+dx_{n-1}$ at t_{n-1}, and so on until x_1 and x_1+dx_1 at t_1.

This is the conditional probability function of a generalized stochastic process. In our work we will be mainly concerned with processes without memory or random processes where the value of the variable x(t) depends only on the value of the variable immediately preceeding t. If our process is discrete in time or if changes occur at fixed times, say $t_1, t_2 \ldots t_n$ where $t_1 < t_2 < t_3 \ldots t_n$, it can be proven that stationary processes have the ergodic attribute for which the time average of a sample component:

$$\bar{x} = \lim_{T \to \infty} 1/T \sum_{i=0}^{T} x(t_i) = \text{sample wear}$$

is a reasonable approximation to the corresponding population average at any time t:

$$\overline{x(t)} = \int_{-\infty}^{\infty} x(t) f(x;t) dx(t)$$

It can also be proven that a condition for ergodicity is that

$$\lim_{T \to \infty} \text{Var}^x (x) = 0$$

6.3 Markov Processes

A Markov process is a stochastic process in which at any given time the subsequent course of the process is affected only by the state at the given time and does not depend on the character of the process at any preceding time. Therefore, the accuracy of predicting the future of our random process is not dependent on any knowledge or the extent of data on the past behavior of the process. The analysis of processes which follow this definition is comparatively easy, and a vast amount of work in the mathematical development of analysis of Markov and Semi-Markov processes has been accomplished. Memory-less processes which can be assumed to be Markov are frequently encountered, especially in reliability theory where the occurrence of failure may be quite unrelated to the history of the component or system. In practice, a large number of systems of engineering interest can be described by a Markov process, by a correct choice of state variables. Like stochastic processes, in general, Markov processes can be divided into discrete parameter and continuous parameter processes. If time is a discrete event, we normally call the Markov process a Markov chain. We, therefore, have discrete or continuous parameter Markov chains, in which time is discrete; and discrete or continuous parameter Markov processes in which time is continuous. If we consider continuous parameter Markov chains first, we define a joint probability density function:

$$w(x_1, t_1; x_2, t_2; \ldots x_n, t_n)$$

The joint probability distribution function:

$$F(x_1,t_1;x_2,t_2;\ldots x_n,t_n) = \int_{-\infty}^{x_1} \cdot \int_{-\infty}^{x_2} \ldots \int_{-\infty}^{x_n} w(x_1,t_1;x_2,t_2;$$

$$\ldots x_n,t_n)dx_n\ldots dx_1 = P[\xi \leq x_1, t_1; \ldots \xi \leq x_n, t_n]$$

= probability that random variable is smaller than x_1, at t_1, smaller than x_2 at t_2, etc.

The Markov condition can then be presented as:

$$P[x_n \leq \xi \leq x_n + dx_n, t_n; x_{n-1} \leq \xi \leq x_{n-1} + dx_{n-1}, t_{n-1}; \ldots, x_1 \leq \xi \leq x_1$$

$$+dx_1,t_1] = P[x_n \leq \xi \leq x_n + dx_n, t_n / x_{n-1} \leq \xi \leq x_{n-1} + dx_{n-1}, t_{n-1}]$$

In reliability theory we will normally be able to deal with discrete variable states when the Markov condition becomes for the conditional probability distribution function:

$$P[\xi \leq x_n, t_n/\xi = x_{n-1}, t_{n-1}; \ldots; \xi=x_1, t_1] = P[\xi \leq x_n, t_n/\xi = x_{n-1}, t_{n-1}]$$

as for the conditional probability function:

$$P[\xi=x_n, t_n/\xi=x_{n-1}, t_{n-1}; \ldots; \xi=x_1, t_1] = P[\xi=x_n, t_n/\xi = x_{n-1}, t_{n-1}]$$

where

$$t_n \geq t_{n-1} \geq \ldots \geq t_2 \geq t_1$$

Similarly, a Markov Process has a conditional probability distribution function:

$$F(x_n; t_n | x_{n-1}, x_{n-2} \ldots x_o; t_{n-1}, t_{n-2} \ldots t_o) = F(x_n; t_n | x_{n-1}; t_{n-1})$$

or

$$P_r[x(t_n) < x_n | x(t_{n-1}) = x_{n-1}, x(t_{n-2}) = x_{n-2} \ldots x(t_o) = x_o] = P[x(t_n) < x_n | x(t_{n-1}) = x_{n-1}]$$

if we denote $x(t_n)$ as the value of variable x at time t_n.

Therefore the accuracy of predicting the future of the random process is not dependent on any knowledge of the past behavior of the process. This assumption remarkably simplifies the mathematics involved in analyzing stochastic processes.

As with general stochastic processes, Markov processes can be divided into those with discrete and continuous random variables and those with discrete and continuous time parameters. Those with discrete time parameters are called Markov chains, and in reliability theory Markov chains with discrete-valued random variables are of primary concern. For these discrete valued Markov chains, the Markov conditions may be expressed as:

$$f(x_n; t_n | x_{n-1}, x_{n-2}, \ldots x_o; t_{n-1}, t_{n-2}, \ldots t_o) =$$

$$f(x_n;t_n|x_{n-1};t_{n-1})$$

or

$$P[x(t_n) = x_n|x(t_{n-1}) = x_{n-1}, x(t_{n-2}) = x_{n-2}, \ldots x(t_o) = x_o] = P[x(t_n) = x_n|x(t_{n-1}) = x_{n-1}]$$

A Markov chain is said to be homogeneous (or to be homogeneous in time or to have stationary transition probabilities), if $f(x_n; t_n|x_{n-1};t_{n-1})$ depends only on the differences $t_n - t_{n-1}$ and is independent of the actual value of t_n. In this case, it can be shown that the limit

$$\lim_{t_n \to \infty} f(x_n;t_n|x_{n-1};t_{n-1})$$

exists, which implies that a homogeneous Markov chain (also process) will always have a steady-state value.

To simplify notation, the following convention will be used for the conditional probability function of a Markov chain. Let

$$P_i(t) \equiv f(x_i;t) \equiv P_r(X_i(t) = X_i)$$

$\equiv$ probability that the random variable X assumes a value (or state) X_i at time t

$$P_{ij}(t,s) \equiv f(X_j;s|X_i,t) = P_r(X(s) = X_j|X(t) = X_i)$$

$\equiv$ transition probability from state i to state j

$\equiv$ probability that the random variable X assumes the value (or state) X_j at time s, given it assumed a value X_i at time t (where $s \geq t$)

In the case when the Markov Chain is homogeneous, then let:

$$P_{ij}(t,s) = P_{ij}(\tau) = f(x_j,s|x_i,t) = f(x_j,t+\tau; x_i,t)$$

where

$$s - t = \tau = \text{stationary transition time.}$$

It should be mentioned that some stochastic processes have a "limited memory" of the past. That is, they may be represented by conditional probability distributions that have the property:

$$P[X_n; t_n | X_{n-1}, X_{n-2} \cdots X_o; t_{n-1}, t_{n-2} \cdots t_o)] = P[(X_n; t_n | X_{n-1}, X_{n-2} \cdots X_{n-r}; t_{n-1}, t_{n-2} \cdots t_{n-r})]$$

or

$$P[X(t_n) = X_n | X_{n-1}(t_{n-1}) = X_{n-1}, \ldots, X_o(t_o) = X_o]$$

$$= P[X(t_n) = X_n | X_{n-1}(t_{n-1}) = X_{n-1}, \ldots, X_{n-r}(t_{n-r}) = X_{n-r}]$$

These processes are called Markov chains of order r because the condition of the variable at the time n depends on its condition during the preceding r time periods. Using this approach, a large number of stochastic processes which are not truly Markov can be analyzed using an expanded Markov Theory.

Using this notation and the Markov property, the joint probability density of a discrete Markov chain can be written in a very simple form:

$$f(X_n, X_{n-1} \cdots, X_o; t_n, t_{n-1}, \cdots t_o) = P[X(t_n) = X_n, X(t_{n-1})$$

$$= X_{n-1}, \ldots X(t_o) = X_o] = P_r[X(t_n) = X_n | X(t_{n-1})$$

$$= X_{n-1}, \ldots, X(t_o) = X_o] \; P[X(t_{n-1}) = X_{n-1}, \ldots X(t_o) = X_o]$$

(using Bayes Theorem)

$$= P[X(t_n) = X_n | X(t_{n-1}) = X_{n-1}] \; P[X(t_{n-1}) = X_{n-1}, \ldots,$$

$$X(t_o) = X_o]$$

(using Markov Property)

$$= P[X(t_n) = X_n | X(t_{n-1}) = X_{n-1}] = P[X(t_{n-1}) = X_{n-1}$$

$$| X(t_{n-2}) = X_{n-2}, \ldots, X(t_o) = X_o] \times P[(X(t_{n-2})$$

$$= X_{n-2}, \ldots, X(t_o) = t_o)]$$

(again using Bayes Theorem)

$$= P[X(t_n) = X_n | X(t_{n-1}) = X_{n-1}] \, P[X(t_{n-1}) = X_{n-1} | X(t_{n-2}) = X_{n-2}] \times P[X(t_{n-2}) = X_{n-2}, \ldots X(t_o) = t_o]$$

(again using Markov property)

In words, this means that the joint probability density of the stochastic variable x when it assumes a value of x_o at t_o, x_1 at t_1,...x_n at t_n is equal to the product of probability that the variable takes on the value X_o at t_o times the probability it takes on the value X_1 at t_1 given that it took on the value X_o at t_o .... times the probability it takes on the value X_n at t_n given that it took on the values X_{n-1} at t_{n-1} . If it happens that the Markov chain is also homogeneous and that the separation between times t_o, t_1 ...t_n is one time unit (or step) so that

$$t_n - t_{n-1} = t_{n-1} - t_{n-2} = \ldots = t_n - t_o = 1$$

then the above expression may be further simplified to:

$$f(X_n, X_{n-1} \ldots X_o; t_n, t_{n-1} \ldots t_o) = P_{n-1,n}[(t_{n-1}, t_n)]$$

$$P_{n-2,n-1}[(t_{n-2}, t_{n-1})] \ldots P_{0,1}[(t_o, t\,) P_o(t_o)]$$

$$- P_{n-1,n}(1) \, P_{n-2,n-1}(1) \, \ldots \, P_{o,1}(1) \, P_o(t_o)$$

where $P_{ij}(t_i, t_j)$ is the probability that the variable x is equal to x_i at t_i given it was equal to x_j at t_j. Therefore, in a stationary Markov Chain:

$$P[\xi = x_i, t_i / \xi = x_{i-1}, t_{i-1}]$$

depends only on $(t_i - t_{i-1})$ and is independent of the actual value of t_i. In this case it can be shown that the

$$\lim_{t_n \to \infty} P[\xi=x_n, t_n | \xi=x_{n-1}, t_{n-1}]$$

exists, which implies that a stationary Markov process (or chain) guarantees that a steady-state solution exists. Similarly, if the transition probability of going from state i to a state j in n steps is represented by $P_{ij}(n)$*, then

$$\sum_{j=0}^{m} P_{ij}(n) = 1 \qquad (6.8)$$

since the process can be in only one of the (m+1) steps after any number of steps. It should be noted that $P_{ij}(n)$ may be also represented in a more expanded form:

$$P_{ij}(n) = \sum_{k=0}^{m} P_{ik}(r) \cdot P_{kj}(n-r) \text{ for any } r \leq n$$

This may be shown by the following proof:

$$P_{ij}(n) = P_r(X(t+n) = X_j, X(t) = X_i)$$

$$= \sum_{k=0}^{m} P_r(X(t+n) = X_j, X(t+r) = X_k | X(t) = X_i) **$$

---

* Note:
$$P_{ij}(0) = \begin{bmatrix} 1 & i=j \\ 0 & i \neq j \end{bmatrix} = \delta_{ij}$$

** Note:
$$\sum_{k=0}^{m} P_r(X(t+N) = X_j, X(t+r) = X_k | X(t) = X_i)$$

$$= \sum_{k=0}^{m} \frac{P_r(X(t+n)=X_j, X(t+r)=X_k, X(t)=X_i)}{P_r(X(t) = X_i)}$$
(Bayes Theorem)

$$= \frac{P_r(X(t+n)=X_j, X(t)=X_i)}{P_r(X(t)=X_i)} = P_r(X(t+n)=X_j | X(t)=X_i) \text{ (again}$$

Bayes Theorem) (Since the summation over all states gives the marginal probability density.)

$$= \sum_{k=0}^{m} P_r[X(t+n)=X_j | X(t+r)=X_k, X(t)=X_i] \, P_r[X(t+r)$$

$$= X_k | X(t) = X_i]$$

(Conditional Form of Bayes Theorem)

$$= \sum_{k=0}^{m} P_r[X(t+n) = X_j | X(t+r) = X_k] \, P_r[X(t+r) = X_k | X(t) = X_i]$$

(From Markov Property)

$$= \sum_{k=0}^{m} \{P_{kj}[(t+n) - (t+r)] \cdot P_{ik}[((t+r) - t)]\}$$

$$= \sum_{k=0}^{m} P_{ik}(r) \, P_{kj}(n-r) \qquad (6.9)$$

This relation is called the Chapman-Kolmogorov equation and in words simply states that the probability of going from state i to state j in n steps is equal to the joint probability of the independent events of going from state i to state k in r steps and from state k to state j in n-r steps, summed over all K.

It should be noted that a similar expression can be derived for non-homogeneous Markov chains with finite states:

$$P_{ij}(m,n) = \sum_{k=0}^{m} P_{ik}(m,u) \, P_{kj}(u,n) \qquad (6.10)$$

Another useful expression is:

$$P_j(n) = \sum_{k=0}^{m} P_{kj}(n-r) \, P_k(r)$$

which may be proven as follows:

$$\sum_{k=0}^{m} P_{kj}(n-r) \, P_k(r) = \sum_{k=0}^{m} P_r(X(t+n) = X_j | X(t+r) = X_k) \cdot$$

$$P_r(X(t+r) = X_k)$$

$$= \sum_{k=0}^{m} P_r(X(t+n) = X_j, X(t+r) = X_k) \quad \text{(Bayes Theorem)}$$

$$= P_r(X(t+n) = X_j) \quad \text{(since summation k over all states gives the marginal density)}$$

$$= P_j(n)$$

Now, a finite state, homogeneous Markov chain can be completely defined by use of the one-step transition probabilities and the Chapman-Kolgomorov equation. It is convenient to organize the one-step transition probabilities in a stochastic matrix:

$$P(1) = \begin{bmatrix} P_{0,0}(1), P_{0,1}(1) \ldots\ldots\ldots\ldots P_{0,m}(1) \\ P_{1,0}(1), P_{1,1}(1) \ldots\ldots\ldots\ldots P_{1,m}(1) \\ , \quad\quad , \quad\quad\quad\quad\quad\quad\quad , \\ , \quad\quad , \quad\quad\quad\quad\quad\quad\quad , \\ P_{m,0}(1), P_{m,1}(1) \ldots\ldots\ldots\ldots P_{m,m}(1) \end{bmatrix}$$

For simplicity the step argument may be omitted:

$$P = \begin{bmatrix} P_{0,0} & P_{0,1} \cdot\quad\cdot\quad\cdot\quad\cdot\quad\cdot\quad\cdot P_{0,m} \\ P_{1,0} & P_{1,1} \cdot\quad\cdot\quad\cdot\quad\cdot\quad\cdot\quad\cdot P_{1,m} \\ , & \quad\quad\quad\quad\quad\quad\quad\quad\quad , \\ , & \quad\quad\quad\quad\quad\quad\quad\quad\quad , \\ P_{m,0} & P_{m,1} \cdot\quad\cdot\quad\cdot\quad\cdot\quad\cdot\quad\cdot P_{m,m} \end{bmatrix}$$

Note that in accordance with equation (6.7) the sum of each row must be equal to one.

To obtain the two-step transition matrix, which presents all the probabilities of transiting from a state i to a state j in two steps, the Chapman-Kolgomorov equation is used with n=2 and r=1:

$$P_{ij}(2) = \sum_{k=0}^{m} P_{ik}(1) P_{kj}(2-1) = \sum_{k=0}^{m} P_{ik} \cdot P_{kj}$$

Quick examination of the above equation results in the discovery that the two-step transition matrix will be just the matrix product of the one-step stochastic matrix.

$$P_{(2)} = p \cdot p = \begin{bmatrix} P_{o,o} & P_{o,1} \cdots P_{o,m} \\ , & , & , \\ , & , & , \\ P_{m,o} \cdots \cdots P_{m,m} \end{bmatrix} \times \begin{bmatrix} P_{o,o} & P_{o,1} \cdots P_{o,m} \\ , & , & , \\ , & , & , \\ P_{m,o} \cdots \cdots P_{m,m} \end{bmatrix}$$

By a similar argument it can be shown that the three-step transition matrix will be the product of three one-step transition matrices:

$$P_{(3)} = P_{(2)} P = P \cdot P \cdot P = P^3$$

It should be noted that similar expressions can be developed for non-homogeneous Markov Chains. In this case, the stochastic matrix will be of the form:

$$P_{(s,t)} = \begin{bmatrix} P_{o,o}(s,t) & P_{o,1}(s,t) & \cdots & P_{o,m}(s,t) \\ P_{1,o}(s,t) & P_{1,1}(s,t) & \cdots & P_{1,m}(s,t) \\ , & , & & , \\ , & , & & , \\ P_{m,o}(s,t) & P_{m,1}(s,t) & & P_{m,m}(s,t) \end{bmatrix}$$

The process may be determined at other time intervals by use of the non-homogeneous Chapman-Kolgomorov equation:

$$P_{(s,t)} = P_{(s,u)} \cdot P_{(u,t)}$$

To make these concepts better understood, consider the homogeneous stochastic process represented by a Markov Chain with two states, 0 and 1, and with the probabilities, P_{ij}, for transiting from one state to another. This process may be pictured as a network, as shown in Figure 6.3.

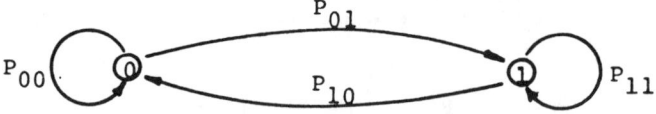

FIGURE 6.3: TWO STATE MARKOV PROCESSES

where the nodes represent the possible states and the arcs the transition probabilities from state to state.

The homogeneous one-step stochastic matrix for the process will be:

$$P = \begin{bmatrix} P_{0,0} & P_{0,1} \\ P_{1,0} & P_{1,1} \end{bmatrix} \begin{matrix} 0 \\ 1 \end{matrix}$$

with states 0, 1 across the top.

The two-step transition matrix will be:

$$P_{(2)} = P \cdot P = \begin{bmatrix} P_{00} & P_{01} \\ P_{10} & P_{11} \end{bmatrix} \begin{bmatrix} P_{00} & P_{01} \\ P_{10} & P_{11} \end{bmatrix} = \begin{bmatrix} P_{00}^2 + P_{01}P_{10} & P_{00}P_{01} + P_{01}P_{11} \\ P_{10}P_{00} + P_{11}P_{10} & P_{10}P_{01} + P_{11}^2 \end{bmatrix}$$

so that, for example, the probability of being in state 1 after two steps when starting from state 0 will be

$$P_{00}P_{01} + P_{01}P_{11}$$

And the n step transition matrix will be:

$$P_{(n)} = P^n = \frac{1}{2 - P_{00} - P_{11}} \begin{bmatrix} 1-P_{1,1} & 1-P_{0,0} \\ 1-P_{1,1} & 1-P_{0,0} \end{bmatrix} + \frac{(P_{0,0} + P_{1,1} - 1)^n}{2 - P_{0,0} - P_{1,1}}$$

$$\begin{bmatrix} 1-P_{0,0} & -(1-P_{0,0}) \\ -(1-P_{1,1}) & 1-P_{1,1} \end{bmatrix}$$

(using the fact that $P_{0,0} + P_{0,1} = 1$ and $P_{1,0} + P_{1,1} = 1$).

If it is not the case that $P_{00} = 1$ and $P_{11} = 1$ so that $P_{00} + P_{11} < 2$, then as n approaches infinity:

$$\lim_{n\to\infty} P(n) = \frac{1}{2-P_{0,0}-P_{1,1}} \begin{bmatrix} 1-P_{1,1} & 1-P_{0,0} \\ \\ 1-P_{1,1} & 1-P_{0,0} \end{bmatrix}$$

$$+ \lim_{n\to\infty}\left[\frac{(P_{0,0}+P_{1,1}-1)^n}{2-P_{0,0}-P_{1,1}}\right] \begin{bmatrix} 1-P_{0,0} & -1-P_{0,0} \\ \\ -(1-P_{1,1}) & 1-P_{1,1} \end{bmatrix}$$

$$- \frac{1}{2-P_{0,0}-P_{1,1}} \begin{bmatrix} 1-P_{1,1} & 1-P_{0,0} \\ \\ 1-P_{1,1} & 1-P_{0,0} \end{bmatrix}$$

In this case the transition probabilities tend to steady state value. So, for example, the steady state probability of being in state 1 after starting in state 0 will approach

$$(1-P_{0,0})/(2-P_{0,0}-P_{1,1})$$

Now, if the unconditional probability state vector is defined as:

$$P(n) = [P_0(n), P_1(n), \ldots P_m(n)] \equiv \text{row vector of the probabilities of being in state } 0, 1, \ldots m \text{ at step } n$$

then $P(0) = [P_0(0), P_1(0), \ldots P_m(0)]$ would represent the initial probability state of the process. For example, $P(0) = [1,0,0..0]$ would simply mean that the process initially starts in state 0 with a probability of 1.

Then it follows that the probability that the process is in state j after one step could be found by the following expression assuming that the initial probability state and one-step transition matrix were known:

$$P_j(1) = \sum_{i=1}^{m} P_i(0) \cdot P_{ij} \qquad (6.11)$$

This is simply the probability of being initially in state i followed by a transition from i to state j, summed over all state i. This expression can be formulated in the matrix form:

$$P(1) = [P_0(1), P_1(1), \ldots, P_m(1)] = [P_0(0), P_1(0), \ldots, P_m(0)] \cdot$$

$$\begin{bmatrix} P_{00} & P_{01} \cdots P_{0m} \\ P_{10} & P_{11} \cdots P_{1m} \\ , & , & , \\ , & , & , \\ P_{m0} & P_{m1} \cdots P_{mm} \end{bmatrix} = P(0) \cdot P$$

In a similar manner, it may be found that:

$$P(2) = P(1) \cdot P = P(0) P^2$$

$$\vdots$$

$$P(n) = P(n-1) \cdot P = P(0) P^n \qquad (6.12)$$

It can be shown that, in general, Markov Chains will tend to steady state values, i.e.,

$$\lim_{n \to \infty} P_{ij}(n) = q_{ij} = \text{constant}$$

only when the Markov chain is ergodic. This occurs when $P_{ii} > 0$ for i=1....m and all states communicate or if there is a positive integer n N so that the n-step probability matrix P(n) has no zero elements. Thus, if Q is defined to be the steady state matrix for an ergodic homogeneous Markov chain, then

$$\lim_{n \to \infty} P_{(n)} = Q$$

the matrix Q may be determined by taking the limit of the identity

$$P_{(n)} = P_{(n-1)} \cdot P$$

and solving for Q

$$\lim_{n\to\infty} [P_{(n)} = P_{(n-1)} \cdot P]$$

$$\lim_{n\to\infty} [P_{(n)} = [\lim_{n\to\infty} P_{(n-1)}] \cdot P]$$

$$Q = Q \cdot P \qquad (6.13)$$

Applying this formula to the general two-state, homogeneous Markov Chain:

$$\begin{bmatrix} q_{0,0} & p_{0,1} \\ q_{1,0} & p_{1,1} \end{bmatrix} = \begin{bmatrix} q_{0,0} & p_{0,1} \\ q_{1,0} & p_{1,1} \end{bmatrix} \begin{bmatrix} q_{0,0} & p_{0,1} \\ q_{1,0} & p_{1,1} \end{bmatrix}$$

$$\begin{bmatrix} q_{0,0} & q_{0,1} \\ q_{1,0} & q_{1,1} \end{bmatrix} = \begin{bmatrix} q_{0,0}\,p_{0,0} + q_{0,1}\,p_{1,0} \,, & q_{0,0}\,p_{0,1} + q_{0,1}\,p_{1,1} \\ q_{1,0}\,p_{0,0} + q_{1,1}\,p_{1,0} \,, & q_{1,0}\,p_{0,1} + q_{1,1}\,p_{1,1} \end{bmatrix}$$

Equivalently,

$$q_{0,0} = q_{0,0}\,p_{0,0} + q_{0,1}\,p_{1,0} \quad \text{In addition } q_{0,0} + q_{0,1} = 1$$

$$q_{0,1} = q_{0,0}\,p_{0,1} + q_{0,1}\,p_{1,1}$$

$$q_{1,0} = q_{1,0}\,p_{0,0} + q_{1,1}\,p_{1,0} \quad \text{In addition } q_{1,0} + q_{1,1} = 1$$

$$q_{1,0} = q_{1,0}\,p_{0,1} + q_{1,1}\,p_{1,1}$$

Each of these triplets of equations may be solved independently (only two or three of each triplet are independent) to yield:

$$q_{0,0} = \frac{p_{1,0}}{1 - p_{0,0} + p_{1,0}}$$

$$q_{0,1} = \frac{p_{0,1}}{1 - p_{1,1} + p_{0,1}}$$

$$q_{1,0} = \frac{p_{1,0}}{1 - p_{0,0} + p_{1,0}}$$

$$q_{1,1} = \frac{p_{0,1}}{1 - p_{1,1} + p_{0,1}}$$

Substituting the identities $p_{o,1} = 1-p_{o,o}$ and $p_{1,o} = 1-p_{1,1}$

$$q_{o,o} = \frac{1-p_{1,1}}{2-p_{o,o}-p_{1,1}} \qquad q_{o,1} = \frac{1-p_{o,o}}{2-p_{o,o}-p_{1,1}}$$

$$q_{1,o} = \frac{1-p_{1,1}}{2-p_{o,o}-p_{1,1}} \qquad q_{1,1} = \frac{1-p_{o,o}}{2-p_{o,o}-p_{1,1}}$$

These values agree with what was found previously. It should be noted that the fact that $q_{o,o} = q_{1,o}$ and $q_{o,1} = q_{1,1}$ is because the equations generated from each matrix row are identical in form. Therefore, each row of Q must be the same. This property is found by induction to be true of all steady state matrices. Thus Q may be more simply represented as:

$$Q = \begin{bmatrix} q_{o,o} & q_{o,1} & \cdots & q_{o,m} \\ \cdot & & & \cdot \\ \cdot & & & \cdot \\ q_{m,o} & q_{m,1} & \cdots & q_{m,m} \end{bmatrix} = \begin{bmatrix} q_o & q_1 & \cdots & q_m \\ \cdot & & & \cdot \\ \cdot & & & \cdot \\ q_o & q_1 & \cdots & q_m \end{bmatrix}$$

If it were given that the initial probability of the process was $P(0) = [P_o(0), P_1(0) \ldots P_m(0)]$, then the steady state probability state of the process would be:

$$\text{steady-state probability state} = \lim_{n \to \infty} P(n) = \lim_{n \to \infty}(P(0)P^n) = \lim_{n \to \infty}(P(0)P(n))$$

$$= P(0)Q \qquad (6.14)$$

$$= [P_o(0), P_1(0), \ldots P_m(0)] \cdot \begin{bmatrix} q_o, q_1, & \cdots & q_m \\ \cdot & & & \cdot \\ \cdot & & & \cdot \\ q_o, q_1, & \cdots & q_m \end{bmatrix}$$

$$\text{So, steady-state probability state} = \sum_{i=0}^{m} P_i(0) \times (q_o, q_1, \ldots q_m)$$

$$= (q_o, q_1, \ldots q_m) \text{ since } \sum_{i=1}^{m} P_i(0) = 1$$

Thus, it is seen that the steady state probability state is independent of the initial probability state.

Using a three-state example, let us define the transition probability matrix P as follows:

$$P = \begin{bmatrix} 1/2 & 1/2 & 0 \\ 1/4 & 1/2 & 1/4 \\ 0 & 1/2 & 1/2 \end{bmatrix} \begin{matrix} 0 \\ 1 \\ 2 \end{matrix} \quad \text{then } P^2 = \begin{bmatrix} 3/8 & 4/8 & 1/8 \\ 2/8 & 4/8 & 2/8 \\ 1/8 & 4/8 & 3/8 \end{bmatrix}$$

with states 0, 1, 2.

if $P(0) = (1, 0, 0)$ then $P(2) = [3/8, 4/8, 1/8]$.

As previously shown in the generalized two-state example,

$$\lim_{n \to \infty} P_{ij}(n) = q_{ij} = \text{constant}$$

It can be shown that this only applies in the case of ergodic Markov chain when $p_{11} > 0$ and all states communicate or

$$P_j = \sum_{i=0}^{\infty} P_i P_{ij} > 0 \text{ for all } j$$

If Q is the matrix q_{ij} and it is possible to go from one state to any other state in a finite number of steps, then

$$P(n) \to Q \text{ in the limit when } n \to \infty$$

If $X = (x_i \ldots x_n)$ = Unique probability state vector when $n \to \infty$, $Q = P(n) = P(n-1)$ and $P_1(n) = P(n-1)$, when P can be written as $XP = X$. In our example then,

$$1/2\ x_1 + 1/4\ x_2 = x_1$$

$$1/2\ x_1 + 1/2\ x_2 + 1/2\ x_3 = x_2$$

$$1/4\ x_2 + 1/2\ x_3 = x_3$$

$$x_1 + x_2 + x_3 = \sum x_i = 1$$

Then we obtain

$$x_1 = 1/4,\ x_2 = 1/2,\ x_3 = 1/4$$

or

$$\lim_{n \to \infty} P(n) = (1/4, 1/2, 1/4)$$

This result is obtained no matter which state process was started and gives us the steady-state probabilities of being in any one of our three states.

To assist in clarifying these concepts, consider the following numerical example of a three-state Markov Chain. The process can be modelled by the network as shown in Figure 6.4.

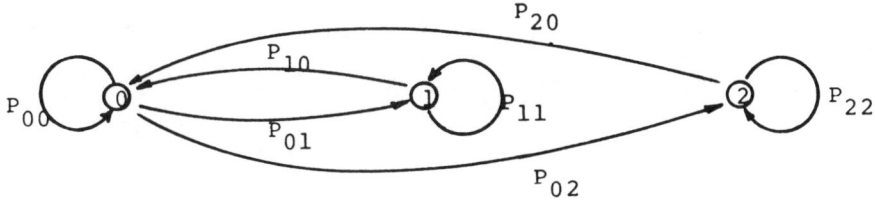

FIGURE 6.4:   Three-State Markov Process

or the stochastic transition matrix:

$$P = \begin{array}{c|ccc} \text{STATE} & 0 & 1 & 2 \\ \hline 0 & 1/2 & 1/4 & 1/4 \\ 1 & 1/2 & 1/2 & 0 \\ 2 & 1/2 & 0 & 1/2 \end{array}$$

The two-step transition matrix will be:

$$P(2) = P \cdot P = \begin{bmatrix} 1/2 & 1/2 & 1/4 \\ 1/2 & 1/2 & 0 \\ 1/2 & 0 & 1/2 \end{bmatrix} \begin{bmatrix} 1/2 & 1/4 & 1/4 \\ 1/2 & 1/2 & 0 \\ 1/2 & 0 & 1/2 \end{bmatrix} = \begin{bmatrix} 1/2 & 1/4 & 1/4 \\ 1/2 & 3/8 & 1/8 \\ 1/2 & 1/8 & 3/8 \end{bmatrix}$$

And, if the initial probability state was $P(0) = (1,0,0)$, then

$$P(2) = P(0)P^2 = [1,0,0] \begin{bmatrix} 1/2 & 1/4 & 1/4 \\ 1/2 & 3/8 & 1/8 \\ 1/2 & 1/8 & 3/8 \end{bmatrix} = [1/2, 1/4, 1/4]$$

The steady state matrix Q may be determined from (6.9) by solving

$$\begin{bmatrix} q_0 & q_1 & q_2 \\ q_0 & q_1 & q_2 \\ q_0 & q_1 & q_2 \end{bmatrix} = \begin{bmatrix} q_0 & q_1 & q_2 \\ q_0 & q_1 & q_2 \\ q_0 & q_1 & q_2 \end{bmatrix} \begin{bmatrix} 1/2 & 1/4 & 1/4 \\ 1/2 & 1/2 & 0 \\ 1/2 & 0 & 1/2 \end{bmatrix}$$

For q_0, q_1 and q_2 (remembering in addition $q_1 + q_2 + q_3 = 1$. The steady state values are found to be:

$$Q = \begin{bmatrix} 1/2 & 1/4 & 1/4 \\ 1/2 & 1/4 & 1/4 \\ 1/2 & 1/4 & 1/4 \end{bmatrix}$$

If it is given that the initial probability state is P(0) = (1,0,0), then the steady state probability state can be found from (6.10) to be:

$$\text{steady state probability state} = P(0)Q = (1,0,0) \begin{vmatrix} 1/2 & 1/4 & 1/4 \\ 1/2 & 1/4 & 1/4 \\ 1/2 & 1/4 & 1/4 \end{vmatrix} = (1/2, 1/4, 1/4)$$

which means that after several steps there will be a 50% chance the process is in state 0, 25% chance it is in state 1, and 25% chance it is in state 2.

Example 6.1

A Markov process is described by a stochastic matrix:

$$\begin{array}{c} \\ 0 \\ 1 \\ 2 \end{array} \begin{array}{ccc} 0 & 1 & 2 \end{array} \\ \begin{bmatrix} 1/4 & 1/2 & 1/4 \\ 1/2 & 1/4 & 1/4 \\ 1/4 & 1/4 & 1/2 \end{bmatrix}$$

If the system is initially in state 0 and is stationary, what is the probability that it is in state 2 after 2 transitions, and what is its steady state probability vector.

    a) From 0 - 2 in two transitions. It can happen in 2 ways only:

$$P_{0-2}(2) = 1/2 \cdot 1/4 + 1/4 \cdot 1/4 + 1/4 \cdot 1/2$$

$$\text{0-1 ; 1-2} \quad \text{0-0 ; 0-2} \quad \text{0-2 ; 2-2}$$

$$P_{0-2}(2) = 2/8 + 1/16$$

$$P_{0-2}(2) = 5/16$$

    b) What is the steady state probability vector?

        1) $P_0 = 1/4\ P_0 + 1/2\ P_1 + 1/4(1-P_0-P_1)$

        2) $P_1 = 1/2\ P_0 + 1/4\ P_1 + 1/4(1-P_0-P_1)$

From (1) $4P_0 - P_0 = 2P_1 + 1 - P_0 - P_1$

              $4P_0 = P_1 + 1$

From (2) $4P_1 = 2P_0 + P_1 + 1 - P_0 - P_1$

$$P_1 = \frac{P_0}{4} + \frac{1}{4}$$

and

$$P_0 = \frac{1}{2}$$

$$P_1 = \frac{1}{3}$$

$$P_2 = \frac{1}{3}$$

6.4 Exercises

    1. A Markov process is described by a stochastic matrix:

$$P = \begin{matrix} & 0 & 1 & 2 & & \text{STATE} \\ & 1/8 & 4/8 & 3/8 & : & 0 \\ & 2/8 & 4/8 & 2/8 & : & 1 \\ & 1/8 & 4/8 & 3/8 & : & 2 \end{matrix}$$

If the system is initially in state 0 and is stationary, what is the probability that it is in state 2 after two transitions, and what is its steady state probability state vector?

2. A system has four states and transition probabilities as shown:

$$\underline{P} = \begin{bmatrix} 0 & 7/8 & 1/16 & 1/16 \\ 0 & 3/4 & 1/8 & 1/8 \\ 0 & 0 & 1/2 & 1/2 \\ 1 & 0 & 0 & 0 \end{bmatrix} \begin{matrix} \text{STATE} \\ 0 \text{ new} \\ 1 \text{ operating} \\ 2 \text{ defective} \\ 3 \text{ failed} \end{matrix}$$

There are three possible decisions that can be made for the system when in a state

Decision No.	Decision
1	Do nothing
2	Repair (return to state 1)
3	Renew (return to state 0)

If the cost of a decision in a particular state is as follows:

		Decision	
State	1	2	3
0	0	0	8
1	1	2	10
2	3	3	12
3	10	12	15

Define three alternative policies and compute the expected cost of the system, given your decisions are unique for each state.

3. Determine the optimum decision matrix for the above problem and compare it to your results.

4. A non-maintained on-line series system with parallel paths requires either path 1 or path 2 and path 3 or all paths to operate.

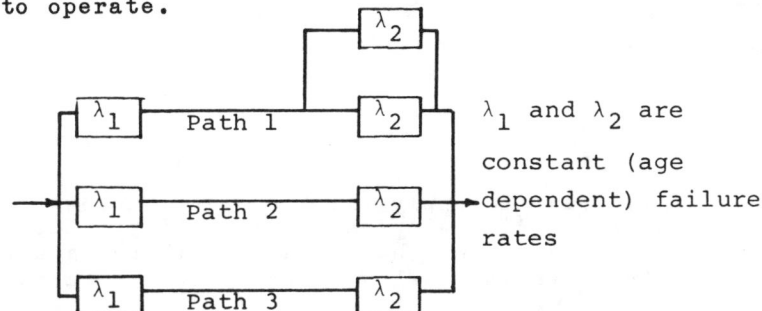

λ_1 and λ_2 are constant (age dependent) failure rates

Determine the number of states of this system. Construct the flow diagram describing the failure processes and establish the transition probability matrix for these processes.

6.5 References

1. Feller, W., "An Introduction to Probability Theory and Its Applications", Vol. I and II, John Wiley and Sons, Inc., New York, 1966.

2. Howard, R.A., "Dynamic Programming and Markov Processes", The M.I.T. Press, Massachusetts Institute of Technology, Cambridge, MA, 1964.

3. Howard, R.A., "Dynamic Probabilistic Systems", Vol. I, Markov Models, John Wiley and Sons, Inc., New York, 1971.

4. Howard, R.A., "Dynamic Probabilistic Systems", Vol. II, Semi-Markov and Decision Processes, John Wiley and Sons, Inc., New York, 1971.

5. Dynkin, E.B., "Theory of Markov Processes", Prentice-Hall Inc., Englewood Cliffs, New Jersey, 1961.

6. Parzen, E., "Stochastic Processes", Holden-Day Inc., San Francisco, 1962.

7. Sittler, R.W., "Systems Analysis of Discrete Markov Processes", I.R.E. Transactions - Circuit Theory, OT-3, No. 1, 1956.

8. Bharucha-Reid, A.T., "Elements of the Theory of Markov Processes and Their Applications", McGraw-Hill Book Company, Inc., New York, 1962.

7.0 THE GENERALIZED FAILURE PROCESS FOR NON-MAINTAINED SYSTEMS

Ernst G. Frankel

Now it is time to apply the theory of Markov processes to systems in order to obtain their reliability. The general approach will be to model the systems' failure as a Poisson process, and then use the Markov matrices to determine system reliability.

First, consider a system of n identical components, each with an age-dependent failure rate $f(t)$. Only one component is required for system operation and only the working component is subject to failure. The states of the system will represent the number of components that have failed. Thus state $k(0 \leq k \leq n)$ represents when k components have failed or equivalently, when (n-k) components still operate. The probability of the system transiting from state k to state k+1 in the time interval t to t+dt is the age-dependent failure rate times the time interval for failure $f(t)\,dt$. If it is assumed that in the limit as $dt \to 0$ the probability of more than one failure or transition to states not adjacent is a second order effect which may be neglected, and if assumed that the transition probability $f(t)\,dt$ is independent of the state of the system, then the system failure behavior may be described by the stochastic matrix:

$P = $ Initial State

FINAL STATE

	0	1	2	n-1	n
0	1-f(t)dt	f(t)dt	0	0	0
1	0	1-f(t)dt	f(t)dt	0	0
2	"	"	"	"	"
"	"	"	"	"	"
n-1	0	0	0	1-f(t)dt	f(t)dt
n	0	0	0	0	1

Note that for each row the probability of transiting to the next higher state (the operating component fails) is $f(t)dt$, which was defined previously. Similarly, the probability that a component in a state will be remaining in the present state is $(1-f(t)dt)$ since the system may either transit to the next higher state or just remain in the present state (remembering that is is assumed that the probability of transiting to non-adjacent higher states is zero). The property that all rows sum to one will be found in all the stochastic matrixes that will be considered later.

Alternatively, the system's failure behavior may be described by a set of difference equations which may be generated by an argument similar to the following. For a system to be in state 0 at time t+dt, it must have been in state 0 at time t and must have remained there during interval dt. Thus, the probability of being in state 0 is the joint (or compound) event of being there initially and not having a transit out of that state. In equation form, this is represented as:

$$P_0(t+dt) = \begin{bmatrix} \text{probability} \\ \text{of being in} \\ \text{state 0 at} \\ \text{time t+dt} \end{bmatrix} = \begin{bmatrix} \text{probability} \\ \text{of being in} \\ \text{state 0 at} \\ \text{time t} \end{bmatrix} \times \begin{bmatrix} \text{probability} \\ \text{of not tran-} \\ \text{siting from} \\ \text{state 0 in} \\ \text{interval dt} \end{bmatrix} = P_0(t)(1-f(t)dt)$$

Similarly,

$$P_1(t+dt) = \begin{bmatrix} \text{probability} \\ \text{of being in} \\ \text{state 1 at} \\ \text{time t+dt} \end{bmatrix} = \begin{bmatrix} \text{probability} \\ \text{of being in} \\ \text{state 0 at} \\ \text{time t} \end{bmatrix} \times \begin{bmatrix} \text{probability} \\ \text{of transit-} \\ \text{ing from} \\ \text{state 0 to} \\ \text{state 1 in} \\ \text{interval dt} \end{bmatrix} + \begin{bmatrix} \text{probability} \\ \text{of being in} \\ \text{state 1 at} \\ \text{time t} \end{bmatrix} \times$$

$$\begin{bmatrix} \text{probability} \\ \text{of not tran-} \\ \text{siting from} \\ \text{state 1 in} \\ \text{interval dt} \end{bmatrix}$$

$$= P_0(t)f(t)dt + P_1(t)(1-f(t)dt)$$

$$\vdots$$

$$P_k(t+dt) = P_{k-1}(t)f(t)dt + P_k(t)(1-f(t)dt)$$

$$\vdots$$

$$P_n(t+dt) = P_{n-1}(t)f(t)dt + P_n(t) \leftarrow \begin{bmatrix} \text{Note: If system is in state n, all} \\ \text{components failed, then it must} \\ \text{remain in that state.} \end{bmatrix}$$

It should be noticed that the coefficients of probability terms, $P_0(t)$, $P_1(t)$...$P_n(t)$, on the right hand side of the expression for $P_k(t+dt)$ correspond to the respective elements of the kth column of the stochastic matrix P.

These difference equations may be reformulated as differential equations. For example:

$$P_k(t+dt) = P_{k-1}(t)f(t)dt + P_k(t)(1-f(t)dt)$$

$$P_k(t+dt) - P_k(t) = (f(t)P_{k-1}(t) - f(t)P_k(t))dt$$

Here the terms $P_k(t)$ are the state probabilities at time t (probability of being in state k at time t). The probability of being in state 0 at time zero, for example, $P_o(t)$ and no failure during the next interval dt, $(1-f(t)dt)$ is equal to the probability that given the system is in state 0 at time t it will remain in this state in the next interval of time dt. The coefficients of the ith state probabilities in the columns of the different equations are the row coefficients in the corresponding row of the transition probability matrix. As a result the set of difference equations could also be represented as a vector

$$[P_o(t+dt), P_1(t+dt), \ldots P_n(t+dt)] = [P_o(t), \ldots P_n(t)]\underline{P}$$

or $\underline{P}(t+dt) = \underline{P}(t) \cdot \underline{P}$ where $\underline{P}(t)$ is the probability state vector.

$$\lim_{dt \to 0} \frac{P_k(t+dt) - P_k(t)}{dt} = \lim_{dt \to 0} (f(t)P_{k-1}(t) - f(t)P_k(t))$$

and

$$P_k'(t) = f(t)P_{k-1}(t) - f(t)P_k(t)$$

Thus the complete set of difference equations results in the set of differential equations:

$$dP_o(t)/dt = P_o'(t) = -f(t)P_o(t)$$

$$dP_1(t)/dt = P_1'(t) = f(t)P_o(t) - f(t)P_1(t)$$

$$\vdots$$

$$dP_k(t)/dt = P_K'(t) = f(t)P_{k-1}(t) - f(t)P_k(t)$$

$$\vdots$$

$$dP_n(t)/dt = P_n'(t) = f(t)P_{n-1}(t)$$

If it is assumed that at t=0 all components are in working order, then the initial conditions for these differential equations are simply:

$$P_0(0) = 1; \quad P_1(0) = P_2(0) = \ldots\ldots\ldots P_n(0) = 0$$

or expressed as the state vector P(0)

$$P(0) = [1, 0, \ldots, 0]$$

These differential equations may be solved to yield the result:

$$P_0(t) = e^{-\int_0^t f(\tau)d\tau}$$

$$P_1(t) = [\int_0^t f(\tau)d\tau] \exp(-\int_0^t f(\tau)d\tau)$$

$$\circ$$
$$\circ$$

$$P_k(t) = (\int_0^t f(\tau)d\tau)^k \exp(-\int_0^t f(\tau)d\tau)/k!$$

$$\circ$$
$$\circ$$

$$P_n(t) = [\int_0^t f(\tau)d\tau]^n \exp(-\int_0^t f(\tau)d\tau)/n!$$

which is a typical Poisson process. Therefore, the probability of having k failures in time (0,t) is:

$$P_k(t) = [\int_0^t f(\tau)d\tau]^k \exp(-\int_0^t f(\tau)d\tau)/k!$$

and the distribution function is obtained simply by summation. Thus, the probability of at most k failure occurring in a time (0,t) is:

$$P(k;t) = \sum_{i=0}^{k} P_k(t) = \exp(-\int_0^t f(\tau)d\tau) \cdot \sum_{i=0}^{k}[\int_0^t f(\tau)d\tau]^i/(i)!$$

The system reliability will be the probability of at most n−1 failures in time t, which is $P(n-1;t)$:

$$R(t) = \exp\left(-\int_0^t f(\tau)d\tau\right) \cdot \sum_{i=0}^{n-1} \left[\int_0^t f(\tau)d\tau\right]^i / (i)!$$

Finally, the probability of exactly k failures at time t will be the probability of the joint event of (k−1) failures in time t and one failure at time t:

$$P_k(t) = \begin{matrix}\text{probability}\\ \text{of exactly}\\ \text{k failures}\\ \text{at time t}\end{matrix} = P_{k-1}(t)f(t) = \frac{\{[\int_0^t f(\tau)d\tau]^{k-1} \exp(-\int_0^t f(\tau)d\tau)\}f(t)}{(k-1)!}$$

This is the density of the Gamma distribution. Now, for the cases when the age-dependent failure rate is constant or a linear function of time, the expressions for $P_k(t)$ and $R(t)$ reduce to:

$$f(t) = \lambda: \quad P_k(t) = \left[\int_0^t \lambda dt\right]^k \exp\left(-\int_0^t \lambda dt\right)/k! = (\lambda t)^k e^{-\lambda t}/k!$$

$$R(t) = \sum_{i=0}^{n-1} P_i(t) = e^{-\lambda t} \sum_{i=0}^{n-1} \frac{(\lambda t)^i}{(i)!}$$

$$f(t) = \lambda t: \quad P_k(t) = \left[\int_0^t \lambda t\, dt\right]^k \exp\left(-\int_0^t \lambda t\, dt\right)/k! = \left(\frac{\lambda t^2}{2}\right)^k \exp\left(-\frac{\lambda t^2}{2}\right)/k!$$

$$R(t) = \sum_{i=0}^{n-1} P_i(t) = e^{-\frac{\lambda t^2}{2}} \sum_{i=0}^{n-1} \left(\frac{\lambda t^2}{2}\right)^i / i!$$

Instead of solving each differential equation separately, the whole system may be solved simultaneously by matrix methods. Let $\underline{A}$ be defined as the differential transition or characteristic matrix, which is a matrix of the transposed coefficients of the differential equations, opbtain by computing the matrix $\underline{P}-\underline{I}$. In the case where $f(t) = \lambda$ and discarding the dt's in the matrix, we obtain the transition probability matrix $\underline{P}$,

$$\underline{P} = \begin{array}{c} \text{STATE} \\ 0 \\ 1 \\ 2 \\ . \\ . \\ . \\ n-1 \\ n \end{array} \begin{array}{c} \begin{array}{cccccc} 0 & 1 & 2 & \cdots & n-1 & n \end{array} \\ \left[\begin{array}{cccccc} 1-\lambda & \lambda & 0 & \cdots & 0 & 0 \\ 0 & 1-\lambda & \lambda & \cdots & 0 & 0 \\ , & , & , & & , & , \\ , & , & , & & , & , \\ , & , & , & & , & , \\ , & , & , & & , & , \\ 0 & 0 & 0 & & 1-\lambda & \lambda \\ 0 & 0 & 0 & \cdots & 0 & 1 \end{array} \right] \end{array}$$

From which after subtracting the identify matrix the characteristic matrix $\underline{A}$ is obtained.

$$\underline{A} = [\underline{P} - I] = \begin{array}{c} \text{STATE} \\ 0 \\ 1 \\ 2 \\ . \\ . \\ . \\ n-1 \\ n \end{array} \begin{array}{c} \begin{array}{cccccc} 0 & 1 & 2 & \cdots & n-1 & n \end{array} \\ \left[\begin{array}{cccccc} -\lambda & \lambda & 0 & \cdots & 0 & 0 \\ 0 & -\lambda & \lambda & \cdots & 0 & 0 \\ , & , & , & & , & , \\ , & , & , & & , & , \\ , & , & , & & , & , \\ , & , & , & & , & , \\ 0 & 0 & 0 & \cdots & -\lambda & \lambda \\ 0 & 0 & 0 & \cdots & 0 & 0 \end{array} \right] \end{array}$$

The set of differential equations may then be expressed as a matrix equation:

$$[(P'_0(t), P'_1(t) \ldots P'_n(t))] = [(P_0(t), P_1(t), \ldots P_n(t))] [\underline{A}] \qquad (7.1)$$

or

$$[P'(t)] = [P(t)] [\underline{A}]$$

where $\underline{P}'(t)$ is the derivative of the probability state vector at time t.

7.1 Solution Using Laplace Transforms

Although the set of differential equations can be solved via a brute force time domain approach to yield $P_k(t)$, Laplace transform methods provide for simple functions for f(t), a far simpler solution technique. This is because Laplace transforms convert the time domain differential equations into a set of s-domain algebraic equations. The Laplace transform on P(t) is defined as follows:

$$P(s) = L\{P(t)\} = \int_0^\infty e^{-st} P(t) dt \qquad (7.2)$$

and

$$P'(s) = L\{P'(t)\} \; sP(s) - P(0) \qquad (7.3)$$

First, (7.2) and (7.3) will be used to solve for $P_0(t)$, then for the remaining states. Applying (7.2) and (7.3) to the first differential equations,

$$P'_0(t) = -f(t) P_0(t)$$

can be written in Laplace transforms as

$$sP_0(s) - P_0(0) = -L\{f(t)P_0(t)\}$$

If $f(t) = \lambda$ then:

$$sP_0(s) - P_0(0) = -\lambda P_0(s)$$

Assuming $P_0(0)$ is defined to be 1; or assuming that we start with the system in state 0 we obtain with $P_0(0) = 1$.

$$sP_0(s) - 1 = -\lambda P_0(s)$$

Solving for $P_0(s)$:

$$P_0(s) = \frac{1}{s+\lambda} \qquad (7.4)$$

To find $P_0(t)$, we must take the inverse transform of (7.4). From a table of Laplace transform pairs it is found that

$$L^{-1}\{P_0(s)\} = P_0(t) = e^{-\lambda t} \qquad (7.5)$$

Thus, the probability of no failures up to time t is simply a negative exponential function in the case for constant age dependent failure rate. With the solution to $P_0(t)$, it is relatively easy to find $P_1(t)$, $P_2(t)$, and so forth. For example, the expression for $P_1'(t)$ or (7.1) with $f(t) =$ is solved as follows:

$$L\{P_1'(t) = \lambda P_0(t) - \lambda P_1(t)\} \qquad \text{(take transform)}$$

$$sP_1(s) - P_1(0) = \lambda P_0(s) - \lambda P_1(s)$$

$$P_1(s) = \lambda \frac{P_0(s)}{s+\lambda} \qquad \text{substitution of } P_1(0) = 0$$

$$\text{and } P_0(s) = \frac{1}{s+\lambda}$$

$$L^{-1}\{P_1(s) = \frac{\lambda}{(s+\lambda)^2} \qquad \text{solving for } P_1(s) \text{ and taking inverse transform}$$

$$P_1(t) = \lambda t \, e^{-\lambda t}$$

Continuing in this fashion, it is seen that in general

$$P_k(s) = \frac{\lambda^k}{(s+\lambda)^{k+1}} \qquad (7.6)$$

whose inverse transform is

$$P_k(t) = \frac{(\lambda t)^k}{k!} e^{-\lambda t}$$

which is the Poisson density function.

To take the Laplace transform of the system simultaneously, it is necessary only to take the transform of 7.1 which yields

$$s[P_0(s), P_1(s), \ldots P_n(s)] - [P_0(0), P_1(0) \ldots P_n(0)]$$
$$= [P_0(s), P_1(s) \ldots P_n(s)][A]$$

or

$$s[P(s)] - [P(0)] = [P(s)][A] \qquad (7.7)$$

where $[P(s)]$ is the Laplace transform of the Probability State Vector $\underline{P}(t)$. Solving (7.7) for $[P(s)]$

$$[P(s)][sI-A] = [P(0)]$$

$$[P(s)] = [P(0)][sI-A]^{-1} \qquad (7.8)$$

In the case where $[P(0)] = [1, 0, 0, \ldots 0]$, (7.8) reduces to:

$$[P(s)] = [1,0,0\ldots 0] \begin{bmatrix} s+\lambda & -\lambda & 0 & \cdots & 0 & 0 \\ 0 & s+\lambda & -\lambda & \cdots & 0 & 0 \\ \vdots & \vdots & \vdots & & \vdots & \vdots \\ 0 & 0 & 0 & & s+\lambda & \lambda \\ 0 & 0 & 0 & & 0 & s \end{bmatrix}^{-1}$$

$$[P(s)] = [\frac{1}{s+\lambda}, \frac{\lambda}{(s+\lambda)^2} \ldots \ldots \frac{\lambda^{n-1}}{(s+\lambda)^n}, \frac{\lambda^n}{(s+\lambda)^{n+1}}]$$

After taking the inverse transform, the same result as before is obtained:

$$L^{-1}\{P(s)\} = [\frac{1}{s+\lambda}, \frac{\lambda}{(s+\lambda)^2}, \ldots \ldots \frac{\lambda^{n-1}}{(S+\lambda)^n}, \frac{\lambda^n}{(S+\lambda)^{n+1}}]$$

$$[P(t)] = [e^{-\lambda t}, \lambda t e^{-\lambda t}, \ldots \frac{(\lambda t)^{n-1} e^{-\lambda t}}{(n-1)!}, \frac{(\lambda t)^n e^{-\lambda t}}{n!}$$

In the following sections, examples of each of these solution methods will be presented.

7.2 Stand-by (Off Line) Redundant System

The generalized Poisson failure process just discussed actually describes a stand-by redundant system where only the single required operating component is subject to failure. The components which stand-by until the immediately preceding component fails are not subject to failure until called upon to take over. The system is only inoperative (failed) if all components fail and the failure event of any component is independent of that of any other component. Switching to stand-by (off-line) components is assumed to be instantaneous, perfect, and failure free. For a system with a constant age dependent failure rate for each of n identical components , it was found that the system reliability is

$$R_T(t) = e^{-\lambda t} \sum_{i=0}^{n-1} (\lambda t)^i \Big/ (i)! \qquad (7.9)$$

Similarly, for a two-component system, the system reliability is

$$R_T(t) = e^{-\lambda t}(1+\lambda t)$$

To make the system somewhat more realistic, assume that the off-line component is a 2-component system that is also subject to failure when off-line, but its failure rate $\lambda_2 < \lambda_1$ where λ_1 is the failure rate of the on-line component.

Define the following system states:

STATE	OPERATING	STAND-BY AVAILABLE	FAILED
0	A	B	---
1	A	---	B
2	B	---	A
3	---	---	A,B

FIGURE 7.1: STATES OF A TWO-COMPONENT STANDBY (OFF-LINE) REDUNDANT SYSTEM

The stochastic transition probability matrix of this system can be derived in a method similar to that used in the Poisson process. In this case it will be

	STATE	0	1	2	3
	0	$1-(\lambda_1+\lambda_2)dt$	$\lambda_2 dt$	$\lambda_1 dt$	0
P =	1	0	$1-\lambda_1 dt$	0	$\lambda_1 dt$
	2	0	0	$1-\lambda_1 dt$	$\lambda_1 dt$
	3	0	0	0	1

and the corresponding difference equations will be:

$P_0(t+dt) = P_0(t)(1-(\lambda_1+\lambda_2)dt)$

$P_1(t+dt) = P_0(t)\lambda_2 dt + P_1(t)(1-\lambda_1 dt)$

$P_2(t+dt) = P_0(t)\lambda_1 dt + P_2(t)(1-\lambda_1 dt)$

Remember, coefficients of $\lambda_1(t)$ and $\lambda_2(t)$, on the right hand side of equations of $R_k(t+dt)$ correspond to the respective

$$P_3(t+dt) = P_1(t)\lambda_1 dt + P_2(t)\lambda_1 dt + P_3(t)$$

entities of the Kth column of P.

The equivalent differential equations will be:

$$P'_0(t) = -(\lambda_1+\lambda_2) P_0(t)$$

$$P'_1(t) = \lambda_2 P_0(t) - \lambda_1 P_1(t)$$

$$P'_2(t) = \lambda_1 P_0(t) \qquad\quad - \lambda_1 P_2(t)$$

$$P'_3(t) = \qquad\qquad + \lambda_1 P_1(t) + \lambda_1 P_2(t)$$

Assuming that the system is initially in State 0, $P_0(0) = 1$; $P_1(0) = P_2(0) = P_3(0) = 0$), then the equivalent Laplace transforms will be:

$$sP_0(s) - P_0(0) = -(\lambda_1+\lambda_2)P_0(s)$$

$$sP_1(s) - P_1(0) = \lambda_2 P_0(s) - \lambda_1 P_1(s)$$

$$sP_2(s) - P_2(0) = \lambda_1 P_0(s) \qquad\quad - \lambda_1 P_2(s)$$

$$sP_3(s) - P_3(0) = \qquad\qquad \lambda_1 P_1(s) + \lambda_1 P_2(s)$$

Solving for $P_0(s)$: $P_0(s) = \dfrac{1}{s+\lambda_1+\lambda_2}$

$$P_1(s) : P_1(s) = \frac{\lambda_2}{s+\lambda_1} P_0(s) = \left(\frac{\lambda_2}{s+\lambda_1}\right)\left(\frac{1}{s+\lambda_1+\lambda_2}\right)$$

$$P_2(s) : P_2(s) = \frac{\lambda_1}{s+\lambda_1} P_0(s) = \left(\frac{\lambda_1}{s+\lambda_1}\right)\left(\frac{1}{s+\lambda_1+\lambda_2}\right)$$

$$P_3(s) : P_3(s) = \frac{\lambda_1}{s}(P_1(s)+P_2(s)) = \frac{\lambda_1}{s}\left(\frac{1}{s+\lambda_1+\lambda_2}\right)\left(\frac{\lambda_1+\lambda_2}{s+\lambda_1}\right)$$

Taking the inverse transforms:

$$P_0(t) = L^{-1}\{\frac{1}{s+\lambda_1+\lambda_2}\} = e^{-(\lambda_1+\lambda_2)t}$$

$$P_1(t) = L^{-1}\{\frac{\lambda_2}{(s+\lambda_1)(s+\lambda_1+\lambda_2)}\} = e^{-\lambda_1} - e^{-(\lambda_1\pm\lambda_2)t}$$

and

$$P_2(t) = \frac{\lambda_1}{\lambda_2}[e^{-\lambda_1 t} - e^{-(\lambda_1+\lambda_2)t}]$$

The reliability of this system is $= \sum_{i=0}^{2} P_i(t)$

$$= e^{-\lambda_1 t} + \frac{\lambda_1}{\lambda_2}[e^{-\lambda_1 t} - e^{-(\lambda_1+\lambda_2)t}]$$

which could also be obtained from $R_T(t) = 1-P_3(t)$. Finally, we solve the system via matrix methods. In this case

$$[A] = [P-I] = \begin{bmatrix} 1-(\lambda_1+\lambda_2) & \lambda & \lambda & 0 \\ 0 & 1-\lambda_1 & 0 & \lambda_1 \\ 0 & 0 & 1-\lambda_1 & \lambda_1 \\ 0 & 0 & 0 & 1 \end{bmatrix} = \begin{bmatrix} 1 & 0 & 0 & 0 \\ 0 & 1 & 0 & 0 \\ 0 & 0 & 1 & 0 \\ 0 & 0 & 0 & 1 \end{bmatrix}$$

$$[A] = \begin{bmatrix} -(\lambda_1+\lambda_2) & \lambda_2 & \lambda_1 & 0 \\ 0 & -\lambda_1 & 0 & \lambda_1 \\ 0 & 0 & -\lambda_1 & \lambda_1 \\ 0 & 0 & 0 & 0 \end{bmatrix}$$

Sp (7.8) becomes

$$[P(s)] = [P(0)][sI-A]^{-1} = [1,0,0,0] \begin{bmatrix} s+\lambda_1+\lambda_2 & -\lambda_2 & -\lambda_1 & 0 \\ 0 & s+\lambda_1 & 0 & -\lambda_1 \\ 0 & 0 & s+\lambda_1 & -\lambda_1 \\ 0 & 0 & 0 & s \end{bmatrix}$$

$$= \frac{[1,0,0,0]}{s(s+\lambda_1)^2(s+\lambda_1+\lambda_2)} \begin{bmatrix} s(s+\lambda_1)^2, & \lambda_2 s(s+\lambda_1)^1, & \lambda_1 s(s+\lambda_1), & (\lambda_1+\lambda_2)(s+\lambda_1) \\ \cdot & \cdot & \cdot & \cdot \\ \cdot & \cdot & \cdot & \cdot \\ \cdot & \cdot & \cdot & \cdot \end{bmatrix}$$

$$= [\frac{1}{s+\lambda_1+\lambda_2}, \frac{\lambda_2}{(s+\lambda_1)(s+\lambda_1+\lambda_2)}, \frac{\lambda_1}{(s+\lambda_1)(s+\lambda_1+s+\lambda_2)}, \frac{\lambda_1(\lambda_1+\lambda_2)}{s(s+\lambda_1)(s+\lambda_1+\lambda_2)}]$$

Taking the inverse transforms of P(s) will yield P(t) identical to that obtained previously.

Similar models can be used for single component systems subject to fatigue or different failure modes where intermediate failure modes result in part-load operation. If state 0 is the full (normal) operating state, state 1 is the state of partial operation due to failure rate λ_1 and state 2 is the state of total failure due to failure rate λ_n; i.e.,

λ_1 = operational wear failure rate

λ_n = catastrophic failure rate

Then the stochastic matrix becomes:

$$P = \begin{bmatrix} 1-(\lambda_1+\lambda_n) & \lambda_1 & \lambda_n \\ 0 & 1-\lambda_n & \lambda_n \\ 0 & 0 & 1 \end{bmatrix}$$

If this component is subjected to repeated "impacts" and the occurrence of each degrades the operation of the component by one state, then

$$P = \begin{bmatrix} 1-(\lambda_1+\lambda_n) & \lambda_1 & & \lambda_n \\ & 1-(\lambda_1+\lambda_n) & \lambda & \lambda_n \\ & & & \lambda_n \\ & & & 1 \end{bmatrix}$$

If the component is considered operable in all but the completely failed state, then the system reliability may be found by analysis similar to that used in the component case:

$$R(t) = e^{-(\lambda_1+\lambda_n)t} \sum_{i=0}^{n-1} \frac{(\lambda_i t)^i}{i!}$$

7.3 Series Systems

Now, it will be demonstrated that the Markov process techniques that have been developed will yield results that agree with what was found by probabilistic and logic arguments previously.

Consider a series system consisting of n-mutually independent components arranged so that the system is assumed to have failed if any component failed. If the n-components are identical and have a constant age dependent failure rate, then the system may be modelled by a two-state process where state 0 is the operating state and state 1 the failed state. The appropriate stochastic matrix and difference equations are:

$$P = \begin{bmatrix} & 0 & 1 \\ 0 & 1-n\lambda & n\lambda \\ 1 & 0 & 1 \end{bmatrix}$$

$$P_0(t+dt) = P_0(t)[1-n\lambda dt] + 0 dt$$

$$P_1(t+dt) = P_0(t)n\lambda dt + P_0(t)$$

then

$$P_0'(t) + n\lambda P_0(t) = 0$$

$$P_0(s) = 1/(s+n\lambda) \text{ and } P_0(t) = e^{-n\lambda t}$$

The reliability of the system is

$$R_t(t) = P_0(t) = e^{-n\lambda_i}$$

If the components are not identical, and λ_i is the failure rate of the ith components, then,

$$R_T(t) = \exp[-\sum_{i=1}^{n} \lambda_i t];$$

and, in general,

$$R_T(t) = \prod_{i=1}^{n} [R_i(t)]$$

Solving the system using matrices......(assuming initially system is operating so $P_0(0) = 1$ and $P_1(0) =$

$$A = P-I = \begin{bmatrix} 1-n\lambda & n\lambda \\ 0 & 1 \end{bmatrix} - \begin{bmatrix} 1 & 0 \\ 0 & 1 \end{bmatrix} = \begin{bmatrix} -n\lambda & \lambda \\ 0 & 0 \end{bmatrix}$$

$$P(s) = P(0)(sI-A)^{-1} = (1,0) \begin{pmatrix} s+n\lambda & -n\lambda \\ 0 & s \end{pmatrix}^{-1}$$

$$= (1,0) \begin{pmatrix} s & n\lambda \\ 0 & s+n\lambda \end{pmatrix} \frac{1}{s(s+n\lambda)}$$

$$(P_0(s), P_1(s)) = (\frac{1}{s+n\lambda}, \frac{n\lambda}{s(s+n\lambda)})$$

$$(P_0(t), P_1(t)) = (e^{-n\lambda t}, 1-e^{-n\lambda t})$$

The reliability of the system is $R_T(t) = P_0(t) = e^{-\lambda n t}$ which agrees exactly with what was obtained previously. If the components are not identical, and λ_i is the age dependent failure rate of the ith component, then the equivalent expression for reliability will be:

$$R_T(t) = e^{-t(\sum_{i=1}^{n} \lambda_i)}$$

and in general

$$R_T(t) = \prod_{i=1}^{n} [R_i(t)]$$

7.4 Redundant (On-Line) Parallel Systems

Consider next a redundant system with on-line stand-by with n identical components, each with $f(t) = \lambda$. The system may be modelled by an n state process where state k represents the system that has k of n components failed. State n will be the system failure state. The appropriate stochastic matrix and difference equations are:

$$P = \begin{bmatrix} 1-n\lambda & n-\lambda & 0 & \cdots & & 0 \\ & 1-(n-1)\lambda & (n-1)\lambda & \cdots & & 0 \\ \cdot & \cdot & \cdot & \cdot & 1-\lambda & \lambda \\ \cdot & \cdot & \cdot & \cdot & \cdot & 1 \end{bmatrix}$$

If we assume n = 2, then:

$$P_0(t+dt) = P_0(t)(1-2\lambda dt) + 0 dt$$

$$P_1(t+dt) = P_0(t)(2\lambda dt) + P_1(t)(1-\lambda dt) + 0 dt$$

$$P_2(t+dt) = P_1(t)\lambda dt + P_2(t) + 0 dt$$

when

$$P_0'(t) = -2\lambda P_0(t)$$

$$P_1'(t) = 2\lambda P_0(t) - \lambda P_1(t)$$

$$P_2'(t) = \lambda P_1(t)$$

Introducing transforms, we obtain:

$$sP_0(s) - P_0(0) = -2\lambda P_0(s)$$

$$sP_1(s) - P_1(0) = 2\lambda P_0(s) - \lambda P_1(s)$$

$$sP_2(s) - P_2(0) = \qquad\qquad -\lambda P_1(s)$$

Solving for $P_0(s)$, $P_1(s)$ and $P_2(s)$:

$$P_0(s) = \frac{1}{s+2\lambda}$$

$$P_1(s) = \frac{2\lambda}{s+\lambda} P_0(s) = \frac{2\lambda}{(s+\lambda)(s+2\lambda)}$$

$$P_2(s) = \frac{\lambda}{s} P_1(s) = \frac{2\lambda^2}{s(s+\lambda)(s+2\lambda)}$$

Taking inverse transforms:

$$P_0(t) = e^{-2\lambda t}$$

$$P_1(t) = 2(e^{-\lambda t} - e^{-2\lambda t})$$

$$P_2(t) = 1 - 2e^{-\lambda t} + e^{-2\lambda t}$$

The reliability of the system, $R_T(t)$, is:

$$R_T(t) = 1 - P_2(t) = P_0(t) + P_1(t) = 2e^{-\lambda t} - e^{-2\lambda t}$$

which is exactly the same as was obtained previously via logic and probabilistic arguments.

For an n-component redundant system with identical components, system reliability

$$R_T(t) = \sum_{i=0}^{n-1} P_i(t) = [1 - (1 - e^{-\lambda t})^n]$$

And, by similar methods to that used above, it may be found for a system for which m out of the total of n components are required that the system reliability is:

$$R_T(t) = \sum_{i=m}^{n} \binom{n}{i} (e^{-\lambda t})^i (1 - e^{-\lambda t})^{n-1} \quad (n > m)$$

And, finally, if the n components are not identical (different λ's) then the system reliability is:

$$R_T(t) = [1 - \prod_{i=1}^{n} (1 - e^{-\lambda_i t})]$$

Comparing a two-component (identical) on-line and off-line system, we find that the off-line system will be better for any value of t. (See Figure 7.2.) It should be noted that perfect, infallible switching is assumed here, which does not apply in reality.

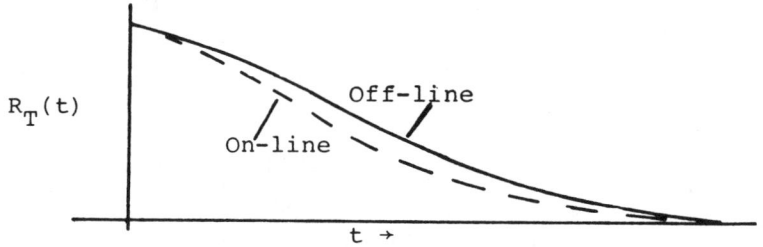

FIGURE 7.2: COMPARISON OF THE RELIABILITY OF TWO-COMPONENT OFF-LINE AND ON-LINE SYSTEMS

Similarly, it can be shown that for non-maintained systems, the lowest level of redundancy is normally best. In General, this is not the best policy for a maintained system.

Although, for the simple series and redundant on-line component systems, it appears that the logic and probabilistic methods previously discussed are simpler, this is not the case for a more complex system. It will be seen that as the systems become more complex, the logic required for solution by probabilistic means becomes intricate, while the Markov process method remains at the level of complexity used in the examples above. Thus, for very complex systems, the Markov process method will always be the method of choice.

7.5 State-Dependent Reliability Models

In many systems, on-line redundancy gives a good simulation of the realistic conditions if the failure rate is assumed to be a function of the state when

$$\underline{P} = \begin{bmatrix} 1-\lambda dt & \lambda_0 dt & \cdots & \cdots & \cdots & \cdots \\ 0 & 1-\lambda_1 dt & \lambda_1 dt & \cdots & \cdots & \cdots \\ \cdot & \cdot & \cdot & \cdot & \cdot & \cdot \\ \cdot & \cdot & \cdot & \cdot & \cdot & \cdot \\ \cdot & \cdot & \cdot & \cdot & \cdot & \cdot \\ \cdot & \cdot & \cdot & \cdots & 1-\lambda_{n-1} dt & \lambda_{n-1} dt \\ \cdot & \cdot & \cdot & \cdots & \cdots & 1 \end{bmatrix}$$

or

$$P_0(t+dt) = P_0(t)(1-\lambda_0 dt) + 0 dt$$

$$P_1(t+dt) = P_0(t) \lambda_0 dt + P_1(t) (1-\lambda_1 dt) + 0 dt$$
$$\vdots$$
$$P_k(t+dt) = P_{k-1}(t) \lambda_{k-1} dt + P_k(t)(1-\lambda_k dt) + 0 dt$$

A very important distribution resulting in a cascade effect is the Yule-Fary distribution where $\lambda_n = (n+1)\lambda$ and

$$P_0(t+dt) = P_0(t) (1-\lambda dt)$$

$$P_1(t+dt) = P_0(t) \lambda dt + P_1(t) (1-2\lambda dt)$$

$$P_k(t+dt) = P_{k-1}(t) k\lambda dt + P_k(t)]1-(1+1)dt]$$

where

$$P_0(s) = \frac{1}{s+\lambda} \qquad P_0(t) = e^{-\lambda t}$$

$$P_1(s) = \frac{P_0(s)\lambda}{s+2\lambda} \qquad P_1(t) = e^{-\lambda t} - e^{-2\lambda t}$$

$$P_k(s) = \frac{P_{k-1}(s)k\lambda}{s+(k+1)\lambda} \qquad P_k(t) = e^{-\lambda t}[1-e^{-\lambda t}]^k$$

from which we obtain the reliability of a system requiring only one of n components

$$R_T(t) = \sum_{i=0}^{(n-1)} P_i(t) = \sum_{i=0}^{(n-1)} e^{-\lambda t}[1-e^{-\lambda t}]^i$$

7.6 Linear Stress Model

Let us next assume that the instantaneous failure rate of each component in an on-line stand-by parallel system is directly proportional to the amount of load carried by the component. If we have n components in the system initially and each component has a failure rate λ, then if one of the components has failed the failure rate of the remaining components becomes $\frac{n}{(n-1)}\lambda$, etc. In such a system

$$P = \begin{bmatrix} 1-n\lambda dt & n\lambda dt & \cdots & \cdots & 0 \\ 0 & 1-n\lambda dt & n\lambda dt & & 0 \\ 0 & & & & \\ 0 & & & & 1 \end{bmatrix}$$

$$P_0(t+dt) = P_0(t)][1-n\lambda dt]$$

$$P_1(t+dt) = P_0(t)\, n\lambda dt + P_1(t)[1-n\lambda dt]$$

$$P_n(t+dt) = P_{n-1}(t) n\lambda dt + P_n(t)$$

whence we obtain that

$$P_0(t) = e^{-n\lambda t}$$

$$P_1(t) = n\lambda t e^{-n\lambda t}$$

$$\frac{[n\lambda t]^i\, e^{-n\lambda t}}{i!}$$

and

$$R_T(t) = \sum_{i=0}^{n-1} e^{-n\lambda t}\, \frac{(n\lambda t)^i}{i!}$$

which is the well-known Erlang distribution.

This kind of distribution is useful where parallel components share the load and where the failure rate of each component is a direct function of the load carried by the component (Figure 7.3).

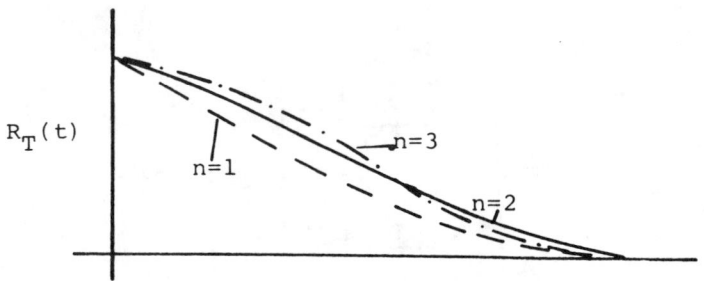

FIGURE 7.3: RELIABILITY OF n-COMPONENT ON-LINE STANDBY PARALLEL SYSTEM

7.7 The Effect of Switching

In discussing off-line stand-by systems, we assumed perfect, instantaneous, and infallible switching. Such conditions do not normally exist in reality; and the switching function must, in actual practice, be assumed to have a failure rate or probability of effective operation. In a simple example, we may assume this failure rate to be constant λ_s although, as shown previously, a time function could be considered.

Let us consider a two-component system (Figure 7.4) where A and B have the same failure rate λ when operating and zero failure rate when standing by. If the switch is only required to put stand-by component on-line when required, but may fail at any time with failure rate λ_s, then:

STATE	OPERATING	STANDING BY	FAILED
0	A	S B	0
1	A	B	S
2	S B	---	A
3	---	B	A S
4	---	S	A B
5	---	---	A S B

$$P = \begin{bmatrix} & 0 & 1 & 2 & 3 & 4 & 5 \\ 0 & 1-(\lambda+\lambda_s) & \lambda_s & \lambda & 0 & 0 & 0 \\ 1 & - & 1-\lambda & 0 & \lambda & 0 & 0 \\ 2 & 0 & 0 & 1-(\lambda+\lambda_s) & \lambda_s & \lambda & 0 \\ 3 & 0 & 0 & 0 & 1 & 0 & 0 \\ 4 & 0 & 0 & 0 & 0 & 1 & 0 \\ 5 & 0 & 0 & 0 & 0 & 0 & 1 \end{bmatrix}$$

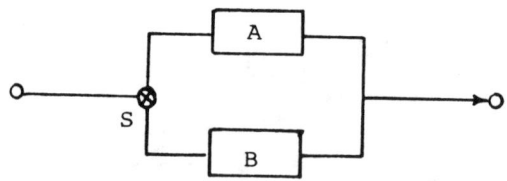

FIGURE 7.4: TWO-COMPONENT SYSTEM WITH SWITCHING

As only the first three states are acceptable, we only compute:

$$P_0(t+dt) = P_0(t)[1-(\lambda+\lambda_2) dt]$$

$$P_1(t+dt) = P_0(t)\lambda_s dt = P_1(t)(1-\lambda dt)$$

$$P_2(t+dt) = P_0(t)\lambda dt + P_2(1-(\lambda+\lambda_s)dt)$$

When assuming initially all components are operative, we obtain

$$P_0'(t) = -(\lambda+\lambda_s) P_0(t) \text{ and } P_0(t) = e^{-(\lambda+\lambda_s)t}$$

$$P_1'(t) = \lambda_s P_0(t) - \lambda P_1(t) \text{ and } P_1(t) = e^{-\lambda t} - e^{-(\lambda+\lambda_s)t}$$

$$P_2'(t) = P_0(t)\lambda - (\lambda+\lambda_s) P_2(t) \text{ and } P_2(t) = \lambda t\, e^{-(\lambda+\lambda_s)t}$$

Therefore the reliability of the system can be expressed as:

$$R_T(t) = \sum_{i=0}^{2} P_i(t) = e^{-\lambda t} + \lambda t\, e^{-(\lambda+\lambda_s)t} = e^{-\lambda t}[1+\lambda t e^{-\lambda_s t}]$$

when $\lambda_s = 0$, then this result is equal to that of a simple two-component redundant system.

If we assume the condition where A and B may fail no matter if they are on-line or off-line, then

$$R_T(t) = \frac{2\lambda+\lambda_s}{\lambda+\lambda_s} e^{-\lambda t} - \frac{\lambda}{\lambda+\lambda_s} e^{-(2\lambda+\lambda_s)t}$$

If in the previous off-line system, we have a total of n components, each with its own switch, and if a single component alone is required for system operation, then the system's reliability is:

$$R_T(t) = \sum_{i=0}^{n-1} \frac{(\lambda t)^i}{i!} e^{-(\lambda+(n-1)\lambda_s)t}$$

Comparing now an ordinary (non-switching) on-line system

with an off-line switching system to compute the acceptable switching failure rate for a two-component system, then for the on-line system to have an equal or better reliability, if

$$2e^{-\lambda t} - e^{-2\lambda t} > e^{-\lambda t} + \lambda t \, e^{-(\lambda+\lambda_s)t}$$

or if

$$\frac{\lambda_s}{\lambda} > -\ln\left\{\frac{1-e^{-\lambda t}}{\lambda t}\right\} \Big/ \lambda t$$

then the on-line (non-switching) system is preferable.

Example 7.1

A problem of a simple series dynamic non-maintained system could consist of a 3-component series system in complete on-line redundancy. If the age dependent failure rates are equal for all components ($\lambda=0.01$/hour)

a. What is the probability that the system is still operative after 10 hours? (λ is independent of the number of components operating and initially all components are operating.)

b. What is the probability that the system is still operative after 10 hours if the redundant series is off-line with instantaneous 100% reliable switching ($\lambda=0$ while components off-line)?

c. What is the MTBF of the two systems?

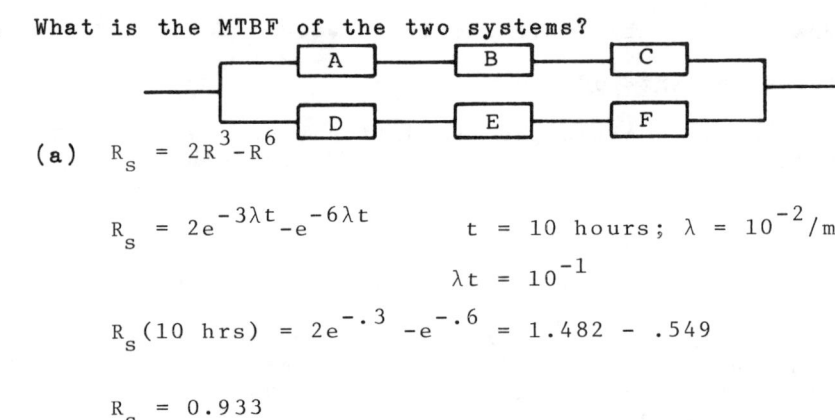

(a) $R_s = 2R^3 - R^6$

$R_s = 2e^{-3\lambda t} - e^{-6\lambda t}$ $t = 10$ hours; $\lambda = 10^{-2}$/m

$\lambda t = 10^{-1}$

$R_s(10 \text{ hrs}) = 2e^{-.3} - e^{-.6} = 1.482 - .549$

$R_s = 0.933$

(b) Off-line (100% reliable switch)

State	Operating	Stand-by	Failed
0	A	B	-
1	B	-	A
2	-	-	A,B

$P(0) = [1, 0, 0]$

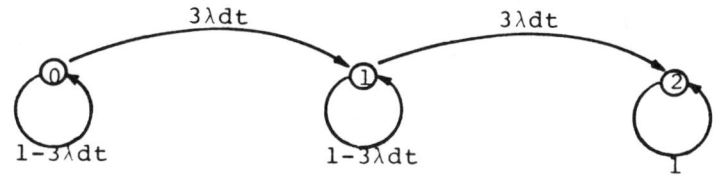

$P_0(t+dt) = P_0(t)[1-3\lambda dt] \implies P_0'(t) = -3\lambda P_0(t)$

Taking transforms:

$sP_0(s) - 1 = -3\lambda P_0(s)$

$P_0(s) = \dfrac{1}{s+3\lambda} \implies P_0(t) = e^{-3\lambda t}$

$P_1(t+dt) = P_0(t)[3\lambda dt] + P_1(t)[1-3\lambda dt]$

$P'(t) = 3\lambda P_0(t) - 3\lambda P_1(t)$

$P_1(s) = \dfrac{3\lambda P_0(s)}{(s+3\lambda)} = \dfrac{3\lambda}{(s+3\lambda)^2} \implies P_1(t) = 3\lambda t e^{-3\lambda t}$

$R(t) = P_0(t) + P_1(t)$

$R(t) = e^{-3\lambda t} + 3\lambda t e^{-3\lambda t}$

$R(10) = e^{-.3} + .3e^{-.3} \approx \underline{.963}$

(c) $L(0) = \int_0^\infty R(t)\,dt$

For system in (a) above:

$L(0) = -2/3\lambda e^{-3\lambda t}\Big|_0^\infty \quad -1/(-6)\lambda e^{-6\lambda t}\Big|_0^\infty$

$L(0) = 2/3\lambda - 1/6\lambda = 1/[2/3 - 1/6]\lambda$

$L(0) = 1/2\lambda = 50$ hr

For system in (b)

$L(0) = \int_0^\infty e^{-3\lambda t}\,dt - 3\lambda \int_0^\infty t e^{-3\lambda t}\,dt$

$L(0) = -1/3 e^{-3\lambda t}\Big|_0^\infty + \int_0^\infty t(-3\lambda)e^{-3\lambda t}\,dt$

$$L(0) = te^{-3\lambda t} \Big|_0^\infty + 1/3\lambda + 1/3\lambda$$

$$L(0) = 2/3\lambda = 66.7 \text{ hr}$$

Example 7.2 Redundant Dual-Parallel System Problem

To show how Markov process techniques can be applied to a more complex system, consider an on-line dual parallel system consisting of four identical components with age dependent failure rate λ.

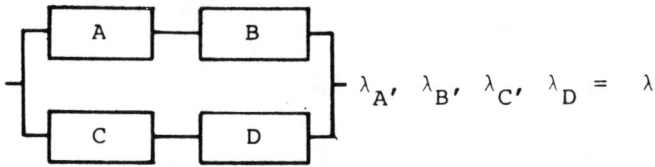

$\lambda_A, \lambda_B, \lambda_C, \lambda_D = \lambda$

The system states may be defined as:

State	Operating	Failure Rate to Next Higher State
0	AB and CD	4λ (4 components x λ)
1	AB or CD	2λ (2 components x λ)
2	None	

The appropriate stochastic matrix for the system will be:

$$P = \begin{array}{c|cccc} & 0 & 1 & 2 \\ \hline 0 & 1-4\lambda & 4\lambda & 0 \\ 1 & 0 & 1-2\lambda & 2\lambda \\ 2 & 0 & 0 & 1 \end{array}$$

with difference equations:

$$P_0(t+dt) = P_0(t)[1-4\lambda dt]$$

$$P_1(t+dt) = P_0(t)\,4\lambda dt + P_1(t)[1-2\lambda dt]$$

$$P_2(t+dt) = \qquad\qquad P_1(t)\,2\lambda\,dt + P_2(t)$$

Solving the associated differential equations (assuming initially the system is in state 0, so $P_0(0) = 1$)

$$P_1(0) = P_2(0) = 0$$

$$P_0'(t) = -4\lambda P_0(t)$$

$$P_1'(t) = 4\lambda P_0(t) - 2\lambda P_1(t)$$

$$P_2'(t) = 2\lambda P_1(t)$$

by Laplace Transform methods yields:

$$\left.\begin{array}{l} sP_0(s) - P_0(0) = -4\lambda P_0(s) \\ sP_1(s) - P_1(0) = 4\lambda P_0(s) - 2\lambda P_1(s) \\ sP_2(s) - P_2(0) = 2\lambda P_1(s) \end{array}\right\} \text{Transform of Differential Equations}$$

$$P_0(s) = \frac{1}{s+4\lambda}$$

$$P_1(s) = \frac{4\lambda P_0(s)}{s+2\lambda} = \frac{4\lambda}{(s+4\lambda)(s+2\lambda)} = \frac{2}{s+2\lambda} - \frac{2}{s+4\lambda}$$

Solving for $P_0(s)$, $P_1(s)$, $P_2(s)$

$$P_2(s) = \frac{2\lambda P_1(s)}{s} = \frac{8\lambda^2}{s(s+4\lambda)(s+2\lambda)} = \frac{1}{s+2\lambda} - \frac{2}{s+2\lambda} + \frac{1}{s}$$

$$\left.\begin{array}{l} P_0(t) = e^{-4\lambda t} \\ P_1(t) = 2e^{-2\lambda t} - 2e^{-4\lambda t} \\ P_2(S) \quad e^{-4\lambda t} - 2e^{-2\lambda t} + 1 \end{array}\right\} \text{Taking inverse transform}$$

The reliability of the system, $R_T(t)$, is $P_0(t) + P_1(t)$,

$$P_T(t) = P_0(t) + P_1(t) = e^{-4\lambda t} + 2e^{-2\lambda t} - 2e^{-4\lambda t} = 2e^{-2\lambda t} - e^{-4\lambda t}$$

7.8 **Exercises**

1. A system consists of two identical components in series, each with a redundancy of one. The stand-by components are off line, and their failure rate is one-half that of the main components as long as they are off line. Use a Markov model to solve this problem and determine the reliability and mean time to failure of this system.

2. Assuming that in the above example there is no perfect switching. In fact, there is a switch for each series component and that switch has a reliability of switching of one-quarter that of the main component. Once the off-line components are switched on the switches are no longer required. (The switch reliability becomes 100%.) What is the reliability and mean time to failure of this system now?

7.9 **References**

1. Barlow, R.E. and Proschan, F., "Mathematical Theory of Reliability", Wiley, New York, 1965.

2. Howard, R.A., "Dynamic Probabilistic Systems - Vol. I: Markov Models - Vol. II: Semi-Markov and Decision Provcesses", John Wiley and Sons, Inc., New York, 1971.

3. Lieberman, G.J., "The Status and Impact of Reliability Methodology", Naval Research Logistics Quarterly 16, 1969.

4. Shooman, M.L., "Probabilistic Reliability - An Engineering Approach", McGraw-Hill, New York, 1968.

5. Ross, S., "Introduction to Probability Models", Academic Press, New York, 1972.

8.0 ANALYSIS OF MAINTAINED SYSTEMS

Ernst G. Frankel

Maintained systems consist of components, some or all of which can be maintained. Similarly the assemblage of components is assumed maintainable. Maintenance comprises different types of actions designed to:

1. monitor performance or conditions of components of systems;
2. adjust and calibrate components or systems;
3. perform preventative repairs;
4. perform scheduled repairs;
5. perform complete overhauls; and,
6. perform casualty repairs.

For the purpose of our analysis it is convenient to divide maintenance into:

a. monitoring and calibration normally done without shutting down a system;
b. preventative repairs done intermittently but not necessarily scheduled;
c. scheduled repairs performed at preplanned intervals and involving planned maintenance actions;
d. overhauls which involve a complete systems repair and may include large-scale component replacement. Overhauls may or may not be scheduled. They can be performed as part of a plan or as a result of unexpected casualties. and,
e. casualty repairs are defined as repairs required to put a system or component back into operation after an unexpected breakdown.

In maintained systems analysis we may have an imposed, scheduled monitoring and calibration as well as scheduled repair plan. There may also be a plan for preventative repairs and overhauls which may be required at predetermined intervals. Conversely the analysis may be designed to determine an effective maintenance plan including some or all the above maintenance actions.

8.1 Systems Availability

A measure of effectiveness of great importance to the analysis of maintained systems is availability. Three types of availability are usually of interest depending on the particular function of a system:

1. **Instantaneous Availability** which can be defined as the probability that the system will be available at any random time t during its life.

2. **Average Up-Time** or **Up-Time Availability** is defined as the proportion of time that the system is available for use during a specific time interval (0, T).

3. **Availability** or **Steady-State Availability** is the proportion of time that the system is available for use when the time interval is very large.

In the limit, the up-time availability approach is the steady-state availability.

The particular availability measure chosen will depend upon the mission requirements of the system. For continuously operated systems, steady-state availability is normally a proper measure. If a duty cycle is defined for a system, then average up-time availability is a satisfactory measure. If the system is required to perform a particular function at any random time (traffic control - guns, etc.) but remains idle between the performance of the functions for very long time periods compared to the duration of the function then instantaneous availability may be usefully employed as an effective performance measure. We will find that in many cases more than one measure of availability is applicable. Considering up-time availability, A_u, first, we define:

$$A_u = \frac{T-T_R-T_S}{T-T_S} = \frac{\text{Actual Operating Time/Repair Period}}{\text{Available Operating Time/Repair Period}} \quad (8.1)$$

T = Average time between scheduled repairs or downtime (overhauls, inspections, etc.)

T_R = Average number of random (unscheduled) failures x average repair time per random failure

T_S = Scheduled downtime per scheduled repair period

In computing availability of a complex system, we must often reduce the plant to an equivalent network of critical components where critical implies that component failure causes system failure or performance degradation. Repair or downtime after failure of a component must include the entire period the plant is off-line following failure of items of interest.

Average repair time is usually hard to obtain, and the scant data available has large variations due to reporting differences, policy differences, labor influence, etc. As critical components in series (reliability standpoint) must often be combined with subsystems, an equivalent failure rate and repair time for groups are usually employed.

Considering next the steady-state availability, A_∞, defined as the percentage of total life time during which the system is available, we obtain:

$$A_\infty = \frac{\text{Life Time} - \text{Total Downtime}}{\text{Life Time}} = A$$

$$= (L-T_S' - T_R')/L = \frac{L-N(T_S+T_R)}{L} \qquad (8.2)$$

where

$\quad$ N $\quad$ = Number of scheduled downtimes during life

$\quad$ L $\quad$ = Expected life time of a system

$\quad$ T_S' = Expected scheduled downtime during $L=NT_S$

$\quad$ T_R' = Expected unscheduled repair downtime during $L=NT_R$

Scheduled downtime includes overhauls, inspection, calibration, required idleness and other time periods during which the system is not available according to a predetermined plan.

Unscheduled (repair) downtime is the expected time loss resulting from casualties and other unscheduled events that require the system to be taken out of operation.

The expected probability, availability, or reliability is a number for the average of the whole sample population but not for all the individual units forming the population. A rigorous representation of reliability or availability of one member of a population should show the effect of the distribution of failures. Such a representation may be made by plotting availability or reliability against the probability of achieving each value of availability or reliability.

Considering a one-component system with an uptime availability requirement, A_u, we can write:

$$N_i = \frac{(1-A_u)T}{r_i} = \text{maximum number of permissible failures of the } i^{th} \text{ component}$$

where

r_i = average repair time of ith component

If $r_i \geq (1-A_u)T$ no failures are allowed. In that case, the <u>probability of achieving A_u</u> or better, such that $(1-A_u)T$ is less than r_i, is equal to the <u>reliability</u> of the system, and

$$P((1-A_u)T \leq r_i) = R(t)$$

t = time between scheduled downtime

The probability of achieving A_u or better for the ith component (assuming $A_u \to A$) is:

$$P_{i_A} = P_{i_A}(0) + \ldots + P_{i_A}(N_i) \tag{8.3}$$

where

$P_{i_A}(x)$ = probability of having exactly x failures in time AT

If we assume an exponential distribution of failure, then the reliability of the component or the probability of zero failures in time AT is

$$P_{i_A}(0) = e^{-f_i(t)AT} = e^{-f_i AT}$$

where

f_i = age dependent failure rate of the ith component.

Similarly,

$$P_{i_A}(x) = \frac{e^{-t_i AT}(f_i AT)^x}{x!}$$

then

$$P_{i_A}(x) = \frac{e^{-f_i AT}(f_i AT)^x}{x!}$$

where N_i is the maximum number of failures permitted within the reliability requirement A. Since N_i will not, in general, be an integer, we may express this as:

$$P_{i_A} = 1 - \frac{e^{-f_i AT}(f_i AT)^{N_i+1}}{(N_i+1)\,\Gamma(N_i+1)} \left[1 + \frac{f_i AT}{(N_i+2)} + \frac{(f_i AT)^2}{(N_i+2)(N_i+3)} + \cdots \right]$$

where $\Gamma(y)$ is the gamma function.

The probability that the ith component has an availability between A and A+dA is

$$\frac{d}{dA}(1-P_{i_A})\,dA;$$

and, therefore, the instantaneous availability or the probability that the ith component will be available at any time during the interval T is:

$$(PA)_i = \int_0^1 A \frac{d}{dA}(1-P_{i_A})\,dA \qquad (8.5)$$

If we require the instantaneous availability of a system with a redundant on-line system working (parallel) with at least $N_i < M_i$ of the components required for system operation, we obtain:

$$(PA_i)_{M_i, N_i} = (PA_i)^{M_i} + \binom{M_i}{M_i-1}(PA_i)^{M_i-1}(1-(PA_i)) + \cdots$$

$$\binom{M_i}{N_i}(PA_i)^{N_i}(1-(PA_i))^{M_i-N_i}$$

Example 8.1

Let us consider a simple problem of a maintained system which consists of a heater (Unit #1) and two identical pumps (Units #2) in active redundancy which is just another way of saying that the two pumps are on line.

The policies set for the system are

1. <u>Operation Policy</u>: Every unit will be operated as long and as soon as operable.

2. <u>Repair Policy</u>: Unit #1 has higher priority than Unit #2 in repair.

3. <u>Number of Repair Crews</u>: One

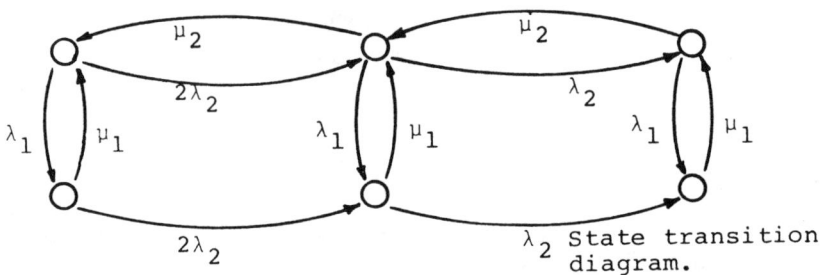

State transition diagram.

The problem may then be to:

a. Find the MTBF of the system.
b. Find the MTTR of the system.
c. Check your expression for MTTR by solving for availability (steady state) and substituting in this expression the MTBF expression found in (a) above.

States	Operating	Failed
0	Both #2, #1	--
1	One #2, #1	One #2
2	Both #2	#1
3	One #2	#2, #1
4	#1	Both #2
5	--	Both #2, #1

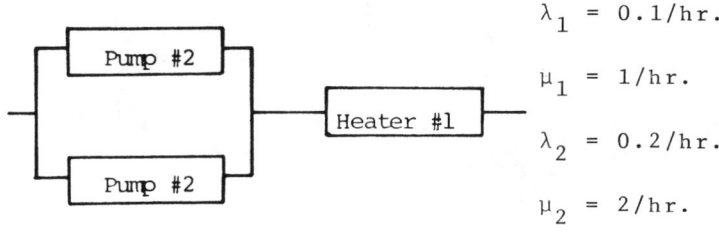

$\lambda_1 = 0.1/\text{hr.}$

$\mu_1 = 1/\text{hr.}$

$\lambda_2 = 0.2/\text{hr.}$

$\mu_2 = 2/\text{hr.}$

a. MTBF = ?

$$Rs(t) = e^{-\lambda_1 t}[2e^{-\lambda_2 t} - e^{-2\lambda_2 t}]$$

and using $\lambda_2 = 2\lambda_1$

$$Rs(t) = e^{-\lambda_1 t}[2e^{-2\lambda_1 t} - e^{-4\lambda_1 t}]$$

$$= 2e^{-3\lambda_1 t} - e^{-5\lambda_1 t}$$

$$L(0) = \int_0^\infty Rs(t)\,dt = \frac{2}{3\lambda_1} - \frac{1}{5\lambda_1}$$

$L(0) \simeq 4.66$ hr.

Note: MTBF as computed here refers to the case without maintenance.

b. MTTR = mean time until first <u>component</u> failure

$$Ps(t) = e^{-(\lambda_1 + 2\lambda_2)t} \quad \leftarrow \text{probability that no component has failed by time } t$$

$$\text{MTTR} = \int_0^\infty t e^{-(\lambda_1 + 2\lambda_2)t}\,dt$$

But we know that $\int_0^\infty \lambda e^{-\lambda t}\,dt = \frac{1}{\lambda} \int_0^\infty e^{-\lambda t}\,dt = \frac{1}{\lambda^2}$

$$\text{MTTR} = \frac{1}{(\lambda_1 + 2\lambda_2)^2} = \frac{1}{(5\lambda_1)^2} \quad \text{But } \lambda_1 = 0.1$$

$$\text{MTTR} = \frac{100}{25} = \underline{4 \text{ hr.}}$$

8.2 Markov Models for Maintained Systems

Let us assume that in a maintained system failures of components are immediately detected and repair is instigated. The probability of completing repair of a component in the time interval T to t+dt is assumed to be equal to $r(t)dt$. If the repair time t follows an exponential distribution, then the instantaneous repair rate $r(t) = \mu$ becomes a constant.

Considering a single component with a failure rate λ and a repair rate μ, we define an operating state 0 and a failed state 1. The stochastic matrix then becomes:

$$P = \begin{bmatrix} 1-\lambda & \lambda \\ \mu & 1-\mu \end{bmatrix}$$

and the birth and death equations

$$P_0(t+dt) = P_0(t)(1-\lambda dt) + P_1(t)\mu dt$$

$$P_1(t+dt) = P_0(t)\lambda dt + P_1(t)(1-\mu dt)$$

which results in the differential equations

$$P_0'(t) = -\lambda P_0(t) + \mu P_1(t)$$

$$P_1'(t) = \lambda P_0(t) - \mu P_1(t)$$

If we assume that our system is initially in the operating state 0, then the probability state vector at time 0 becomes

$$P(0) = [P_0(0), P_1(0)] = [1,0]$$

Transforming the above equations

$$(s+\lambda)P_0(s) - \mu P_1(s) = 1$$

$$-\lambda P_0(s) + (s+\mu)P_1(s) = 0$$

when

$$P_0(s) = \frac{s+\mu}{s(s+\lambda+\mu)} = \frac{\mu}{(\lambda+\mu)}\frac{1}{s} + \frac{\lambda}{(\lambda+\mu)}\frac{1}{(s+\lambda+\mu)}$$

$$P_0(t) = L^{-1}[P_0(s)] = \frac{\mu}{(\lambda+\mu)} + \frac{\lambda}{(\lambda+\mu)} e^{-(\lambda+\mu)t}$$

As zero is the only operating state, the instantaneous availability of the system is the probability that we find our system in state zero at any time t. Therefore,

$$A(t) = P_0(t) = \frac{\mu}{(\lambda+\mu)} + \frac{\lambda}{(\lambda+\mu)} e^{-(\lambda+\mu)t} \qquad (8.6)$$

Similarly,

$$[1-A(t)] = \text{Unavailability} = P_1(t) = \frac{\lambda}{(\mu+\lambda)} - \frac{\lambda}{(\mu+\lambda)} e^{-(\lambda+\mu)t}$$

$$= (1-\underline{P}(t))$$

If our state probability vector is $P(0) = [0,1]$ or the system is initially in the failed state, then it is

$$A(t) = P_0(t) = \frac{\mu}{\lambda+\mu} - \frac{\lambda}{\lambda+\mu} e^{-(\lambda+\mu)t}$$

and

$$P_1(t) = \frac{\lambda}{\lambda+\mu} - \frac{\lambda}{\lambda+\mu} e^{-(\lambda+\mu)t}$$

If we are interested in the percentage of time our system is in the operative state in a period T, then the up-time availability

$$A(T) = \frac{1}{T}\int_0^T A(t)\,dt \qquad (8.7)$$

becomes

$$A(T) = \frac{\mu}{(\lambda+\mu)} + \frac{\lambda}{(\lambda+\mu)^2 T} - \frac{\lambda e^{-(\lambda+\mu)t}}{(\lambda+\mu)^2 T} \quad \text{when } P(0) = [1,0]$$

$$A(T) = \frac{\mu}{(\lambda+\mu)} - \frac{\lambda}{(\lambda+\mu)^2 T} + \frac{\lambda\, e^{-(\lambda+\mu)T}}{(\lambda+\mu)^2 T} \quad \text{when } P(0) = [0,1]$$

In the limit when $T \to \infty$, the up-time availability becomes the steady-state availability at $\lim_{T \to \infty} A(T) = A(\infty) = \frac{\mu}{(\lambda+\mu)}$ independent of $P(0)$. This result can also be obtained by solving the birth and death equations and making the derivative of the state probabilities zero, which implies that we assume a steady state exists in the limit when $t \to \infty$. We then obtain:

$$P_0'(\infty) = -\lambda P_0(\infty) + \mu P_1(\infty) = 0$$

$$P_1'(\infty) = \lambda P_0(\infty) - P_1(\infty) = 0$$

and

$$P_0(\infty) + P_1(\infty) = 1 \quad \text{when } P_0(\infty) = A(\infty) = \frac{\mu}{(\lambda+\mu)}$$

An interesting problem is the consideration of partial failure and make-shift repair. If our component is new or completely overhauled, we may apply a failure rate λ_1 to it which leads to the requirement of a make-shift repair and a failure rate λ_2 leading to full overhaul need with repair rates μ_1 and μ_2 respectively associated with them. Let us assume our component after a make-shift repair has a failure rate λ_3 leading to make-shift repairs and λ_4 leading to overhauls. We consider that an overhaul brings it back to an "as-good-as-new" condition. Let:

State

0   system after complete repair (or new)
1   system undergoing make-shift repair
2   system after make-shift repair
3   system undergoing overhaul

$$P_0(t+dt) = P_0(t)[1-(\lambda_1+\lambda_2)dt] + P_3(t)\mu_2 dt$$

$$P_1(t+dt) = P_0(t)\lambda_1 dt + P_1(t)[1-\mu_1 dt] + P_2(t)\lambda_3 dt$$

$$P_2(t+dt) = P_1(t)\mu_1 dt + P_1(t)[1-\lambda_3+\lambda_4)dt]$$

$$P_3(t+dt) = P_0(t)\lambda_2 dt + P_3(t)[1-\mu_2 dt] + P_2(t)\lambda_4 dt$$

when the stochastic matrix

$$P = \begin{bmatrix} (1-\lambda_1-\lambda_2) & \lambda & 0 & \lambda_2 \\ 0 & (1-\mu_1) & \mu_1 & 0 \\ 0 & \lambda_3 & (1-\lambda_3-\lambda_4) & \lambda_4 \\ \mu_2 & 0 & 0 & (1-\mu_2) \end{bmatrix}$$

Assuming we require the steady-state availability of the system (omitting the arguments),

$$P_0' = -(\lambda_1+\lambda_2)P_0 + \mu_2 P_3 = 0$$

$$P_1' = \lambda_1 P_0 - \mu_1 P_1 + \lambda_3 P_2 = 0$$

$$P_2' = \mu_1 P_1 - (\lambda_3+\lambda_4)P_2 = 0$$

$$P_3' = \lambda_2 P_0 - \mu_2 P_3 + \lambda_4 P_2 = 0$$

when

$$A(\infty) = P_0 + P_2 = \frac{\mu_2 \mu_1 [\lambda_1+\lambda_4]}{\lambda_1 \lambda_4 (\mu_1 \mu_2) + \mu_1 \mu_2 (\lambda_1 \lambda_4) + \mu_2 \lambda_1}$$

8.2.1 Maintained Series Systems

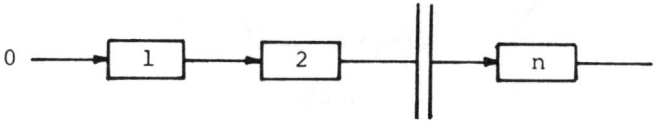

FIGURE 8.1

Assume n identical components are placed in series, each of which has $f(t) = \lambda$ and $r(t) = \mu$. If state one is the operating state and state two is the state that any component has failed and is being repaired:

$$P = \begin{bmatrix} 1-n\lambda & n\lambda \\ \mu & 1-\mu \end{bmatrix}$$

and

$$P_0'(t) = -n\lambda P_0(t) + \mu P_1(t)$$

$$P_1'(t) = n\lambda P_0(t) - \mu P_1(t)$$

and if $P_0(0) = [1,0]$

$$sP_0(s) + n\lambda P_0(s) - \mu P_1(s) = 1$$

$$sP_1(s) + \mu P_1(s) - n\lambda P_0(s) = 0$$

and

$$A(t) = P_0(t) = \frac{\mu}{n\lambda+\mu} + \frac{n\lambda}{(n\lambda+\mu)} e^{-(n\lambda+\mu)t}$$

$$A(T) = \frac{\mu}{n\lambda+\mu} + \frac{n\lambda}{(n\lambda+\mu)^2 T} [1-e^{-(n\lambda+\mu)T}]$$

If the n components are not identical and have failure rates λ_i i=1,2... yet their respective repair rates are equal, then

$$A(t) = P_0(t) = \frac{\mu}{(\sum_{i=1}^{n} \lambda_i) + \mu} + \frac{\sum_{i=1}^{n}(\lambda_i)}{(\sum_{i=1}^{n} \lambda_i) + \mu} e^{-(\lambda + \sum_{i=1}^{n} \lambda_i)t}$$

If the repair rates vary for the different components as well, then different repair states have to be defined (one for each component) or assuming a two-component system, for example:

State
- 0   Both components operate
- 1   Component A has failed and is repaired at rate
- 2   Component B has failed and is repaired at rate

$$\begin{bmatrix} 1-(\lambda_1+\lambda_2) & \lambda_1 & \lambda_2 \\ \mu_2 & (1-\mu_1) & 0 \\ \mu_2 & 0 & (1-\mu_2) \end{bmatrix}$$

and

$$P_0'(\infty) = -(\lambda_1+\lambda_2)P_0 + \mu_1 P_1 + \mu_2 P_2 = 0$$

$$P_1'(\infty) = \lambda_1 P_0 - \mu_1 P_1 = 0$$

$$P_2'(\infty) = \lambda_2 P_0 - \mu_2 P_2 = 0$$

when

$$P_0 = \frac{\mu_1 P_1}{\lambda_1} = \frac{\mu_2 P_2}{\lambda_2} = \frac{\mu_1 \mu_2}{(\mu_1 \mu_2 + \lambda_1 \mu_2 + \lambda_2 \mu_1)}$$

and, in general, in an n component system:

$$P_0 = \frac{\mu_i}{\lambda_i} P_i = \frac{\prod_{i=1}^{n} \mu_i}{\prod_{j=1}^{n} \mu_j + \sum_{i=1}^{n} \lambda_i \prod_{\substack{j=1 \\ j \neq i}}^{n} \mu_j}$$

8.2.2 Maintained Parallel Systems

Let us assume that two components are placed in an on-line parallel arrangement with each of the components having an age dependent failure rate of $f(t) = \lambda$ and repair rate of $r(t) = \mu$. Assuming that state zero is the state when both components operate, state one when either of the two components has failed and is being repaired, and state two is the state when both components have failed and one of the failed components is being repaired, then:

$$P = \begin{bmatrix} 1-2\lambda & 2\lambda & \\ \mu & 1-\lambda-\mu & \lambda \\ & \mu & 1-\mu \end{bmatrix}$$

and

$$P_0'(t) = -2\lambda P_0(t) + \mu P_1(t)$$

$$P_1'(t) = 2\lambda P_0(t) - (\lambda+\mu)P_1(t) + \mu P_2(t)$$

$$P_2'(t) = \lambda P_1(t) - \mu P_2(t)$$

and given $P(0) = [1, 0, 0]$

$$sP_0(s) + 2\lambda P_0(s) - \mu P_1(s) = 1$$

$$sP_1(s) + (\lambda+\mu)P_1(s) = -2\lambda P_0(s) - \mu P_2(s) = 0$$

$$sP_2(s) + \mu P_2(s) - \lambda P_1(s) = 0$$

$$P_0(s) = \frac{1+\mu P_1(s)}{s+2\lambda}$$

$$P_1(s) = \frac{2\lambda P_0(s)}{s+\lambda+\mu} + \frac{\mu P_2(s)}{s+\lambda+\mu}$$

$$P_2(s) = \frac{\lambda \underline{P}_1(s)}{s+\mu}$$

$$A(t) = P_0(t) + P_1(t) = L^{-1}[P_0(s) + P_1(s)]$$

Similarly if we have a two-component off-line redundant parallel system with perfect switching and two repairmen, then state one is the state when one component operates and the second component stands by, state two is the state when the first component fails and the second component operates, while the first component is being repaired, while state three is the state when both components are failed and under repair. Assuming that a repairman can only work on one failed component and that the probability of both failed components repair being completed in an increment of time is zero,

$$P = \begin{bmatrix} 1-\lambda & \lambda & 0 \\ \mu & 1-\lambda-\mu & \lambda \\ 0 & 2\mu & 1-2\mu \end{bmatrix}$$

Example 8.2

A two component on-line stand-by system is fully operative if one or two components are on line. If the failure rate is λ_1, if one is operating and λ_2 if both are operating, and if the repair rate is μ_1 if one has failed, but μ_2 if both have failed, what is:

    a. steady state availability of system
    b. reliability of system for an interoverhaul period T
    c. instant availability of system.

a.

STATE	OP	FAIL
0	2	0
1	A	B
2	B	A
3	0	2

$$P_0' = sP_0 - 1 = -2\lambda_2 P_0 + \mu_1 P_1 + \mu_1 P_2$$

$$P_1' = sP_1 = \lambda_s P_0 - (\lambda_1 + \mu_1) P_1 + \mu_2 P_3$$

$$P_2' = sP_2 = \lambda_2 P_0 - (\lambda_1 + \mu_1) P_2 + \mu_2 P_3$$

$$P_3' = sP_3 = \lambda_1 P_1 + \lambda_1 P_2 - \mu_2 P_3$$

$$1 = P_0 + P_1 + P_2 + P_3$$

$$P_0 = \frac{2\mu_1}{s+2\lambda_2} P_1$$

$$P_2 = P_1$$

$$(s+\lambda+\mu_1)P_1 = \frac{2\mu_1\lambda_2}{s+2\lambda_2}P_1 + \mu_2 P_3$$

$$P_3 = \left[\frac{s+\lambda_1+\mu_1}{\mu_2} - \frac{2\mu_1\lambda_2}{\mu_2(s+2\lambda_2)}\right]P_1$$

$P_3(t)$ solve as above. So,

$$A(\infty) = P_0+P_1+P_2 \mid s=0 \qquad P_1 = \frac{\lambda_2\mu_2}{\mu_1\mu_2+2\mu_2\lambda_2+\lambda_1\lambda_2}$$

$$A(\infty) = \frac{\mu_1\mu_2+2\lambda_2\mu_2}{\mu_1\mu_2+2\mu_2\lambda_2+\lambda_1\lambda_2} \qquad P_2 = P_1$$

$$P_0 = \frac{\mu_1\mu_2}{\mu_1\mu_2+\mu_2\mu_2+\lambda_1\lambda_2}$$

b. $$A(T) = \frac{1}{T}\int_0^T A(t)\,dt$$

$$= \frac{1}{T}\int_0^T (P_0(t)+P_1(t)+P_2(t))\,dt$$

c. $$A(t) = P_0(t)+P_1(t)+P_2(t)$$

Example 8.3

A single component operates 80% at half load and 20% at full load. The catastrophic failure rate at full load is λ_1 and at half load is $\lambda_1/4$. If a catastrophic failure occurs, the repair rate is μ. At full load the component also has a minor failure rate λ_2 and at half load $\lambda_2/2$. If the rate at which minor failures are repaired is 10μ, what is the instant and steady state availability of the component?

STATE

0   Operating
1   Catastrophic Failure
2   Minor Failure

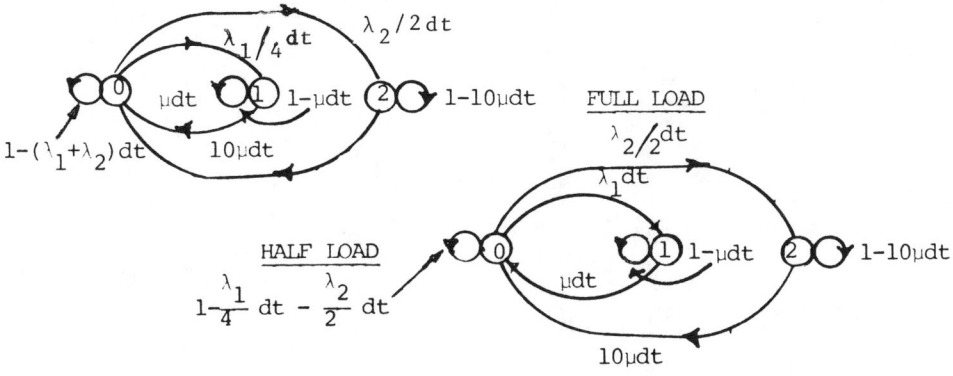

Full Load

$$sP_0 - 1 = P_0(-\lambda_1 - \lambda_2) + \mu P_1 + 10\mu P_2$$

$$sP_1 = P_0 \lambda_1 - \mu P_1$$

$$sP_2 = P_2 \lambda_2 - 10\mu P_2$$

Half Load

Same for Half Load except

$$\lambda_1 = \lambda_1/4$$

$$\lambda_2 = \lambda_2/2$$

Solve for steady state, $\frac{d}{dt} = 0$;

$$P_2 = \frac{\lambda_2}{10\mu} P_0$$

$$P_1 = \frac{\lambda_1}{\mu} P_0$$

$$P_0 + \frac{\lambda_1}{\mu} P_0 + \frac{\lambda_2}{10\mu} P_0 = 1$$

$$P_0 = \frac{1}{1 + \frac{\lambda_1}{\mu} + \frac{\lambda_2}{10\mu}} = \frac{10\mu}{10\mu + 10\lambda_1 + \lambda_2}$$

$$P_0' = \frac{20\mu}{20\mu + 5\lambda_1 + \lambda_2}$$

$$P_1 = \frac{10\lambda_1}{10\mu + 10\lambda_1 + \lambda_2}$$

$$P_2 = \frac{\lambda_2}{10\mu + 10\lambda_1 + \lambda_2}$$

Steady State Availability

$$A(\infty) = \frac{2\mu}{10\mu + 10\lambda_1 + \lambda_2} + \frac{16\mu}{20\mu + 5\lambda_1 + \lambda_2}$$

Time Variation:

$$P_1 = P_0 \frac{\lambda_1}{s+\mu}$$

$$P_2 = P_0 \frac{\lambda_2}{s+10\mu}$$

$$P_0 \left(1 + \frac{\lambda_1}{s+\mu} + \frac{\lambda_2}{s+10\mu}\right) = 1/s$$

$$P_0 = \frac{(s+\mu)(s+10\mu)}{s[s^2 + s(11\mu + \lambda_1 + \lambda_2) + 10\mu^2 + 10\mu\lambda_1 + \mu\lambda_2]}$$

$$= \frac{(s+\mu)(s+10\mu)}{s(s^2 + FS + G)}$$

$$= \frac{A}{s-B} + \frac{C}{s-D} + \frac{L}{S}$$

$$B = \frac{-F+\sqrt{F^2-4G}}{2} \qquad D = \frac{-F-\sqrt{F^2-4G}}{2}$$

$$A = \frac{(B+\mu)(B+10\mu)}{B(B-D)} \qquad C = \frac{(D+\mu)(D+10\mu)}{D(D-B)}, \quad L = \frac{10\mu^2}{BD}$$

$$P_0(t) = Ae^{Bt} + Ce^{Dt} + L$$

$P_1(t)$, $P_2(t)$, $P_0'(t)$, $P_1'(t)$, $P_2'(t)$ solved similarly

Instantaneous Availability

$$A(t) = 0.2\, P_0(t) + .8\, P_0'(t)$$

Solution is mechanically very tedious.

Example 8.4

a. A maintained two-component on-line system in which each component has a failure rate of $\lambda = .001$/day independent of how many components are on line, is initially in state zero when both components are operative. If the system is to be in this state at least 50% of the time and if it is to be in an operative state (at least one component working) 95% of the time, what are the required repair rate(s)?

b. Assuming you could only work on one component at a time and the repair rate was therefore independent of the number of failed components, what is the required repair rate now, assuming the above availabilities?

c. If it costs $1.00 for each percent decrease in failure rate and $2.00 for each percent of increase in repair rate, what is the best policy?

a.

STATE	OPERATING	FAIL
0	2	0
1	1	1
2	0	2

$$P_0 \geq .5 \qquad P_0 = .5$$

$$P_0 + P_1 \geq .95 \quad P_1 = .45 \qquad P_2 = .05$$

$$\underline{\mu_1 = .00111}$$

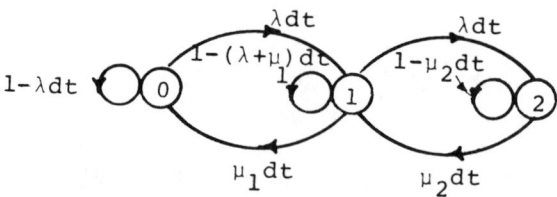

Steady State

$$\lambda P_0 = \mu_1 P_1 \qquad (.001)(.05) = \mu_1(.45)$$

$$\mu P_1 = \mu_2 P_2 \qquad (.001)(.45) = \mu_2(.05)$$

$$P_0 + P_1 + P_2 = 1$$

b) $\mu_1 = \mu_2 = \mu$

$$P_0 + P_1 = .95 = \frac{1}{1 + \frac{\lambda}{\mu} + \frac{\lambda^2}{\mu^2}} + \frac{\lambda/\mu}{1 + \lambda/\mu + \lambda^2/\mu^2} \rightarrow \lambda/\mu = .258$$

check:

$P_0 = .755 \qquad\qquad P_0 > .5$

$P_1 = \frac{\lambda}{\mu} P_0 = .195 \qquad P_0 + P_1 \geq .95$

$P_2 = \frac{\lambda}{\mu} P_1 = .05 \qquad P_0 + P_1 + P_2 = 1$

c) Minimize $P_2 = 1 - P_0 - P_1 = 1 - \frac{\mu^2 + \lambda\mu}{\mu^2 + \lambda\mu + \lambda^2}$

Cost	μ	λ	P_2	Improved Percentage
0	.0039	.001	.05	--
$10	.0039	.0001	.001	4.9% decrease
$10	.0195	.001	.003	4.7% decrease

For same cost, P_2 is decreased most by reducing λ rather than improving μ.

Example 8.5

A two component redundant system as shown, consists of a switch (including a sensor), a main component 1 (with high efficiency) and an auxiliary component 2 (low efficiency). The time to failure of both components and the switch is exponential as is also their time to complete a repair.

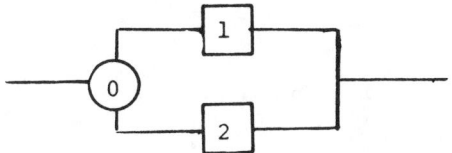

If λ_1, λ_2, λ_s and μ_1, μ_2, μ_s are respectively (assuming repair and failure rate are independent) the failure and repair rates of the components of the system and if component 1 will always work when it is available (independent of the availability of component 2) what is:

    a. availability of the system (steady state)?
    b. availability at time t?
    c. percentage time component 1 is operating?
    d. MTBF of system?
    e. number of times component 1 will fail in time T?
    f. if the efficiency of component 1 is e_1 and component 2 is e_2 and component s is e_s, what is the aggregate operating efficiency of the system? (average efficiency while operating)

    a. Since all failure and repair rates are independent, the problem can be separated into a series of two systems: one composed of the switch and the other of components 1 and 2.

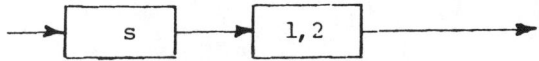

First determine the availability of the switch.

$$P_0(t+dt) = (1-\lambda_s dt)P_0(t) + \mu_s dt(1-P_0(t))$$

$$P_0'(t) = (\lambda_s + \mu_s)P_0(t) = \mu_2$$

Assume operating at t=0

$$sP_0(s) - 1 + (\lambda_s + \mu_s)P_0 = \frac{\mu_s}{s}$$

$$P_0(t) = e^{-(\lambda_s+\mu_s)t} + \frac{\mu}{\lambda_s+\mu_s}(1 - e^{-(\lambda_s+\mu_s)t})$$

Now determine availability of Parallel System

State	Up	Down
0	1,2	0
I	1	2
II	2	1
III	0	1,2

We assume that the failure rates don't change no matter which component is on line.

$$P_0(t+dt) = P_0(1-\lambda_1-\lambda_2) + \mu_2 P_I + \mu_1 P_{II}$$

$$P_I(t+dt) = \lambda_2 P_0 + (1-\mu_2-\lambda_1)P_I + \mu_1 P_{III}$$

$$P_{II}(t+dt) = \lambda_1 P_0 + (1-\mu_1-\lambda_2)P_{II} + \mu_2 P_{III}$$

$$P_{III}(t+dt) = \lambda_1 P_L + \lambda_2 P_{II} + (1-\mu_1-\mu_2)P_{III}$$

$$P_0' + (\lambda_1+\lambda_2)P_0 = \mu_2 P_I + \mu_1 P_{II}$$

$$P_I' + (\lambda_1+\mu_2)P_I = \lambda_2 P_0 + \mu_1 P_{III}$$

$$P_{II}' + (\lambda_2+\mu_1)P_{II} = \lambda_1 P_0 + \mu_2 P_{III}$$

$$P_{III}' + (\mu_1+\mu_2)P_{III} = \lambda_1 P_1 + \lambda_2 P_{II}$$

It is simpler to combine the states I and II in which one component is up.

State	Up	Down
A	1,2	0
B	1 or 2	1 or 2
C	ϕ	1 and 2

$$\begin{array}{c|c}
& \text{Average failure} \\
& \text{and repair rates} \\
& \lambda_a \quad \lambda_b \quad \mu_a \quad \mu_b \\
\hline
\lambda_a = \dfrac{\lambda_1+\lambda_2}{2} & \\
\hline
\mu_a = P_r(1 \text{ fails, not } 2)\mu_1 + P_r(2 \text{ fails, not } 1)\mu_2 & \\
\hline
\mu_a = \dfrac{\lambda_1\mu_1+\lambda_2\mu_2}{\lambda_1+\lambda_2} & \\
\hline
\lambda_b = P_r(1 \text{ failed, not } 2)\lambda_2 + P_r(2 \text{ failed, not } 1)\lambda_1 & \\
\hline
\lambda_b = \dfrac{2\lambda_1\lambda_2}{\lambda_1+\lambda_2} & \\
\hline
\mu_b = \dfrac{\mu_1+\mu_2}{2} & \\
\hline
\end{array}$$

$$P_A' + \lambda_a P_A = \mu_a P_B \qquad 1 \qquad P_A(s+\lambda_a) = \mu_a P_B + 1$$

$$P_B' + (\mu_a+\lambda_b)P_B = \lambda_a P_A + \mu_b P_C \qquad 2 \qquad P_B(s+\lambda_a+\mu_b) = \lambda_a P_A + \mu_b P_C$$

$$P_C' + \mu_b P_C = \lambda_b P_B \qquad 3 \qquad P_C(s+\mu_b) = \lambda_b P_B$$

Solve for P_C = probability of being in failed state.

$$P_C \dfrac{(s+\mu_b)}{\lambda_b}(s+\mu_a+\lambda_b) = \lambda_a \dfrac{(\mu_a P_B - 1)}{s+\lambda_a} + \mu_b P_C$$

$$P_C \dfrac{(s+\mu_b)}{\lambda_b}(s+\mu_a+\lambda_b) = \dfrac{\lambda_a \mu_a (s+\mu_b)}{\lambda_b (s+\lambda_a)} P_C + \mu_b P_C + \dfrac{\lambda_a}{s+\lambda_a}$$

$$P_C((s+\mu_b)(s+\mu_a+\lambda_b) - \frac{\lambda_a\lambda_a(s+\mu_b)}{s+\lambda_a} - \mu_b\lambda_b) = \frac{\lambda_a\lambda_b}{s+\lambda_a}$$

$$P_C(s+\mu_b)(s+\lambda_a)(s+\mu_a+\lambda_b) - \lambda_a\mu_a(s+\mu_b) - \mu_b\lambda_b(s+\lambda_a) = \lambda_a\lambda_b$$

$$P_C(s) = \frac{\lambda_a\lambda_b}{(s+\mu_b)(s+\lambda_a)(s+\mu_a+\lambda_b) - \lambda_a\mu_a(s+\mu_b) - \mu_b\lambda_b(s+\lambda_a)}$$

$$P_A(s) + P_B(s) = \frac{1}{s} - P_C(s)$$

So the Instantaneous Availability of the system is

$$A_{switch} \cdot A_{12} = A_I$$

$$A_I = (e^{-(\lambda_s+\mu_s)t} + \frac{\mu_s}{\lambda+\mu_s}(1-e^{-(\lambda_s+\mu_a)t})(1-\zeta^{-1}(P_C(s)))$$

b. **Steady State Availability**

Let $s \to 0$

$$P_C(s) \to \frac{\lambda_a\lambda_b}{\lambda_a\mu_b(\mu_a+\lambda_b) - \lambda_a\mu_a\mu_b - \mu_b\lambda_b\lambda_a}$$

This is equivalent to setting derivatives equal to zero and solving 1, 2 & 3 as follows:

$$\lambda_a P_A = \mu_a P_B$$

$$(\mu_a+\lambda_b)P_B = \lambda_a P_A + \mu_b P_C$$

$$\mu_b P_C = \lambda_b P_B$$

Solve for P_C, $P_A + P_B + P_C = 1$

$$P_C + \frac{\mu_b}{\lambda_b} P_C + \frac{\mu_a \mu_b}{\lambda_a \lambda_b} P_C = 1$$

$$P_C = \frac{\lambda_a \lambda_b}{\lambda_a \lambda_b + \lambda_a \mu_b + \mu_a \mu_b}$$

Steady State Availability = $(1-P_C) \cdot A_{switch}^{ss}$

$$A_{s.s.} = \frac{\mu_s}{\lambda_s \mu_s} \frac{\lambda_a \lambda_b + \lambda_a \mu_b + \mu_a \mu_b - 1}{\lambda_a \lambda_b + \lambda_a \mu_b + \mu_a \mu_b}$$

c. **Percent time 1 is operating**

This equals the steady state availability of 1 (since it always operates when it is available) times the availability of the switch.

$$\% = \frac{\mu_1 \mu_s}{(\mu_1 + \lambda_1)(\mu_s + \lambda_s)}$$

d. <u>MTBF</u>

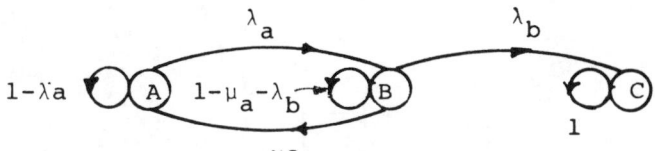

$$L(0) = \int_0^\infty R_S \cdot R_p(t) = \int R_{TOTAL} dt$$

$$R_S = e^{-\lambda st} \rightarrow L(0) = \frac{1}{\lambda_s}$$

R_p must be calculated by letting state c (both failed, 1 & 2) be an <u>absorbing</u> state.

1) $P_A' + \lambda_a P_A = \mu_a P_B \rightarrow P_A(s+\lambda_a) = \mu_a P_B + 1$

2) $P_B' + (\mu_a + \lambda_b) P_B = \lambda_a P_A \rightarrow P_B(s + \mu_a + \lambda_b) = \lambda_a P_A$

3) $P_C' = \lambda_b P_B \rightarrow sP_C = \lambda_b P_B$

eliminate $P_C(s)$ 1 & 2

$$P_B(s) = \frac{\lambda_a}{(s+\lambda_a)(s+\mu_a+\lambda_b) - \lambda_a \mu_a}$$

$$P_A(s) = \frac{s+\mu_a+\lambda_b}{(s+\lambda_a)(s+\mu_a+\lambda_b) - \lambda_a \mu_a}$$

$$\frac{1}{\lambda_s} \int_0^\infty (P_A(t) + P_B(t)) dt = \int R_{TOTAL}(t)$$

$$= [P_A(s+\lambda_s) + P_B(s+\lambda_s)]_{s=0}$$

$$\int_0^\infty R = \frac{\lambda_a}{\lambda_a \mu_a + \lambda_a \lambda_b - \lambda_a \mu_a} + \frac{\mu_a \lambda_a}{\lambda_a \mu_a + \lambda_a \lambda_b}$$

$$\int_0^\infty R\, dt = \frac{1}{\lambda_s} (\frac{1}{\lambda_b} + \frac{1}{1+\lambda_a \lambda_b})$$

$$L(0) = \frac{1}{\lambda_s \lambda_b} + \frac{1}{\lambda_s + \lambda_s \lambda_a \lambda_b}$$

e. Number of times 1 will fail in time T. This is a Poisson process with rate λ_1.

$$\therefore n = \lambda_1 T$$

f.

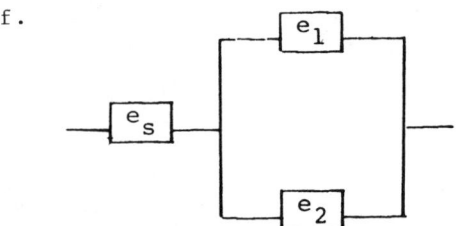

First find the equivalent efficiency of the parallel system, e_p.

$e_p = e_1$ (% of operating time 1 is on line) $+ e_2$ (% of operating time 2 is on line)

$$= e_1 \frac{\mu_1}{\lambda_1+\mu_1} + e_2 \frac{\lambda_1}{\mu_1+\lambda_1}$$

$$e = \frac{e_s(e_1\mu_1+e_2\lambda_1)}{\lambda_1+\mu_1}$$

8.3 Development of the General Expression for the Mean Time to Failure of a Markov Chain

If we consider a general type of birth and death equation of the form

$$P_0(t+dt) = (1-\lambda_0 dt)P_0(t) + \mu_1 dt P_1(t)$$

$$P_1(t+dt) = \lambda_0 dt P_0(t) + (1-\lambda_1 dt - \mu_1 dt)\underline{P}_1(t) + \mu_2 dt P_2(t)$$

$$P_i(t+dt) = \lambda_{i-1} dt P_{i-1}(t) + (1-\lambda_i dt - \mu_i dt)P_i(t) + \mu_{i+1} dt P_{i+1}(t)$$

and now consider the simple case, where all but the last or n^{th} state are operating states, then the differential equations of the probabilities of being in any of the n operating states are:

$$P_0' = -\lambda_0 P_0 + \mu_1 P_1$$

$$P_1' = \lambda_0 P_0 - (\lambda_1+\mu_1)P_1 + \mu_2 P_2$$

$$\vdots$$

$$P_{n-1}' = \lambda_{n-2}P_{n-2} - (\lambda_{n-1}+\mu_{n-1})P_{n-1} + \mu_n P_n$$

where the argument has been left out for simplicity of notation.

This set of simultaneous equations can be written in matrix form:

$$[X'] = [P_0', P_1', \ldots P_{n-1}'] = [P_0, P_1, \ldots P_n] \begin{bmatrix} -\lambda_0 & \lambda_0 & \cdots \\ \mu_1 & -(\lambda_1' + \mu_1) & \cdots \\ & \mu_2 & \cdots \\ & & \cdots \mu_n \end{bmatrix}$$

or $[X'] = [X][P-\delta_{ij}]$ where P = Stochastic Matrix

$$\delta_{ij} = \begin{cases} 1 & j = 1 \\ 0 & j \neq 1 \end{cases}$$

and

$$[P-\delta_{ij}] = [A] = \text{Fundamental Matrix}$$

Taking the Laplace Transform of this equation

$$L[X] = [sP_0(s) - P_0(0), sP_1(s) - P_1(0), \ldots]$$

and if

$$P_0(0) = [1, 0, \ldots]$$

$$= [X(s)] \, s\{[I] - P(0)\} = [X(s)][A]$$

It should be noted that $[X]$ is a row vector (matrix). Rewriting the equation

$$[X(s)] \, s \, [I] - [X(s)][A] = P(0) = [1, 0, \ldots, 0]$$

$$[X(s)] = P(0) \, [sI-A]^{-1}$$

and

$$[X(t)] = L^{-1}\{P(0)[sI-A]^{-1}\}$$

Integrating both sides we obtain:

$$\int_0^\infty [X(t)]\,dt = \lim_{t\to\infty} \int_o^t L^{-1}\{P(0)[sI-A]^{-1}\}\,dt$$

$$= P(0) \lim_{t\to\infty} L^{-1}\{[I-\tfrac{A}{s}]^{-1}\}$$

$$= [\tfrac{D_{ij}}{-A} \;\; \ddots\;]$$

where

$$[-A]^{-1} = \frac{(\text{transpose of cofactor } a_{ij})}{-A}$$

if the cofactor is defined as

$$(-1)^{i+j} D_{ij}$$

then its transpose is

$$(-1)^{j+i} D_{ji}$$

where D_{ji} is the minor of a_{ji}, which is formed by taking the determinant of the matrix which is left after the ith row and j^{th} column have been deleted. (NOTE: $(-1)^{i+j}$ is always equal to +1 for the main diagonal.)

To compute the MTBF, assuming all but the last state n are operating states, we obtain:

$$L(0) = \bar{t} = \int_0^\infty t\, f(t) R(t)\,dt = \int_0^\infty R(t)\,dt$$

$$= \int_0^\infty \sum_{i=0}^{n-1} P_i(t)\,dt = \sum_{i=0}^{n-1} \int_0^\infty P_i(t)\,dt = \sum_{i=0}^{n-1} \frac{D_{io}}{|-A|}$$

$$= \sum_{i=0}^{n-1} \frac{|D_{io}|}{|D_n|}$$

as it can be seen that $|-A|$ determinant is equal to the determinant of I-P with the nth row and column deleted.

Similarly $D_{io} = D_i$ = Determinant $|I-P|$ with the ith and n^{th} rows and columns deleted.

8.3.1 Mean Time to Failure and Variance of Time to Failure of Non-Maintained and Maintained Systems

The mean time to failure $\bar{t} = L(0)$ of a system is by definition the first moment of the probability density of the failure event of the system over the origin of time or as defined below:

$$MTBF = \bar{t} = L(0) = -\int_0^\infty t f(t) dt = \int_0^\infty R(t) dt$$

We have seen in our previous discussion that the MTBF is easily obtained from the reliability $R(t)$ for series redundant or complex systems consisting of identical components. All we had to do then was to integrate the $R(t)$ expression or to evaluate its Laplace transform at $s=0$. Considering a simple series system for instance

$$R_T(t) = e^{-n\int_0^t f(t)dt} = e^{-n\lambda t} \text{ if } f(t) = \lambda$$

and

$$L(0) = \int_0^\infty R_T(t) dt = \frac{1}{n\lambda}$$

if $\lambda_i \neq \lambda_j$ and i is not eqyal to j then

$$L(0) = \frac{1}{\sum_{i=1}^n \lambda_i}$$

Using this transform method we obtain:

$$L[R_T(t)]_{s=0} = L[e^{-n\lambda t}]_{s=0} = \frac{1}{s+n\lambda}\bigg|_{s=0} = \frac{1}{n\lambda} \text{ as before.}$$

If we next consider redundant on-line systems with identical components, then

$$R_T(t) = [1-(1-R(t))^n] = [1-(1-e^{-\lambda t})^n]$$

if a single component only is required for system operation. For n=2 we can write that

$$R_T(t) = 2e^{-\lambda t} - e^{-2\lambda t}$$

and

$$L(0) = \bar{t} = \frac{2}{\lambda} - \frac{1}{2\lambda} = \frac{3}{2\lambda} \text{ as before}$$

This result can also be obtained using the Laplace transform:

$$L(0) = \bar{t} = L[2e^{-\lambda t} - e^{-2\lambda t}]\Big|_{s=0} \quad [\frac{2}{s+\lambda} - \frac{1}{s+2\lambda}]\Big|_{s=0} = \frac{3}{2\lambda}$$

When dealing with complex systems, consisting of non-identical components, or if the system is maintainable, then the above methods of deriving the MTBF are very arduous. In general, we will be required to solve a set of "Birth" or "Birth & Death" equations to obtain the $R(t)$, or we may choose to establish the stochastic matrix of the system and then solve for $R(t)$ by matrix or transform methods. It is, therefore, advisable to use these same tools for the computation of the MTBF directly.

Let P, the stochastic matrix, be such that it satisfies only one absorbing state (i.e., all other absorbing states are cancelled or combined into one state), then we may establish the fundamental matrix $[P-I] = [A]$ and derive the determinant D_n which is the determinant $[I-P]$ with the last row and column cancelled. If we define D_{io} as the determinant of $[I-P]$ in which in addition to the last row and column also the ith row and the oth column are cancelled, then the MTBF in state i is equal to

$$\frac{|D_{io}|}{|D_n|} = \text{Expected time spent in state i given initially in state o}$$

The system MTBF is then

$$\bar{t} = L(0) = \sum_i \frac{|D_{io}|}{|D_n|}$$

If we consider a series system for instance, we find that

$$P = \begin{bmatrix} 1-n\lambda & n\lambda \\ 0 & 1 \end{bmatrix} \text{ and } D_n = n\lambda$$

when $D_{ii} = D_{oo} = 0$, and the MTBF becomes zero.

This approach is therefore only applicable to systems with at least two operating states. If we assume an On-Line Redundant Non-Maintained system with identical components which require but one component for system operation, then

$$[I-P] = \begin{bmatrix} n\lambda & \cdots & -n\lambda & \cdots & 0 & \cdots & 0 \\ & \cdots & (n-1)\lambda & \cdots & -(n-1)\lambda & \cdots & 0 \\ & & & \cdot & & & \\ & & & \cdot & & & \\ & & & \cdot & \lambda & \cdots & -\lambda \\ 0 & & & & & & 0 \end{bmatrix}$$

and

$$D_n = n!\,\lambda^n$$

$$D_{oo} = (n-1)!\,\lambda^{n-1}$$

$$D_{11} = n(n-2)!\,\lambda^{n-1}$$

and

$$\bar{t} = L(0) = \sum_{i=1}^{n} \frac{1}{i\lambda}$$

If more than one component is required to operate this system, then we obtain as many ones on the diagonal of P as the minimum number of components required. In that case the matrix P is reduced to retain only the first absorbing state $n-m+1$.

As another example of a non-maintained system let us investigate an off-line, perfectly switching stand-by system requiring only one component for operation:

$$[I-P] = \begin{bmatrix} \lambda & -\lambda & & & 0 \\ & \lambda & -\lambda & & \\ & & \cdot & & \\ & & & \lambda & -\lambda \\ & & & & 0 \end{bmatrix}$$

and

$$D_n = \lambda^n; \quad D_{10} = D_{20} = \ldots D_{(n-1)(n-0)} = \lambda^{n-1}$$

therefore

$$L(0) = \bar{t} = \sum_{i=0}^{n-1} \left|\frac{D_{io}}{D_n}\right|$$

In case $m < n$ components are required for operation then

$$L(0) = \bar{t} \text{ (on-line)} = \sum_{i=m}^{n} \frac{1}{i\lambda}$$

and

$$L(0) = \bar{t} \text{ (off-line)} = \sum_{i=0}^{n-m} \frac{D_{ii}}{D_n} = \frac{n-m+1}{\lambda}$$

If we are concerned with maintained systems, then the reliability $R(t)$ of the system is obtained by making the first failed state an absorbing state (and ignoring the transition probabilities of return to operating states from all failed states).

This implies, similarly to the procedure adopted in the non-maintained systems, that we only consider the first failed state and ignore or cancel all subsequent failed states, as, for example, in the case where more than one component is required to operate the system. The resulting stochastic matrix for an on-line redundant system with identical components and a single repairman with service rate μ is:

$$P = \begin{bmatrix} 1-n\lambda, & n\lambda, & 0 & \ldots & & & & \cdot \\ \mu, & 1-[(n-1)\lambda+\mu], & (n-1)\lambda, & 0 & \ldots & & & \cdot \\ 0, & \mu, & 1-[(n-2)\lambda+\mu], & (n-2)\lambda, & 0 & \ldots & & \cdot \\ \multicolumn{7}{c}{\dotfill} & 1 \end{bmatrix}$$

and

$$[I-P] = \begin{bmatrix} n\lambda & -n\lambda & 0 & 0 & 0 & \cdots & 0 \\ -\mu & (n-1)\lambda+\mu & -(n-1)\lambda & 0 & 0 & \cdots & 0 \\ 0 & -\mu & (n-2)\lambda+\mu & -(n-2)\lambda & 0 & \cdots & 0 \\ \multicolumn{7}{c}{\dotfill 0} \end{bmatrix}$$

if $n=2$, $D_n = 2\lambda^2$, $D_{oo} = \lambda+\mu$, $D_{11} = 2\lambda$

and

$$\bar{t} = L(0) = \sum_{i=0}^{n-1} \frac{|D_{io}|}{|D_n|} = \frac{3\lambda+\mu}{2\lambda^2}$$

Considering the case of r=n repairmen (assuming one repairman per component)

$$[I-P] = \begin{bmatrix} n\lambda_1 & -n\lambda_1 & 0 & 0 & \cdots \\ -\mu & (n-1)\lambda+\mu & -(n-1)\lambda & 0 & \cdots \\ 0 & -2\mu & (n-2)\lambda+2\mu & -(n-2)\lambda & \cdots \\ & & & & 0 \end{bmatrix}$$

If $n=2$, $\bar{t} = L(0) = \dfrac{3\lambda+\mu}{2\lambda^2}$

For the general case of n=n and r=n

$$\bar{t} = L(0) = \frac{1}{\lambda} \sum_{i=0}^{n-1} \frac{(1+\mu/\lambda)^i}{(i+1)}$$

If we are concerned with an off-line redundant system where r=1, then

$$[I-P) = \begin{bmatrix} \lambda & -\lambda & 0 \cdots \cdots \cdots \cdots 0 \\ -\mu & \lambda+\mu & -\lambda \cdots \cdots \cdots \cdots 0 \\ 0 & -\mu & \lambda+\mu & -\lambda \cdots \cdots \cdots 0 \\ \multicolumn{4}{c}{\dotfill 0} \end{bmatrix}$$

If $n=2$, $D_n = \lambda^2$; $D_{oo} = \lambda+\mu$; $D_{11} = \lambda$

and

$$\bar{t} = L(0) = \frac{2\lambda+\mu}{\lambda^2}$$

$n=3$, $D_n = \lambda(\lambda+\mu)^2 - \mu\lambda^2 - \lambda(\mu\lambda+\mu^2)$

$$= \lambda^3$$

$$D_{oo} = \lambda^2 + 2\mu\lambda + \mu^2 - \mu\lambda = \lambda^2 + \mu\lambda + \mu^2$$

$$D_{11} = \lambda^2 + \lambda\mu$$

$$D_{22} = \lambda^2$$

therefore $t = L(0) = \dfrac{3\lambda^2 + 2\lambda\mu + \mu^2}{\lambda^3}$

if $r=n$,

$$[I-P] = \begin{bmatrix} \lambda & -\lambda & 0 & 0 & 0\ldots\ldots 0 \\ -\mu & \lambda+\mu & -\lambda & 0 & 0\ldots\ldots 0 \\ 0 & -2\mu & \lambda+\mu & -\lambda & 0\ldots\ldots 0 \\ 0 & 0 & -3\mu & \lambda+3\mu & -\lambda\ldots\ldots 0 \\ \multicolumn{5}{c}{\cdots\cdots\cdots\cdots\cdots\cdots\cdots\cdots\cdots\cdots} \\ \multicolumn{5}{c}{\cdots\cdots\cdots\cdots\cdots\cdots\cdots\cdots\cdots 1} \end{bmatrix}$$

If $n=2$

$$\bar{t} = L(0) = \frac{2\lambda+\mu}{\lambda^2}$$

If $n=3$

$$D_n = \lambda[\lambda+\mu][\lambda+2\mu] - 2\mu\lambda^2 - \lambda[\mu][\lambda+2\mu] = \lambda^3$$

$$D_{oo} = (\lambda+\mu)(\lambda+2\mu) - 2\lambda\mu = \lambda^2 + \lambda\mu + 2\mu^2$$

$$D_{11} = \lambda^2 + 2\lambda\mu$$

$$D_{22} = \lambda^2$$

$$\bar{t} = L(0) = \frac{3\lambda^2 + 3\lambda\mu + 2\mu^2}{\lambda^2}$$

In general, if r=n

$$\bar{t} = L(0) = \frac{1}{\lambda} \sum_{i=0}^{n-1} \frac{n!}{(i+1)!\,(n-i+1)!} \left[\frac{\mu}{\lambda}\right]^i$$

Variance of Time to Failure

To investigate the statistical parameters of the distribution of time to failure we need, besides the mean time to failure, the standard deviation of the time to failure. The variance which is the square of the standard deviation is defined as

$$\sigma_t^2 = \int_0^\infty t^2 f(t) R(t)\,dt - \bar{t}^2, \text{ where } \bar{t}^2 = L(0)$$

using the matrix method discussed we can compute the variance for all systems with more than one operating state again ignoring all but the first absorbing (failed state)

$$[\text{Var } t] = \sigma_t^2 = \sum_{i=0}^{n-1} \frac{D_{ii}^2}{D_n^2}$$

where the definition of D_{ii} and D_n is as before. Considering a two-component, non-maintained, on-line, redundant system

$$[I-P] = \begin{bmatrix} 2\lambda & -2\lambda & 0 \\ 0 & \lambda & -\lambda \\ 0 & 0 & 0 \end{bmatrix}$$

we obtain

$$[I-P]^2 = \begin{bmatrix} 4\lambda^2 & -6\lambda^2 & 2\lambda^2 \\ 0 & \lambda^2 & -\lambda^2 \\ 0 & 0 & 0 \end{bmatrix}$$

and

$$D_2^2 = 4\lambda^4 \qquad D_{11}^2 = 4\lambda^2 \qquad D_{oo}^2 = \lambda^2$$

whereupon

$$\sigma_t^2 = \frac{5}{4\lambda^2}$$

and for the same off-line system

$$\sigma_t^2 = \frac{n}{\lambda^2}$$

In both cases it is assumed that only one component is required for system operation. Similarly considering maintained systems the variance of n=2, r=1 on-line system is derived:

$$[I-P]^2 = \begin{bmatrix} 2\lambda(2\lambda+\mu) & -2\lambda(\lambda+\lambda+\mu) & -2\lambda^2 \\ -2\lambda\mu-\lambda\mu-\mu^2 & 2\lambda\mu+(\lambda+\mu)^2 & -\lambda(\lambda+\mu) \\ 0 & 0 & 0 \end{bmatrix}$$

and

$$D_n^2 = 4\lambda^4$$

$$D_{oo}^2 = 2\lambda\mu+(\lambda+\mu)^2$$

$$D_{11}^2 = 4\lambda^2+2\lambda\mu$$

therefore

$$\sigma_t^2 = \frac{5\lambda^2+6\lambda\mu+\mu^2}{4\lambda^4}$$

For the n=2, r=1 off-line system we obtain

$$\sigma_t^2 = \frac{2\lambda^2+4\lambda\mu+\mu^2}{\lambda^4}$$

from

$$[I-P]^2 = \begin{bmatrix} \lambda(\lambda+\mu) & -\lambda(2\lambda+\mu) & \lambda^2 \\ -\mu(2\lambda+\mu) & \lambda\mu+(\lambda+\mu)^2 & -\lambda(\lambda+\mu) \\ 0 & 0 & 0 \end{bmatrix}$$

general expressions for these cases can be found by similar means as those adopted in deriving the formulas for the MTBF.

Recurrence

In maintained systems we are normally concerned with ergodic processes in which all states (except the failed state) communicate. This means that each state can be reached from any other state, which implies that if, in a redundant system, only one component is required for operation when the 0 to (n-1) states communicate.

In order to compute the mean time the process is in any state j before entering the absorbing (failed) state given it was initially in state i we truncate the matrix P=P and define the inverse of the matrix $[I-P_T] = M$. If we consider again a n=2 component, r=1, maintained on-line system, then

$$[I-P_T] = \begin{bmatrix} 2\lambda & -2\lambda \\ -\mu & \lambda+\mu \end{bmatrix} = [-A_T]$$

where P_T and A_A stand for the truncated matrix $\underline{P}$ and A respectively with last row and column cancelled and

$$M = [I-P_T]^{-1} = \begin{bmatrix} \dfrac{\lambda+\mu}{2\lambda^2} & \dfrac{2\lambda}{2\lambda^2} \\ \\ \dfrac{\mu}{2\lambda^2} & \dfrac{2\lambda}{2\lambda^2} \end{bmatrix}$$

where

$$m_{ij} = (-1)^{(i+j)} \dfrac{D_{ji}}{D_n} = \text{expected time system spends in state i given it started in state j before failure}$$

The expected time the system spends in state 0 before failing giving initially in state 0 is then

$$\dfrac{\lambda+\mu}{2\lambda^2}$$

and the expected time in state i is then

$$\frac{2\lambda}{2\lambda^2} = \frac{1}{\lambda}$$

The total expected time before failure is therefore

$$\frac{\lambda+\mu}{2\lambda^2} + \frac{2\lambda}{2\lambda^2} = \frac{3\lambda+\mu}{2\lambda^2} = \bar{t} = L(0) \text{ given start in state 0}$$

as obtained before.

Similarly, had we started in state 1, then the expected time in state 0 is

$$\frac{\mu}{2\lambda^2}$$

and in state 1 is

$$\frac{2\lambda}{2\lambda^2}$$

and the total expected time before failing becomes

$$\frac{2\lambda+\mu}{2\lambda^2} < \frac{3\lambda+\mu}{2\lambda^2}$$

Therefore, the expected time before failure (MTBF) is larger if the system is initially in the zero state.

8.4 Models of Maintained Systems with Redundant Off-Line Components

MODEL 1

Assumptions

1. Only one component is on-line at a time.

2. Upon failure perfect switching occurs to the next off-line, stand-by component.

3. The repairmen concentrate their effort on bringing the component which failed last to working condition.

4. The repaired component is brought on-line again while the working component is dropped off-line (to a state of stand-by).

5. No failure occurs when components are off-line.

Notation

i = state number   i = 1, 2, ..., n+1

In i = 1 to n, the system is in working condition.

In i = n + 1, the system is in fail condition.

$P_i = P_i(t=\infty)$ = steady-state probability of being in state i

= steady-state probability of having (i-1) components out of action, the ith component on line and working (provided 1 ≠ n + 1) and the $(i-1)^{st}$ component under repair.

λ_i = failure rate of the ith component when on line

μ_i = rate of repair at which the ith component is brought back on line

2-Component System (Figure 8.2)

The three required equations are:

$$0 = -\lambda_1 P_1 + \mu_1 P_2 \qquad\qquad P_2 = \lambda_1/\mu_1 \, P_1$$

$$0 = -(\mu_1+\lambda_2) P_2 + \lambda_1 P_1 + \mu_2 P_3 \qquad P_3 = \frac{\lambda_1 \lambda_1}{\mu_1 \mu_2} P_1$$

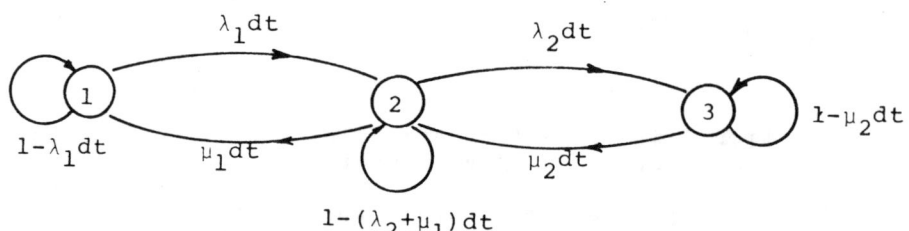

FIGURE 8.2

$$1 = P_1 + P_2 + P_3 \qquad\qquad 1 = P_1[1+\lambda_1/\mu_1+\lambda_2\lambda_1/\mu_1\mu_2]$$

from which

$$P_1 = \frac{1}{1+\lambda_1/\mu_1+\lambda_1\lambda_2/\mu_1\mu_2}$$

$$P_2 = \frac{\lambda_1/\mu_1}{1+\lambda_1/\mu_1+\lambda_1\lambda_2/\mu_1\mu_2}$$

$$P_3 = \frac{\lambda_1\lambda_2/\mu_1\mu_2}{1+\lambda_1/\mu_1+\lambda_1\lambda_2/\mu_1\mu_2}$$

3-Component System (Figure 8.3)

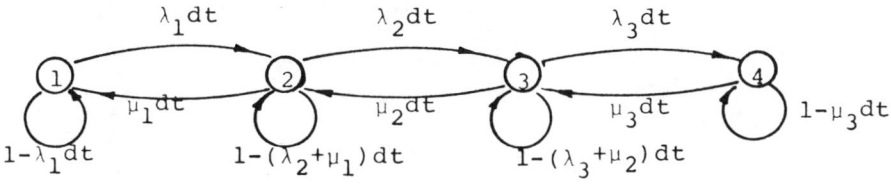

FIGURE 8.3

The four required equations are:

$0 = -\lambda_1 P_1 + \mu_1 P_2$, from which $P_2 = \lambda_1/\mu_1 \; P_1$

$0 = -(\lambda_2+\mu_1)P_2 + \lambda_1 P_1 + \mu_2 P_3$, from which $P_3 = \dfrac{\lambda_1\lambda_2}{\mu_1\mu_2} P_1$

$0 = -(\mu_2+\lambda_3)P_3 + P_2\lambda_2 + P_4\mu_3$, from which $P_4 = \dfrac{\lambda_1\lambda_2\lambda_3}{\mu_1\mu_2\mu_3} P_1$

and

$1 = P_1+P_2+P_3+P_4$

from which

$$P_1 = \frac{1}{1+\lambda_1/\mu_1 + \dfrac{\lambda_1\mu_2}{\lambda_1\mu_2} + \dfrac{\lambda_1\lambda_2\lambda_3}{\mu_1\mu_2\mu_3}}$$

$$P_2 = \frac{\lambda_1/\mu_1}{1+\lambda_1/\mu_1 + \dfrac{\lambda_1\lambda_2}{\mu_1\mu_2} + \dfrac{\lambda_1\lambda_2\lambda_3}{\mu_1\mu_2\mu_3}}$$

$$P_3 = \frac{\dfrac{\lambda_1 \lambda_2}{\mu_1 \mu_2}}{1 + \dfrac{\lambda_1}{\mu_1} + \dfrac{\lambda_1 \lambda_2}{\mu_1 \mu_2} + \dfrac{\lambda_1 \lambda_2 \lambda_3}{\mu_1 \mu_2 \mu_3}}$$

and

$$A(\infty) = \frac{1 + \dfrac{\lambda_1}{\mu_1} + \dfrac{\lambda_1 \lambda_2}{\mu_1 \mu_2}}{1 + \dfrac{\lambda_1}{\mu_1} + \dfrac{\lambda_1 \lambda_2}{\mu_1 \mu_2} + \dfrac{\lambda_1 \lambda_2 \lambda_3}{\mu_1 \mu_2 \mu_3}}$$

N Parallel Redundant Off-Line System (1 Component Required for System Operation) - Figure 8.4)

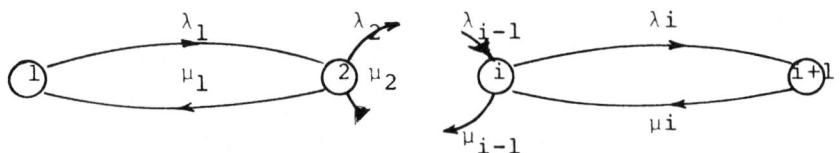

FIGURE 8.4

λ_i = failure rate of ith component when on line (λ_i =0 when stand-by)

μ_i = repair rate bringing the ith component from failure to state of work

$$A(\infty) = \frac{1 + \sum_{i=1}^{n} \left[\prod_{j=1}^{j=1} P_j \right]}{1 + \sum_{i=1}^{n} \left[\prod_{j=1}^{j=i} P_j \right]}$$

where

$P_j = \lambda_j / \mu_j$

MODEL 2

Assumptions

1. Same as 1 and
2. Model 1
3. Maintenance work is carried out at the same time on all the components which failed. As a result, the system when in state i, can return only to state 1 or deteriorate further to state i+1.
4. If the ith component is at present on line and <u>all</u> the (i-1) components which were under repair have been repaired, component one (1) is brought back on line and component i becomes a stand-by unit.
5. Same as 5 in Model 1.

Notation

In an n component system, i = state number, i = 1, 2, ..., n+1. In i=1 to n, the system is in working condition. In i = n+1, the system is in fail condition.

P_i)
) are as defined in Model 1
λ_i)

$\bar{\mu}_i$ is the rate of repair at which all i components which are under repair are brought back into state 1 stand-by simultaneously

2-Component System (Figure 8.5)

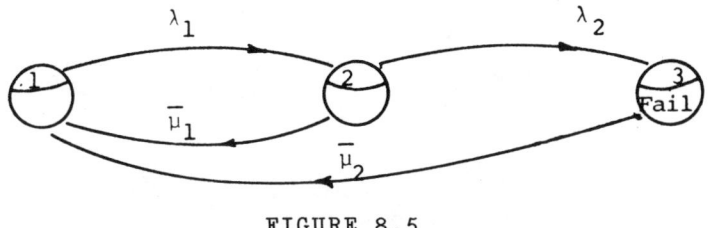

FIGURE 8.5

The three required equations are:

$$0 = -\lambda_1 P_1 + \bar{\mu}_1 P_2 + \bar{\mu}_2 P_3$$

$$0 = -(\lambda_2 + \bar{\mu}_1) P_2 + \lambda_1 P_1$$

$$1 = P_1 + P_2 + P_3$$

from which

$$P_2 = \frac{\lambda_1}{\lambda_2 + \bar{\mu}_1} P_1$$

$$P_3 = \frac{\lambda_1 \lambda_2}{\bar{\mu}_2 (\lambda_2 + \bar{\mu}_1)} P_1$$

and

$$P_1 = \frac{1}{1 + \dfrac{\lambda_1}{\lambda_2 + \bar{\mu}_1} + \dfrac{\lambda_1 \lambda_2}{\bar{\mu}_2 (\lambda_2 + \bar{\mu}_1)}} = 1/\Delta$$

$$P_2 = \frac{\dfrac{\lambda_1}{\lambda_2 + \bar{\mu}_1}}{\Delta}$$

$$P_3 = \frac{\dfrac{\lambda_1 \lambda_2}{\mu_2 (\lambda_2 \bar{\mu}_1)}}{\Delta}$$

and

$$A(\infty) = P_1 + P_2$$

$$= \frac{1 + \dfrac{\lambda_1}{\lambda_2 + \bar{\mu}_1}}{\Delta}$$

3-Component System (Figure 8.6)

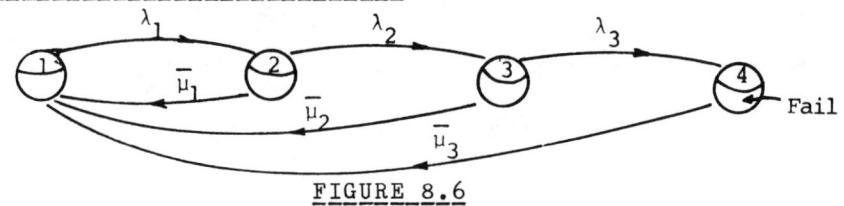

FIGURE 8.6

The four required equations are:

$$0 = -\lambda_1 P_1 + \bar{\mu}_1 P_2 + \bar{\mu}_2 P_3 + \bar{\mu}_3 P_4$$

$$0 = \lambda_1 P_1 - (\lambda_2 + \bar{\mu}_1) P_2$$

$$0 = \lambda_2 P_2 - (\bar{\mu}_2 + \lambda_3) P_3$$

$$1 = P_1 + P_2 + P_3 + P_4$$

from which

$$P_2 = \frac{\lambda_1}{\lambda_2 + \bar{\mu}_1} P_1$$

$$P_3 = \frac{\lambda_2}{(\bar{\mu}_2 + \lambda_3)} P_2 = \frac{\lambda_1 \lambda_2}{(\lambda_2 + \bar{\mu}_1)(\lambda_3 + \bar{\mu}_2)} P_1$$

$$P_4 = \frac{\lambda_1}{\bar{\mu}_3} P_1 - \frac{\bar{\mu}_1}{\bar{\mu}_3} P_2 - \frac{\bar{\mu}_2}{\bar{\mu}_3} P_3$$

$$= \left\{ \frac{\lambda_1}{\bar{\mu}_3} - \frac{\bar{\mu}_1}{\bar{\mu}_3} \frac{\lambda_1}{(\lambda_2 + \bar{\mu}_1)} - \frac{\bar{\mu}_2 \lambda_1 \lambda_2}{\bar{\mu}_3 (\lambda_2 + \bar{\mu}_1)(\lambda_3 + \bar{\mu}_2)} \right\} P_1$$

$$= \frac{\lambda_1 \lambda_2 \lambda_3}{\bar{\mu}_3 (\lambda_2 + \bar{\mu}_1)(\lambda_3 + \bar{\mu}_2)} P_1$$

from which

$$P_1 = \frac{1}{1 + \left(\frac{\lambda_1}{\lambda_2 + \mu_1}\right) + \frac{\lambda_1 \lambda_2}{(\lambda_2 + \mu_1)(\lambda_3 + \mu_2)} + \frac{\lambda_1 \lambda_2 \lambda_3}{(\lambda_2 + \mu_1)(\lambda_3 + \mu_2)\mu_3}}$$

$$= 1/\Delta$$

$$P_2 = \frac{\frac{\lambda_1}{\lambda_2 + \mu_1}}{\Delta}$$

$$P_3 = \frac{\frac{\lambda_1 \lambda_2}{(\lambda_2+\mu_1)(\lambda_3+\mu_2)}}{\Delta}$$

$$P_4 = \frac{\frac{\lambda_1 \lambda_2 \lambda_3}{(\lambda_2+\mu_1)(\lambda_3+\mu_2)\mu_3}}{\Delta}$$

and

$$A(\Sigma) = \sum_{i=1}^{3} P_i$$

N-Component System (Figure 8.7)

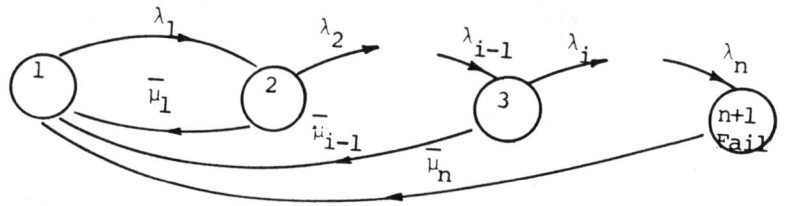

FIGURE 8.7

$$A(\infty) = \sum_{i=1}^{n-1} P_i$$

$$= \frac{1 + \sum_{i=1}^{n-1} \frac{\prod_{j=1}^{i} \lambda_j}{\prod_{j=1}^{i} [\lambda_{j+1}+\bar{\mu}_j]}}{1 + \sum_{i=1}^{n} \frac{\prod_{j=1}^{i} \lambda_i}{\prod_{j=1}^{i} [\lambda_{j+1}+\bar{\mu}_j]}}$$

where

$\lambda_{j+1} = 0$, where $j+1 = n+1$.

MODEL 3

Assumptions

    1. Two components in parallel are needed to keep the system operating.

    2. Upon failure of an on-line component perfect switching brings one of the stand-by components on-line.

    3. Repair work commences on any component upon failure, each component havings its own failure rate and repair rate.

    4. If the stand-by maintenance work on component ξ is completed when component i is on-line, then if $\xi < i$, then ξ is brought on-line and i is reduced to a stand-by unit. If $\xi > i$, then ξ becomes a stand-by unit and i continues its operation as an on-line unit.

    5. As 5 in Model 1.

Notation

The notation used will distinguish between the state number and the state of the component in the system. (See Figure 8.8.)

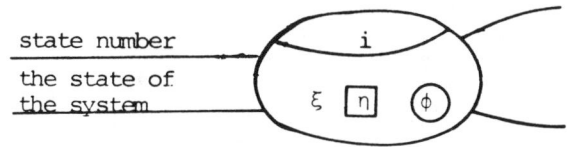

FIGURE 8.8

Let us take component ξ

    $\boxed{\xi}$ means that component ξ is out of action (fail) and being repaired

    $\widehat{\xi}$ means that component ξ is in a state of stand-by

    ξ means that component ξ is on-line

Therefore, in the flow graph, we will have the description of each state as shown in Figure 8.8.

    Also,

    λ_ξ = failure rate of component ξ

    μ_ξ = repair rate of component ξ

and

    P_i = steady-state probability that the system will be in state i

in state i

2-Components System (Figure 8.9)

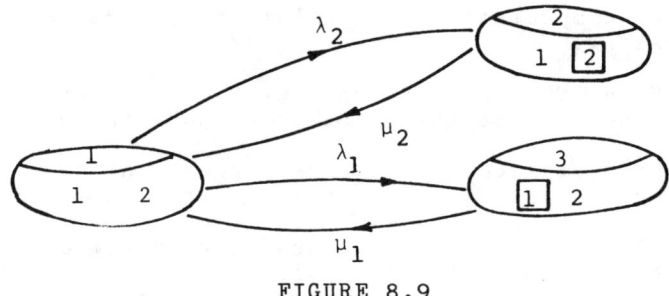

FIGURE 8.9

The flow graph is

$\eta = 2$

$\gamma = 2$

$S_o b_o = 0$

Then

$0 = -(\lambda_1 + \lambda_2)P_1 + \mu_2 P_2 + \mu_1 P_3$

$0 = -\mu_2 P_2 + \lambda_2 P_1$

$1 = P_1 + P_2 + P_3$

from which we get

$P_2 = \lambda_2/\mu_2 \; P_1$

$P_3 = \lambda_1/\mu_1 \; P_1$

so;

$P_1 = \dfrac{1}{1 + \dfrac{\lambda_1}{\mu_1} + \dfrac{\lambda_2}{\mu_2}} = A(\infty)$

3-Component System (Figure 8.10)

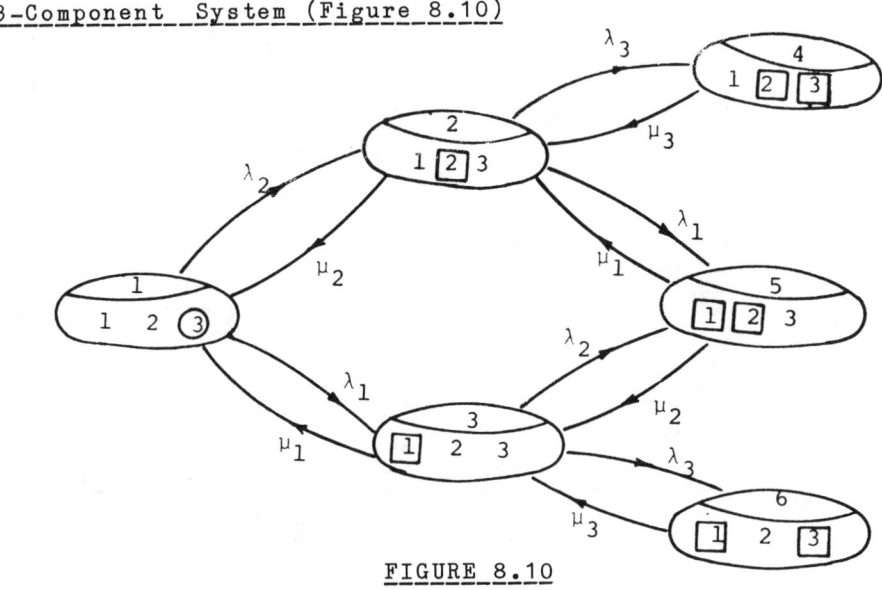

FIGURE 8.10

ξ on-line condition

(ξ) stand-by condition

$\boxed{\xi}$ fail

The required equations are:

$$0 = -(\lambda_1+\lambda_2)P_1 + \mu_2 P_2 + \mu_1 P_3$$

$$0 = -(\mu_2+\lambda_3+\lambda_1)P_2 + \mu_3 P_4 + \lambda_2 P_1 + \mu_1 P_5$$

$$0 = -\mu_3 P_4 + \lambda_3 P_2$$

$$0 = -(\lambda_2+\mu_1+\lambda_3)P_3 + \lambda_1 P_1 + \mu_2 P_5$$

$$0 = -\mu_3 P_6 + \lambda_3 P_3$$

$$1 = P_1 + P_2 + P_3 + P_4 + P_5 + P_6$$

from which:

$$P_2 = P_1 \frac{1}{\mu_2} \frac{(\lambda_1+\lambda_2) - \frac{\lambda_1\mu_1-\lambda_2\mu_2}{\lambda_2+\mu_1+\lambda_3}}{1 + \frac{\lambda_1+\mu_2}{\lambda_2+\mu_1+\lambda_3}}$$

$$P_3 = P_1 \frac{1}{\mu_1} \left[\frac{\mu_2+\lambda_1}{\lambda_2+\mu_1+\lambda_3} \frac{(\lambda_1+\lambda_2) - \frac{\lambda_1\mu_1-\lambda_2\mu_2}{\lambda_2+\mu_1+\lambda_3}}{1 - \frac{\lambda_1+\mu_2}{\lambda_2+\mu_1+\lambda_3}} + \frac{\mu_1\lambda_1-\mu_2\lambda_2}{\lambda_2+\mu_1+\lambda_3} \right]$$

$$P_4 = \frac{\lambda_3}{\mu_3} P_2 = \frac{\lambda_3 P_1}{\lambda_2 \mu_3} \frac{(\lambda_1+\lambda_2) - \frac{\lambda_1\mu_1-\lambda_2\mu_2}{\lambda_2+\mu_1+\lambda_3}}{1 + \frac{\lambda_1=\mu_2}{\lambda_2+\mu_1+\lambda_3}}$$

$$P_5 = \frac{\mu_2+\lambda_1}{\mu_1} P_2 - \frac{\lambda_2}{\mu_1} P_1 = \left[\frac{\mu_2+\lambda_1}{\mu_1\mu_2} \frac{(\lambda_1+\lambda_2) - \frac{\lambda_1\mu_1-\lambda_2\mu_2}{\lambda_2+\mu_1+\lambda_3}}{\frac{\lambda_1+\mu_2}{\lambda_2+\mu_1+\lambda_3}} - \frac{\lambda_2}{\mu_1} \right] P_1$$

$$P_6 = \frac{\lambda_3}{\mu_3} P_3 = P_1 \left[\frac{\lambda_3}{\mu_1\mu_3} \frac{\mu_2-\lambda_1}{\lambda_2+\mu_1+\lambda_3} \frac{(\lambda_1+\lambda_2) - \frac{\lambda_1\mu_1-\lambda_2\mu_2}{\lambda_1+\mu_1+\lambda_3}}{1 + \frac{\lambda_1+\mu_3}{\lambda_2+\mu_1+\lambda_3}} \right.$$
$$\left. + \frac{\lambda_1\mu_1+\lambda_2\mu_2}{\lambda_2+\mu_1+\lambda_3} \right]$$

and from

$$1 = \sum_{i=1}^{6} P_i \text{ we get}$$

$$A(\infty) = \sum_{i=1}^{3} P_i$$

The expression for $A(\infty)$ is cumbersome but can easily be derived for the above expressions.

Let us increase the number of stand-by components in the previous case to two and examine the corresponding increase in the number of states the reliability model will experience. See Figure 8.11.

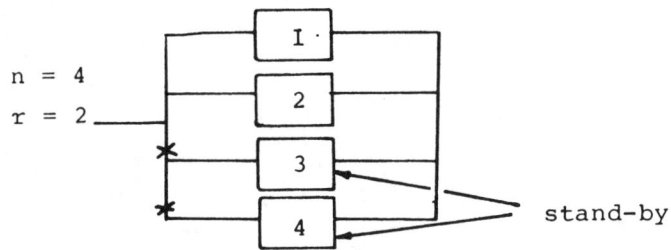

FIGURE 8.11:   TWO-COMPONENT PARALLEL SYSTEM WITH TWO
              STAND-BY COMPONENTS

As before, we will have the convention of

  $\boxed{\xi}$ ξ is in fail condition

  $\circled{\xi}$ ξ in in stand-by condition

  ξ ξ is in on-line condition

We can use the decision tree to mark all possible events in
our system. On each of the nodes, we shall write the state
number, while the branch will carry the event causing the
system to go to the particular state. The reader is reminded
that the complication arises because of the different failure
and maintenance rate each component has which necessitates
tracing all possible outcomes with each of the n components.

The "failed" states can be seen in Figure 8.12 (on the
following page). A flow graph and the necessary equations
can be derived from Figure 8.12.

MODEL 4

Assumptions

1. Same as in Model 1.

2. Same as in Model 1.

3. Each component has its own repairmen and specific
   rate of repair and rate of failure.

4. Whenever component i is on-line and component ξ
   where $\xi < i$ was repaired, perfect switching occurs
   as a result of which "ξ" is brought on-line and "i"
   becomes a stand-by component. If $\xi > i$, component i
   remains on-line while component ξ becomes a stand-by
   component.

5. Same as in Model 1.

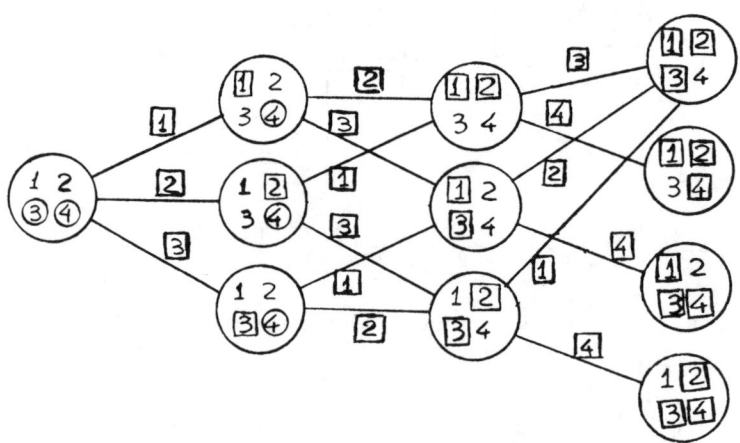

FIGURE 8.12: STATES OF TWO-COMPONENT SYSTEM WITH TWO STAND-BY COMPONENTS (MODEL 3)

Notation

Care must be taken in this model (as in Model 3) to separate the state number from the number of the components currently on-line. Therefore, in the flow graph, each state has its number in addition to symbolic description of the state in which the system is.

Thus,

$\boxed{\xi}$
$\textcircled{\xi}$ have the same meaning as in Model 3.
ξ

P_i = steady-state probability that the system will be in state i

μ_ξ = repair rate of component no.

λ_ξ = failure rate of component no.

2-Components System

The flow graph is shown in Figure 8.13 and the necessary equations are:

$$0 = -\lambda_1 P_1 + \mu_1 P_2 + \lambda_2 P_4$$

$$0 = -(\mu_1+\lambda_2)P_2 + \lambda_1 P_1 + \mu_2 P_3$$

$$0 = -(\mu_2+\mu_1)P_3 + \lambda_1 P_4 + \lambda_2 P_2$$

$$1 = P_1 + P_2 + P_3 + P_4$$

from which we get:

$$P_2 = P_1 \frac{\lambda_1(\lambda_1+\mu_1+\mu_2)}{(\mu_1+\mu_2)(\mu_1+\lambda_2) + (\lambda_1\mu_1-\lambda_2\mu_2)}$$

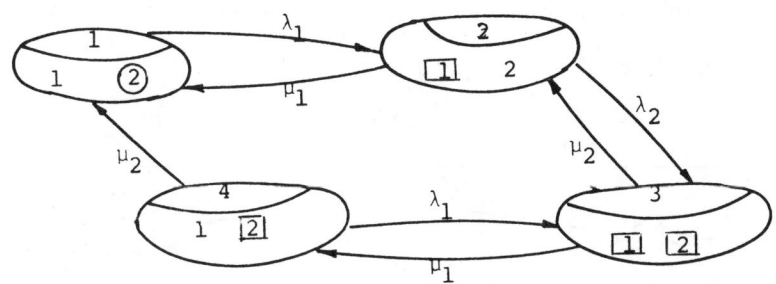

FIGURE 8.13: STATE TRANSITION DIAGRAM OF ONE-COMPONENT SYSTEM WITH ONE STAND-BY COMPONENT (MODEL 4)

Let $\Delta_1 = (\mu_1+\mu_2)(\mu_1+\lambda_2) + (\lambda_1\mu_1-\lambda_2\mu_2)$

then,

$$P_3 = \frac{\frac{\lambda_1\lambda_2}{\mu_2}(\lambda_1+\mu_2)}{\Delta_1} P_1$$

$$P_4 = \frac{\frac{\lambda_1\lambda_2}{\mu_2}\mu_1}{\Delta_1} P_1$$

and so

$$P_1 = \frac{1}{1 + \frac{\lambda_1\mu_2}{\mu_2}\frac{(\lambda_1+\mu_1+\mu_2)}{\Delta_1} + \frac{\lambda_1\lambda_2}{\mu_2}\frac{(\lambda_1+\lambda_2)}{\Delta_1} + \frac{\lambda_1\lambda_2}{\mu_2}\frac{\mu_1}{\Delta_1}} = 1/\Delta_2$$

$$P_2 = \frac{\lambda_1(\lambda_1+\mu_1+\mu_2)}{\Delta_1\Delta_2}$$

$$P_3 = \frac{\lambda_1\lambda_2}{\mu_2} \frac{(\lambda_1+\mu_2)}{\Delta_1\Delta_2}$$

$$P_4 = \frac{\lambda_1\lambda_2}{\mu_2} \frac{\mu_1}{\Delta_1\Delta_2}$$

and

$$A(\infty) = P_1 + P_2 + P_4$$

Now, let us have a look on the 3-component system.

3-Component System

The flow graph is as shown in Figure 8.14.

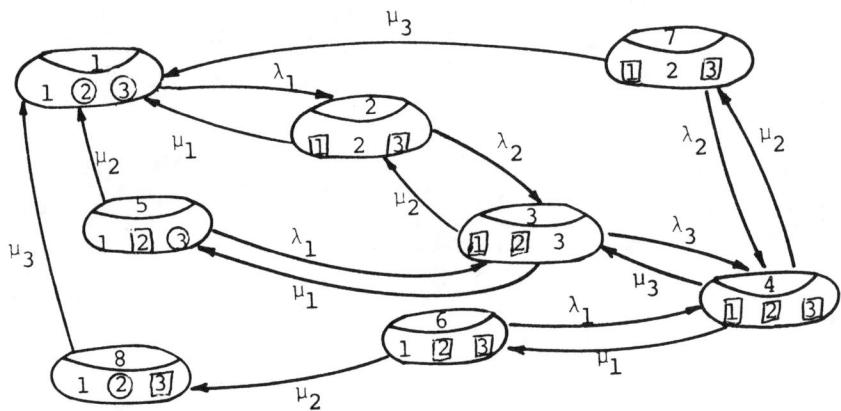

FIGURE 8.14: STATE TRANSITION DIAGRAM OF ONE-COMPONENT SYSTEM WITH TWO STAND-BY COMPONENTS (MODEL 4)

The necessary equations are:

$$0 = -\lambda_1 P_1 + \mu_1 P_2 + \mu_2 P_5 + \mu_3 P_8$$

$$0 = 1(\lambda_2+\mu_1)P_2 + \lambda_1 P_1 + \mu_2 P_3 + \mu_3 P_7$$

$$0 = 1(\mu_1+\lambda_3+\mu_2)P_3 + \lambda_2 P_2 + \mu_3 P_4 + \lambda_1 P_5$$

$$0 = -(\mu_3+\mu_2+\mu_1)P_4 + \lambda_3 P_3 + \lambda_2 P_7 + \lambda_1 P_6$$

$$0 = -(\lambda_1+\mu_2)P_5 + \mu_1 P_3 + \mu_3 P_6$$

$$0 = -(\lambda_1+\mu_2+\mu_3)P_6 + \mu_1 P_4$$

$$0 = -(\mu_3+\lambda_2)P_7 + \mu_2 P_4$$

$$0 = -\mu_3 P_8 + \mu_2 P_6$$

or

$$1 = P_1 + P_2 + P_3 + P_4 + P_5 + P_6 + P_7 + P_8$$

The reader can note that the usual

$$\sum_{i=1}^{8} P_i = 1$$

is not used in this case if we are going to use the stochastic matrix to find the P_i's.

From the state equations, we can derive the stochastic matrix of the system. As discussed before, matrix inversion can become quite difficult, both in the time and transform domain. Many different approaches can be used to get around this problem. In this case, because the inversion of this matrix using transform techniques is somewhat cumbersome, the equations are solved using a simultaneous equation approach as follows. Letting:

$$\phi = (\lambda_2+\mu_1)$$
$$\eta = (\mu_3+\mu_2+\mu_1)$$
$$\beta = (\lambda_1+\mu_2+\mu_3)$$
$$\xi = (\mu_1+\lambda_3+\mu_2)$$
$$\alpha = (\lambda_1+\mu_2)$$

$$\gamma = (\mu_3 + \lambda_2)$$

and letting

$$X = (\xi\alpha - \lambda_1\mu_1)(\eta\gamma\beta - \lambda_2\mu_2^{\beta-\lambda_1}\mu_1\gamma) - \lambda_3\mu_3\gamma(\alpha\beta + \lambda_1\mu_1)$$

$$= (\xi\alpha - \lambda_1\mu_1)\beta\gamma Y - \lambda_3\mu_3(\alpha\beta + \lambda_1\mu_1)$$

$$Y = [\eta - \frac{\lambda_2\mu_2}{\gamma} - \frac{\lambda_1\mu_1}{\beta}]$$

and

$$Z = \phi\gamma - \alpha\beta\gamma\lambda_2\mu_2(\gamma + \lambda_3\mu_3)$$

Now, we can express P_2 to P_8 in terms of P_1 and the above factors:

$$P_2 = \frac{\lambda_1 \gamma \times P_1}{Z} = C_2 P_1$$

$$P_3 = \frac{\lambda_1 \lambda_2 \alpha\beta \gamma^2 Y}{Z} P_1 = C_3 P_1$$

$$P_4 = \frac{\lambda_1 \lambda_2 \lambda_3 \alpha\beta \gamma^2}{Z} P_1 = C_4 P_1$$

$$P_5 = \frac{\lambda_1 \mu_1 \lambda_2 \gamma^2 [\beta Y + \mu_3 \lambda_3]}{Z} P_1 = C_5 P_1$$

$$P_6 = \frac{\lambda_1 \mu_1 \lambda_2 \lambda_3 \alpha \gamma^2}{Z} P_1 = C_6 P_1$$

$$P_7 = \frac{\lambda_1 \lambda_2 \mu_2 \lambda_3 \alpha\beta \gamma}{Z} P_1 = C_7 P_1$$

$$P_8 = \frac{\lambda_1 \mu_1 \lambda_2 \mu_2 \lambda_3 \alpha \gamma^2}{Z} P_1 = C_8 P_1$$

now,

$$P_1 = \frac{1}{\sum_{i=1}^{8} C_i} \quad \text{where } C_1 = 1$$

and

$$P_i = \frac{C_i}{\sum_{i=1}^{8} C_i} \quad i = 1, 2, \ldots, 8$$

So,

$$A_\infty = 1 - P_4 = 1 - \frac{C_4}{\sum_{i=1}^{8} C_i}$$

8.5 Exercises

1. The reliabilities of the components of a steam cycle are as listed (for time between overhauls). If the cycle consists of a series of boilers, turbines, condenser, condensate pumps, (one-stage) feed heaters, and feed pumps, and if the numbers of each component installed and required are as listed,

Component	No. Installed (Parallel)	No. Required for Operation)	Reliability
Boiler	1	1	0.96
Turbine	1	1	0.99
Condenser	1	1	0.995
Condensate Pump	2	1	0.98
Feed Heater	2	1	0.98
Feed Pump	3	2	0.97

then:

    a. Find the total systems reliability.

    b. Assume that the overhaul period is two years. What is the corresponding MTBF of the system, and if the age-dependent failure rates are constants for all components, what are these age-dependent failure rates?

c. What is the probability that the total system will have 1, 2, ... n failures during the first year after overhaul?

2. There are systems which, for safety hazards or other reasons, have limited repair capability while "on line". Such a system, shown as a functional diagram below, is a set of pumps and heat exchangers for a liquid-sodium heat exchanger system for a nuclear powered ship.

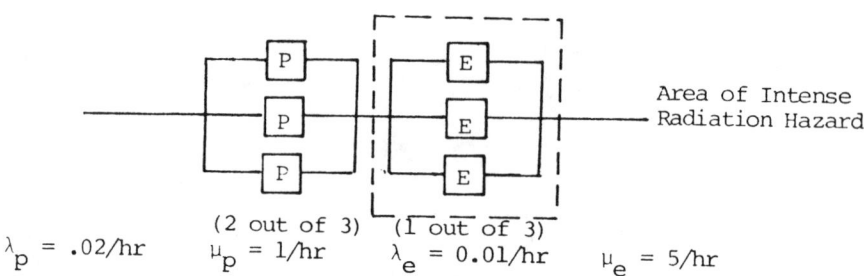

(2 out of 3)   (1 out of 3)
$\lambda_p = .02/hr$ $\mu_p = 1/hr$ $\lambda_e = 0.01/hr$ $\mu_e = 5/hr$

The valve and drain system permits easy and safe draining of pumps so that they may be serviced while on-line. The intense heat and radiation hazard of the heat exchanger require that all three exchangers be shut down and drained into a reserve tank before servicing. There are two servicemen, but only one serviceman can work on a unit (pump or exchanger) at a time.

It is desired to establish a service and an operational policy which will maximize up-time.

$$A_\infty = \frac{MTBF}{MTBF + MTTR} = \frac{1}{1 + MTTR/MTBF}$$

MTTR = mean time to repair

a. List possible repair policies.

b. Should the service policy be to service the pumps as they fail, or wait until two fail, and repair all units once a fail state has been reached, or resume operation as soon as possible?

c. Should the policy be to wait until one, two, or three exchangers fail before servicing, and repair all, or resume operations as soon as possible? (Prove answers numerically.)

3. Mean time to repair (MTR) is defined as the mean time a system spends in a non-operating state before returning to an operating state. Find the MTR for a system consisting of two identical components in parallel arranged as _ideal_ (perfect switching) _off-line_ standby. A single

repairman providing a repair rate is available. What is the percentage time both operate?

4. What is the MTBF and MTR of a two identical component off-line system which requires a switch and sensor for each component as shown?

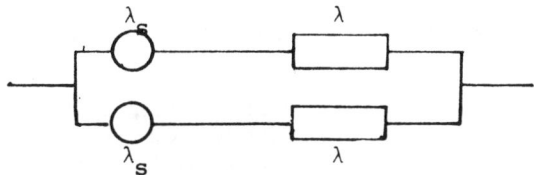

The repair rate is μ for the components independent of the number of components failed. The switches similarly have a repair rate of μ_s independent of the number of switch failures. μ and μ_s are mutually independent.

5. A ship steam turbine plant has 3 identical forced draught fans with constant age-dependent failure rate λ (in on-line stand-by), of which 2 out of 3 are needed for operation. There are two repairmen, each providing a repair rate and both able to work on one fan at a time. (Total repair rate constant = 2.) Repair policy is first-come, first-served and initially one fan is down. Find:

    a. steady state availability

    b. steady state availability if only 1 out of 3 are needed for operation

    c. repeat b. assuming repair cannot start until all 3 fans are down and the system is not restarted until all 3 fans are again repaired.

6. A pumping station has three identical pumps, two of which are required to pump full loads. A single pump can pump 60% of the load. Switching is instantaneous and perfect and two pumps, when available, will usually work in parallel with the third on off-line stand-by or under repair. If the failure rate of a pump is equal to $\lambda = 1 \times 10^{-4}$/hr independent of the number of (1 or 2) pumps working and if the repair rate of a pump is $\mu_1 = 2 \times 10^{-2}$ with one pump failed and $\mu_2 = 4 \times 10^{-2}$ with two pumps down, what is:

    a. the MTBF of the system. (The mean time before all three pumps fail).

    b. the percentage of the MTBF that one or two pumps are working and the resulting average percentage output of the system before the first failure.

c. the mean time before the failure of the first pump after start

d. the variance of the MTBF and the resulting probability that the system will not fail without 90% of the MTBF.

Discuss your assumption of the distribution of the time to failure.

7. A two-component <u>off-line</u> system with perfect switching is composed of components whose age-dependent failure rates are λ_1 and λ_2 respectively. Each requires a different repair process after failure resulting in a repair state of μ_1 and μ_2 respectively if component 1 or 2 is under repair. If both components have failed they are repaired simultaneously with a repair rate μ which brings both of them back into operation. Find:

a. the MTBF if both components are operative initially

b. the MTBF if component 2 alone was operative initially

c. the percentage time both componenbts are operative before system failure given system starts with both components operative.

8.6 References

1. Shooman, M.L., "Probabilistic Reliability: An Engineering Approach". McGraw-Hill, New York, 1968.

2. Jardine, A.K.S., "Maintenance, Replacement, and Reliability". Wiley and Sons, New York, 1973.

3. Sandler, G.H., "Systems Reliability Engineering". Prentice-Hall, Englewood Cliffs, New Jersey, 1963.

4. Smith, C.O., "Introduction to Reliability in Design". McGraw-Hill, New York, 1976.

5. Kapur, K.C. and Lamberson, L.R., "Reliability in Engineering Design". John Wiley & Sons, New York, 1977.

6. Bazovsky, I., "Reliability Theory and Practice". Prentice-Hall, Englewood Cliffs, New Jersey, 1963.

7. Hart, G.C., "Uncertainty Analysis, Loads, and Safety in Structural Engineering". Prentice-Hall Inc.,

Englewood Cliffs, New Jersey, 1982.

8. Singh, C. and Billintou, R., "System Reliability Modelling and Evaluation", Hutchinson, London, 1977.

9. Kapur, K.C. and Lamberson, L.R., "Reliability in Engineeringd Design". John Wiley, New York, 1977.

10. Smith, D.J., "Reliability Engineering", Pitman, New York, 1973.

11. Smith, D.J. and Bagg, A.H., "Maintainability Engineering", Pitman, New York, 1973.

12. Zelen, M., "Statistical Theory of Reliability", University of Wisconsin Press, Madison, Wisc., 1963.

9.0 STRATEGIES FOR REPAIR POLICIES

Ernst G. Frankel

Before discussing optimization of system operation with respect to resources (weight, volume, cost, etc.), it might be interesting to consider some of the aspects of the variables affecting operational inputs. When discussing the expected time m_{ij}, a system finds itself in a certain state j given it was initially in state i then we may define a ratio

$\dfrac{m_{ij}}{t}$ = proportion of time before absorption system finds itself in state j given initially in state i

where t = MTBF.

Such a ratio is of importance for the analysis of utilization and cost of the repair effort, if the reliability of the system is the performance criterion.

For instance, considering a two-component on-line and off-line system,

		State	
		0	1
Proportion of time system spends in state before entering failed state given starts in state 0.	On-line	$\dfrac{\lambda+\mu}{3\lambda+\mu}$	$\dfrac{2\lambda}{3\lambda+\mu}$
	Off-line	$\dfrac{\lambda+\mu}{2\lambda+\mu}$	$\dfrac{\lambda}{2\lambda+\mu}$

The above approach is useful for systems which are not allowed to fail and where therefore ($A\infty$) is not a maximizing variable.

If, on the other hand, we require to maximize the availability with a given amount of repair resources or a fixed amount of systems cost, then we are interested in the proportion of total time spent in each state, including the failed states. Returning to our two-component system we therefore find the steady state probabilities of being in state j given we were initially in state i (i is usually taken as 0). These probabilities can be shown to tend to the proportion of time (limit → ∞) that our system finds itself in any of the states. Considering a two-component system again,

			State 0	State 1	State 2
Proportion of time in State j given initially in State o	On-line	r=1	$\dfrac{\mu^2}{\mu^2+2\lambda\mu+2\lambda^2}$	$\dfrac{2\lambda\mu}{\mu^2+2\lambda\mu+2\lambda^2}$	$\dfrac{2\lambda^2}{\mu^2+2\lambda\mu+2\lambda^2}$
		r=2	$\dfrac{\mu^2}{\mu^2+2\lambda\mu+\lambda^2}$	$\dfrac{2\lambda\mu}{\mu^2+2\lambda\mu+\lambda^2}$	$\dfrac{\lambda^2}{\mu^2+2\lambda\mu+\lambda^2}$
	Off-line	r=1	$\dfrac{\mu^2}{\mu^2+\lambda\mu+\lambda^2}$	$\dfrac{\lambda\mu}{\mu^2+\lambda\mu+\lambda^2}$	$\dfrac{\lambda^2}{\mu^2+\lambda\mu+\lambda^2}$
		r=2	$\dfrac{2\mu^2}{2\mu^2+2\lambda\mu+\lambda^2}$	$\dfrac{2\lambda\mu}{2\mu^2+2\lambda\mu+\lambda^2}$	$\dfrac{\lambda^2}{2\mu^2+2\lambda\mu+\lambda^2}$

If the lifetime T of our system is comparatively short (in other words, if we expect transient effects not to have died down before T), then instead of the above steady state expressions, we obtain formulas for $P_j(t)$ and integrate those to obtain the proportion of time in state j during T or

$$P_j(T) = \frac{1}{T} \int_0^T P_j(t)\,dt$$

For instance, a two-component, parallel, on-line maintained system initially in state 0 has a

$$P_2(t) = \frac{\lambda^2}{(\lambda+\mu)^2} + \frac{\lambda^2 e^{-2(\lambda+\mu)t}}{(\lambda+\mu)^2} - \frac{2\lambda^2 e^{-(\lambda+\mu)t}}{(\lambda+\mu)^2}$$

when r=2, which gives us the proportion of time T spent in state 2 (failed state).

$$P_2(T) = \frac{1}{T}\int_0^T P_2(t)\,dt = \frac{\lambda^2}{(\lambda+\mu)^2}\left[1 - \frac{3}{2(\lambda+\mu)T} + \frac{e^{-2(\lambda+\mu)T}}{2(\lambda+\mu)T} + \frac{2e^{-(\lambda+\mu)T}}{(\lambda+\mu)T}\right]$$

In a similar fashion $P_1(T)$ and $P_2(T)$ could be found and then be used in optimizing the allocation of system parameters to achieve the desired performance goal. The above procedure can be applied to any state system and be used to evaluate utilization of applied repair resources.

It should be noted that the proportion of time a system spends in a state before entering another or the failed state is different from the expected time the system spends in a state or in a set of states before first systems failure. For example, a two-component on-line redundant system with one repair channel would expect to spend

$$\frac{\mu+\lambda}{2\lambda^2}$$

units of time in state 0 and $1/\lambda$ units of time in state 1 before first systems failure, given the system started operating in state 0 at the start of time 0.

9.0.1 General Repair Strategy Determination

In the most general sense we can divide systems costs into operating costs of individual units and downtime costs of the total system. Assuming we deal with a parallel system consisting of n identical on-line components where

K = steady state probability of one component failing

K^n = steady state probability of all n components failing

if
C_1 = total operating and maintenance cost per component per unit time

and
C_2 = cost of downtime of total system per unit time

Then total cost for n component system

$$C_n = C_1 n + C_2 K^n \text{ as } K < 1$$

Then

$$C_n - C_{n-1} = C_1(n-(n-1)) + C_2(K^n - K^{n-1})$$

Therefore

$$\frac{C_1}{C_2} = K^{n-1} - K^n$$

and

$$n = \log\left(\frac{C_1 K}{C_2(1-K)}\right) / \log K$$

To solve these equations we must solve n Markov transition matrices. A solution for larger problems, with 4 or more states, is preferably solved using dynamic programming or other optimization techniques.

9.0.2 Cost of Scheduled Overhauls and Inspections

Various models have been developed to determine the optimum overhaul and inspection interval for a system subject to failure. If we assume that the system is subject to preventative surveys at intervals of t and that the ith item is surveyed every $K_i t$ interval, then

$K_i t$ = time between preventative surveys of the i item inspected every $K_i t$ period

where

K_i = 1, 2,....n (where n = number of items)

and

C_{1i} = cost of preventative surveys

P_i = probability of preventative repair of item i performed given preventative survey of the item is performed

and

C_{2i} = cost of a preventative repair, than

$(\frac{P_i}{K_i t}) C_{2i}$ = expected annual cost of preventative repair attributable to i^{th} item

E_i = expected number of failues of i item between preventative surveys

C_{3i} = cost of unscheduled repairs

C_i = total annual M&R cost of component i is then

$$C_i = \frac{1}{K_i t} C_{1i} + \frac{P_i}{K_i t} C_{2i} + \frac{E_i}{K_i t} C_{3i}$$

Assuming that

C_o = cost of setting up periodic surveys

and

$(1/t)C_o$ = annual cost of periodic inspections

C_m = annual cost of repair person

M = number of repair persons

then

$$C_T = \text{total cost} = (1/t)C_o + C_m M + \sum_i C_i$$

The problem boils down to finding the pair (t,M) which minimizes the total cost of C. Assuming C_o and C_m are independent of K_i, M, and T and $C_i = f(C_{1i}, C_{2i}, C_{3i}, P_i, E_i)$, we may next define the various action strategies possible on the basis of the different states of particular components.

State of Component	Action Under One Given Strategy
Good	None
Deteriorated	Repair During Next Scheduled Inspection
Failed	Repair Immediately

These three states can next be related by their transition probabilities

		FINAL STATE		
		1	2	3
	1	p	p	p
Initial State	2	0	p	p
	3	1	0	0

If we define $P(k)$ as the probability state vector at time k, then $P(n)P = P(n+1)$, and assuming $P(0) = [1,0,0]$ for system in state one (1) at time zero, then

$$P(n) = P(0)P^n$$

If we let P_1 = probability of preventative repair after N time intervals where

$$N_i = \frac{K_i t}{\Delta T}$$

and $P_i = P_2(N_i)$ = probability in 2nd state at end of N_i^{th} time interval (component i), then we can write that

$$E_i = \sum_{j=1}^{N_i} P_3(j) = \text{expected number of failures of i in } N_1 \text{ time intervals}$$

Using this approach we can now obtain the total repair cost including the cost of downtime. Obviously many different states and strategies, which define sets of proposed actions for each of the states, could be developed and tested for minimum total repair cost.

Numerous models have been developed for the determination of minimum repair costs. The above model assumes linear cost functions. Other models allow the use of parabolic or higher order non-linear, but monotonically increasing cost functions. Mathematical programming techniques such as linear, non-linear or dynamic programming are usually applied for the determination of the optimum or minimum repair cost strategy. In some models repair and operating costs are combined and the strategy which will provide the lowest cost per unit time of effective operation is determined, subject to particular operating requirements. When the model becomes complex or the cost functions discontinuous, mathematical simulation is usually found to be the only effective method of approach.

A large number of models have been developed in recent years which are designed to permit determination of the optimum number of spare parts required or money spent for spare parts for a particular time period, assuming a certain reliability or availability of the system is desired. Repair strategy models sometimes incorporate spare part provisioning and purchasing. The problem of optimum spare part provisioning though may not only be a function of spare part purchase and holding costs, but may also be constrained by spare part weight and volume. This is particularly the case when transport or other mobile systems are considered whose primary function is their payload-carrying ability in terms of payload weight and/or volume.

9.0.3 Spare Part Inventory Provisioning

Most spare part inventory provisioning models attempt to minimize spare part shortages over time between overhauls, mission time, or other objective measure. Given

- C = total cost of spare part kits provided for mission
- S_i = number of spare parts of type i stocked in kit
- C_i = cost of one unit of spare of type i
- m = number of different parts which must be stocked
- T = time of mission
- K_i = essentiality factor for the part i in the ordered set of M different parts
- $f_i(x_i)$ = probability of a demand for x during mission time
- x_i = parts i demanded in time t

we can then proceed using the <u>Geisler and Kurr</u> model by defining as the objective function:

$$\text{MIN } E(S_i) = \sum_{i=1}^{m} K_i \int_{S_i}^{\infty} (x_i - S_i) f_i(x_i) dx_i, \quad S_i \geq 0$$

from which we determine the value of S_i for all parts i subject to the cost constraint

$$\sum_{i=1}^{m} S_i C_i \leq C$$

Using a Lagrange multiplier approach, we obtain the equation

$$\lambda = K_i / C_i \int_{S_i}^{\infty} f_i(x) dt$$

which must be satisfied for all i to obtain if $S_i = 0$.

$$\int_{S_i}^{\infty} f(x_i) dx = 1$$

In general we obtain

$$\lambda = K_i / C_i$$

and therefore if $K_i / C_i \leq \lambda$ no spare part is stocked for i, for minimum cost of spare part kit.

A similar model was proposed by Gouray.

$$\text{Expected Demand} = E(x_i) = \int_{o}^{\infty} x_i f_i(x_i) dx_i \quad i=1,2,\ldots m$$

$$\text{Expected Supply} = E(S_i) = \int_o^{S_i} x_i f_i(x_i) dx_i + S_i \int_{S_i}^{\infty} f_i(x_i) dx_i$$

and

$$\text{Expected Unfilled Demand} = E(x_i) = E(S_i)$$

To maximize the utility of the spare part kit we now maximize

$$\sum_{i=1}^{m} K_i E(S_i)$$

subject to the kit cost or volume constraints.

$$\text{Volume Constraint } V = \sum_{i=1}^{m} S_i V_i \text{ by Lagrange}$$

$$\text{Cost Constraint } C = \sum_{i=1}^{m} S_i C_i$$

K_i = essentiality factor

to obtain now the optimum values S_i we solve

$$1 = \frac{\lambda V_i}{K_i} \int_o^{S_i} f_i(x_i) dx_i, \text{ for all i, if the volume constraint applies, and}$$

$$1 = \frac{\lambda C_i}{K_i} \int_o^{S_i} f_i(x_i) dx_i, \text{ for all i, if the cost constraint applies.}$$

In the Black and Proschan Model the ratio of marginal increase in reliability of performance assurance over marginal increase in spare part cost is used. The addition of spares is adjusted to the increase in the marginal assurance of not running out of spares as follows:

$$r_i = \frac{f_i(x_i+1) - f_i(x_i)}{F_i(x_i+1) - C_i(x_i)} = \Delta P_1 / \Delta C_i = \frac{\text{Marginal Assurance}}{\text{Marginal Cost}}$$

For kit to be optimum this ratio must be equal for all parts i or

$$r_i = r_j = r_k = \text{constant}$$

$f_i(x_i)$ is the probability of demand for x_i in time T while

$C_i(x_i)$ is the cost of x_i units of parts i.

9.1 Use of Dynamic Programming in Systems Reliability

In systems reliability we are inveriently confronted with an optimization problem in which we require maximization of the reliability, availability, or maintainability of a system subject to certain constraints. We may similarly desire the attainment of a limiting value of reliability while minimizing the allocation of certain resources. The problem of optimizing system reliability then consists of maximizing a function such as

$$R(s_1, s_2 \ldots s_n) = h_1(s_1) \cdot h_2(s_2) \ldots h_n(s_n) = \text{Series System Reliability}$$

over the region of available resources s_i where $i = 1, 2, \ldots n$ and

$$\sum_{i=1}^{n} s_1 = s$$

with the further constraint that $s_i \geq 0$.

In the series system above, $h_i(s_i)$ is the reliability of ith component with an allocation constraint s_i implying that some allocation is to be made for each stage in the process. The total allocation s in this case may be fixed.

If, instead of considering the process as static, we imagine allocations s_i to be made for each system component one at a time, the process assumes a dynamic character; hence the name dynamic programming or dynamic allocation. We first assign an allocation s_n to the nth component, then s_{n-1} to the $(n-i)^{th}$ component, and so on. The maximum reliability of the system Max $R(s_i \ldots s_n)$ over the total region of allocations depends on S and n; and we consequently can write

$$f_n(s) = \underset{s_i}{\text{Max}} \; R(s_1 \ldots s_n) = \text{Optimal Achievable Reliability for a Total Allocation of Resources S}$$

where

$$s_i > 0, \quad \sum_{i=1}^{n} s_i = s.$$

It is clear that for s = 0, $f_n(0) = 0$. Similarly, for a one component system with an allocation s,

$$f_1(s) = h_1(s) \qquad s \geq 0$$

To obtain a recurrence equation connecting two stages of the allocation process $f_n(s)$ and $f_{n-1}(s)$ we let $0 \leq s_n \leq s$ be the allocation for the nth component. Then regardless of the value of s_n allocated for the nth component with a resulting return or component reliability $h_n(s_n)$, the remaining quantity of resources for the other (n-1) components will be $(s-s_n)$. By definition the reliability of the (n-1) component system with an allocation of $(s-s_n)$ is:

$$f_{n-1}(s-s_n)$$

and, therefore, the reliability of the n component system is:

$$h_n(s_n) \cdot f_{n-1}(s-s_n)$$

The optimal choice of s_n is obviously that one which maximizes the system reliability; and we, therefore, obtain the basic functional equation:

$$f_n(s) = \text{Max} [h_n(s_n) \cdot f_{n-1}(s-s_n)]$$
$$n = 1, 2, \ldots \qquad s \geq 0 \qquad \text{with } f_1(s) = h_1(s_1)$$

as determined before.

The above recurrence equation is a mathematical statement of the "Principle of Optimality" developed by R. Bellman (3) which states:

> "An optimal policy has the property that whatever the initial state and initial decisions are, the remaining decisions must constitute an optimal policy with regard to the state resulting from the first decision."

Under certain conditions, we may desire to maximize the mean time before failure of a system as a measure of the system reliability. As will be seen subsequently, this implies maximizing the sum of the inverse functions of the component allocations such as

$$T_f(s_1, s_2 \ldots s_n) = [q_1(s_1) + g_2(s_2) + \ldots + g_n(s_n)],$$

whereupon the recurrence equation then becomes:

$$f_n(s) = \underset{s_i}{\text{Max}}\, T_f(s_1 \ldots s_n) = \text{Max MTBF within allocation } s_i$$
$$= \text{Max}\, [g_n(s_n) + f_{n-1}(s-s_n)]$$

where again

$$s \geq 0 \quad \sum_{i=1}^{n} s_i = s$$

Let us assume that our system consists of n independent components arranged in series. System reliability is then:

$$R_s(t) = \prod_{i=1}^{n} R_i(t)$$

If the components have a constant failure rate,

$$R_s(t) = \prod_{i=1}^{n} R_i(t) = \prod_{i=1}^{n} e^{-\lambda_i t}$$

$$\lambda_i = f_i(t) = \text{Instant Failure Rate of Component } i$$

Similarly, if m components are placed in parallel to form a system, system reliability:

$$R_s(t) = 1 - \prod_{j=1}^{m} [1-R_j(t)] = 1 - \prod_{j=1}^{m} [1-e^{-\lambda_j t}]$$

If a complex system consists of n components in series and if redundancy of degree m_i is introduced at component i to increase system reliability, then

$$R_s(t) = \prod_{i=1}^{n} 1 - \prod_{j=1}^{m_i} [1-R_{ij}(t)]$$

Assuming the redundant components are identical, we can write that

$$R_s(t) = \left[\prod_{i=1}^{n} 1 - \prod_{j=1}^{m_i} (1-R_{ij}(t)) \right]$$

$$= \prod_{i=1}^{n} \left[1-(1-R_i(t))^{m_i}\right] = \prod_{i=1}^{n} \left[1-(1-e^{-\lambda_i t})^{m_i}\right]$$

(for constant failure rate components)

If the total constraint allocation for the system (weight, cost, etc.) is equal to s and

$$\sum_{i=1}^{n} s_i = s$$

where s_i is the allocation for the ith component which has a redundancy $1 \leq m_i$, then m_i is a function of s_i, and the maximum system reliability achievable within the constraint s can be written as:

$$f_n(s) = \text{Max } R_s(t) = \text{Max } h_1(s_1) \cdot f_{n-1}(s-s_1)$$

where

$h_n(s_n) = $ Reliability of nth component with Constraint Allocation s_n

and

$$f_n(s) = \text{Max } \{[1-(1-e^{-\lambda_n t})^{m_n(s_n)}] \cdot f_{n-1}(s-s_n)\}$$

To compute $f_n(s)$, we can tabulate the possible values of

$$1-[1-e^{-\lambda_n t}]^{m_n(s_n)} = h_n(s_n)$$

for all possible values of allocation of $0 < s_n < s$ for a single-component system consisting of component n with a redundancy m_n. The maximum value of $h_n(s_n)$ is equal to $f_1(s_n)$ for each choice of s_n (discrete steps). We next compute

$$f_2(s) = \text{Max } [h_n(s_n) \cdot f_1(s-s_n)]$$

which is the maximum of the product of the reliability of the nth component with an allocation s_n multiplied by the maximum reliability achievable by component (n-1) with an allocation

$(s-s_n)$ for all values of s_n. The process is continued until we arrive at

$$f_n(s) = \text{Max } [h_n(s_n) f_{n-1}(s-s_n)]$$

9.1.1 Complex System Reliability Analysis Under Constraints

Here we are concerned with the application of the functional equation of dynamic programming as a statement of the principle of optimality to the solution of complex system reliability problems under constraints. Generally, we consider catastrophic component failure with some attempt made to include operation under/out of tolerance conditions. Consequently, we assume that components or "events" have either failed or operate successfully; and only their probability of failure will affect the reliability of the system. In a later section the distribution of tolerances and their effect on the probability of failure distributions will be introduced, and first costs as well as operating costs will be obtained as a probability distribution for a given required value of system reliability. We will then also be able to test the sensitivity of component and system reliability to variations in tolerances.

The probability of x failures of a component in time t is usually described by a Poisson distribution when $R(t)$ is also equal to $1-p(x \geq 1, t)$ where

$$P(x, t) = \frac{m^x e^{-m}}{x!} = \text{probability of x failures in time t}$$

The relation of age dependent failure rate and reliability is expressed by:

$$f(t) = \frac{d \ln R(t)}{dt}$$

and if we assume the failure rate $f(t)$ to be constant during the lifetime of the component, which is justified if the lifetime only comprises the time of random failures only (excluding initial failure and wearout periods), then $f(t)=\lambda$, and $R(t) = e^{-\lambda t}$, and the probability of x failures in time t then becomes:

$$P(x, t) = \frac{(Ct)^x e^{-Ct}}{x!} = \frac{(\lambda t)^x e^{-\lambda t}}{x!}$$

Constant failure rate is an acceptable assumption for most electrical, electronic, and other control elements. For certain mechanical components, failure rate is more realistically represented by a linear function of time, and the Weibull distribution

$$f(t) = m \, c^1 t^{(m-1)}$$

is used when m=2 and

$$R(t) = e^{-c^1 t^2} = e^{-c^1 t^m}$$

In this section though we will use a constant failure rate for all components and events on the understanding that a time-dependent failure rate may be more applicable to some and should be introduced if required. The results and methods will be similar although the mathematical work would be more complicated.

9.1.2 Optimization of Multistage Decision Processes

The functional equation of dynamic programming is a mathematical statement of the principle of optimality as defined previously, which states "that an optimal policy has the property that whatever the initial state and decision, remaining decisions must constitute an optimal policy with regard to the state resulting from the first decision". In other words, in dynamic programming, we first consider the last stage of a process on the understanding that whatever the policy decisions of all but the last stage, the process will only result in an optimum if the last stage is working at its optimum with respect to the feeder of all but the last stage. We, therefore, consider all the possible leads to the last stage and determine its optimal operational parameters with respect to each possible lead. We next consider the last two stages of the process and the possible leads from all but the last two stages and again find the optimum operating parameters of the combination of these two stages with respect to the leads. There will always be a combined optimal policy for these two last stages for any given lead to the first of the two; the first stage of the pair will not necessarily operate at its optimal (one-stage) policy with regard to the lead, but the second stage must use optimal policy with respect to the lead it receives from the first under these conditions. This process is then continued until all stages of the process are included. The great advantage of the functional equation approach is that it not only reduces the computational work of finding an optimum among a large number of possible combinations of leads to stages in a process, but also allows us to investigate the sensitivity of the system to changes in lead and staging. Putting this into mathematical form for a discrete deterministic process, let us consider a one-stage process with an initial state p in which a decision q will result in an outcome $S(p, q)$. If the process has several stages, which may be in a time or space coordinate, and if p_1 is the initial state of the first stage and a decision q_1 is made resulting in a new state p_2, then we can write:

$$p_2 = S(p_1, q_1)$$

and a second decision q_2 will then result in a state

$$p_3 = S(p_2, q_2)$$

and in general

$$p_N = S(p_{N-1}, q_{N-1})$$

An N stage process can normally be described by a function of p_i and q_i in the form:

$$F(p_1, p_2, \ldots \ldots p_N, q_1, q_2, \ldots \ldots q_N)$$

which is then used to formulate an optimum decision policy. The function to be optimized is often expressed as the sum of the return or response of the individual stages.

$$F(p_i, q_i) = g(p_1, q_1) + g(p_2, q_2) + \ldots + g(p_N, q_N)$$

$$= \sum_{i=1}^{N} g(p_i, q_i)$$

for all values of q with $p_i = S(p_{i-1}, q_{i-1})$, $(i=2,\ldots N)$. In considering maximization of our system function by letting the initial state p_1 vary over all possible values and the number of stages N in the process assume values from 1 to N, we generate functions:

$$f_N(p_i) = \text{Max} \ [g(p_1, q_1) + \ldots g(p_N, q_N)]$$

Starting with

$$f_1(p_1) = \underset{q_1}{\text{Max}} \ g(p_1, q_1)$$

which is simple to obtain as a maximization only over q_1, we continue the process. The result of the first decision q_1 is to transform p_1 into $p_2 = S(p_1, q_1)$. Consequently, the contribution of state variable from the new last (N-1) stages will be:

$$f_{N-1}[S(p_1, q_1)]$$

and

$$f_N(p_1) = g(p_1, q_1) + f_{N-1}[S(p_1, q_1)]$$

and, if this decision is q_1 is chosen to maximize the process,

$$f_N(p_1) = \max_{q_1} [g(p_1, q_1) + f_{N-1}\{S(p_1, q_1)\}]$$

This recurrence equation can also be obtained by considering

$$f_N(p_1) = \max_{q_i}, \max_{q_2} \ldots \max [g(p_1, q_1) + g(p_2, q_2) + \ldots + g(p_N, q_N)]$$

$$= \max_{q_1} [g(p_1, q_1)] + \max_{q_2} [g(p_2, q_2)] + \max_{q_3} [g(p_3, q_3)] + \ldots + \max_{q_N} [g(p_N, q_N)]$$

from which

$$f_{N-1}(p_2) = \max_{q_2} [g(p_2, q_2)] + \ldots \max_{q_N} [g(p_N, q_N)]$$

and

$$f_N(p_1) = \max_{q_1} [g(p_1, q_1) + f_{N-1}(p_2)]$$

The function to be maximized is also often expressed as the product of the response of the stages when the recurrence equation assumes the form

$$f_N(p_1) = \max_{q_1} [g(p_1, q_1) \cdot f_{N-1}(p_2)]$$

9.1.3 Complex System with Component Stand-by

If m independent components are arranged in series to form a system, the reliability of the system is expressed by:

$$R(t)_{System} = \prod_{i=1}^{m} R(t)_i = \prod_{i=1}^{m} e^{-C_i t}$$

Similarly, if n identical components are placed in parallel to form a single-component system of redundancy degree n, the reliability of the system is equal to:

$$R(t)_{System} = [1-(1-R(t))^n] = [1-(1-e^{-Ct})^n],$$

if the system requires only the operation of a single component at a time. Considering a complex system with component stand-by, by combining the two preceeding equations, we obtain:

$$R(t)_{System} = \prod_{i=1}^{m} \{1-(1-R(t))^{n_i}\} = \prod_{i=1}^{m} \{1-(1-e^{-C_i t})^{n_i}\}$$

If the constraint of our system is defined as

$$Z = \sum_{i=1}^{m} Z_i$$

where Z_i is the allocation of our constraint parameter to component i which has a degree of redundancy n_i, then, using the recurrence equation of dynamic programming, optimum allocation policy, or allocation for maximum system reliability within the allocation constraint Z, given individual component reliability is fixed, can be expressed in the form:

$$f_m(Z) = \text{Max } \{[1-(1-e^{-C_i t})^{n_1(Z_1)}] \cdot f_{m-1}(Z-Z_1)]\}$$

where n_i is a function of the allocation for the ith stage Z_i.

9.1.4 Complex System with Switching

If this system is to be automatically controlled, then a failure detection and switching device must be incorporated in series with all redundant components (Figure 9.1).

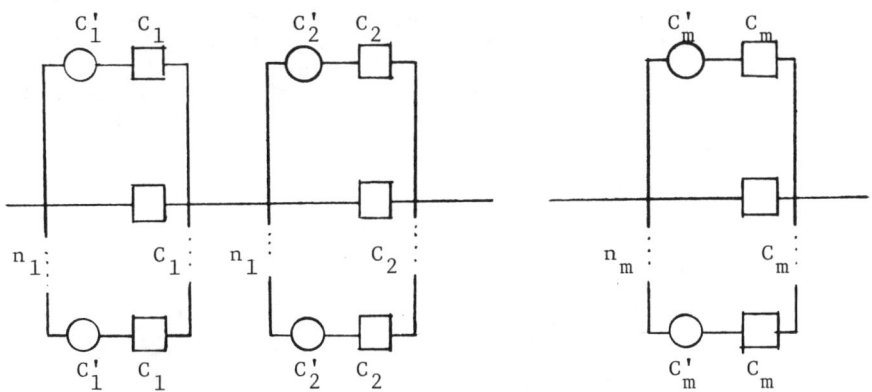

FIGURE 9.1: SERIES SYSTEM WITH REDUNDANT COMPONENTS

The reliability of these devices will affect the overall system reliability. If the probability that these devices will detect faults in the operating component and switch is equal to their reliability

$$R^1(t)_i = e^{-C^1_i t}$$

then

$$R(t)_{System} = \prod_{i=1}^{m} \{1-(1-e^{-C_i t})(1-e^{-C_i t} e^{-C'_i t})^{n_i - 1}\}$$

and the functional equation becomes:

$$f_m(Z) = \text{Max } \{[1-(1-e^{-C_i t})(1-e^{-(C_i + C_1)t})^{n_1}(Z_1)] \cdot f_{m-1}(Z-Z_1)\}$$

Next, we may generalize the problem and consider a redundant series system made up of non-identical subsystems, or systems where stand-by units are dissimilar from the main as well as other stand-by units. This method is also applicable for a simple analysis of repairable systems. By this, we imply that repair and consequent non-availability of stand-by units is allowed. Reduced reliability or probability of non-failure is then a representation of the availability of the redundant part. As availability can be expressed as a probability density function, which for many maintenance schedules assumes a Poisson distribution as well, it can be introduced as an additional series component in each branch and, in fact, be incorporated into the probability of non-failure of the switching device. Furthermore, if we allow redundancy of the non-identical stand-by as well as main components, the general system becomes as shown in Figure 9.2

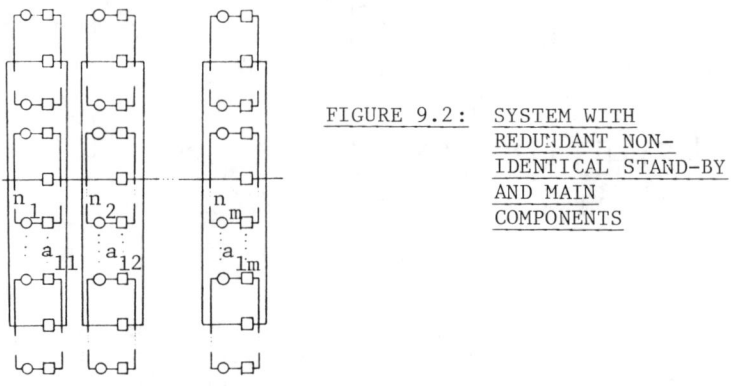

FIGURE 9.2: SYSTEM WITH REDUNDANT NON-IDENTICAL STAND-BY AND MAIN COMPONENTS

and has a reliability of:

$$R(t)_{System} = \prod_{i=1}^{m} \{1-(1-R_{1i}(t))^{a_{1i}} \prod_{k=2}^{n_i} (1-r^1_{ki}(t))$$

$$R_{ki}(t)^{a_{ki}}\} = \prod_{i=1}^{m} \{1-(1-e^{-C_{11}t})^{a_{11}}$$

$$\prod_{k=2}^{n_i} (1-e^{-(C^1_{ki}+C_{ki})t})^{a_{ki}}\}$$

and the recurrence equation then becomes:

$$f_m(Z) = Max \{[1-(1-e^{-C_{11}t})^{a_{11}} \prod_{k=2}^{n_i} (1-e^{-(C^1_{k1}+C_{k1})t})^{a_{k1}}] \cdot f_{m-1}(Z-Z_1)\}$$

In many systems we find it advantageous to sustain normal operations with multiple parallel components. If K_i is now the number of ith components required for normal operation, then the reliability of a complex system with identical redundant components is:

$$R(t)_{System} = \prod_{i=1}^{m} \{1-(1-e^{-C_i t})^{k_i}(1-e^{-(C^1_i+C_i)t})^{n_i-2k_i+1}\}$$

with results in the functional equation of the form:

$$f_m(Z) = \text{Max } \{[1-(1-e^{-C_1 t})^{k_1}(1-e^{-(C^1_1+C_1)t})^{n_1-2k_1+1}] \cdot f_{m-1}(Z-Z_1)\}$$

9.1.5 Reliability of Complex System with Component Maintenance

The problem of maintenance scheduling and spare-part inventory control has been the subject of many intense studies. In the following notes, the author only attempts a simple and rather confined analysis of the effect of rigid repair policies on the availability or reliability of a complex system using the dynamic programming approach. The assumption will be made that total repair resources consisting of facilities, parts, and manpower are strictly divided among the different component parts of the system, resulting in both a constant repair rate for each component as well as a maximum number of a particular component that can be repaired simultaneously. Use of resources once allocated will not be allowed for the repair of other components. It is admitted that this approach is nonrealistic, but if allocated resources were to be used for several different components concurrently, the intersection of the statistical distribution of failure events would have to be introduced. In our present study we consider the case of repairing components subject to random failure only in order to obtain maximum system reliability within the constraints of total available repair resources. The case of preventative maintenance, which can obviously be staggered to use resources optimally, is not considered. This is a problem in which game theory and linear programming will be found useful as well. If we again assume a constant failure rate C_i for component i and a constant repair rate $r_i(p_i)$, we can say that the number of failures of components i occurring in unit time after x_i of components i have failed is

$$(n_i - x_i) C_i = C_{x_i}$$

Similarly, r_{x_i}, the repair number of components i, will be equal to $x_i r_i$ up to the limit of simultaneously repairable components i which is denoted by $X_i(p_i)$. Let us again assume the K_i parallel components are required for normal operation.

Therefore, the largest possible number of concurrent failures is $(n_i - K_i)$. If we now consider the probability of having x_i failures in time $(t+\Delta t)$, then this is equal to the sum of the probability of having:

a) x_i failures in time t and no failures or repairs in the time between t and $(t+\Delta t)$,

b) $(x_i - 1)$ failures in time t and one failure in the time between t and $(t+\Delta t)$,

c) $(x_i + 1)$ failures in time t and one repair in the time between t and $(t+t)$,

or

$$\overline{R}_{x_i}(t+\Delta t) = \overline{R}_{x_i}(t)[1-(C_{x_i}\Delta t][1-r_{x_i}\Delta t] + \overline{R}_{x_i-1}(t)[C_{x_i-1}\Delta t]$$
$$+ \overline{R}_{x_i+1}(t)[r_{x_i+1}\Delta t]$$

This difference equation can be solved by iteration procedures or by converting it into a differential equation and, thereafter, transforming it using Laplace Transforms and solving the set of resulting linear equations by the use of the Gauss-Jordan Method. We then obtain values of $R_{x_i}(t)$ for all possible values of x_i in terms of r_{x_i}.

The reliability of the system of m series subsystems of components each with a redundancy i is then:

$$R_{system}(t) = \prod_{i=1}^{m}\left(\sum_{x_i=0}^{n_i-k_i} \overline{R}_{x_i}(t)\right) = \prod_{i=1}^{m}[1-\overline{R}_{n_i-k_i+1}(t)]$$

where

$$\sum_{x_i=0}^{n_i-k_i+1} \overline{R}_{x_i}(t) = 1$$

and

$$\overline{R}_{n_i-k_i+1}(t) = (1-R(t))^{n_i-k_i+1}$$

also
$$\bar{R}_o(0) = 1 \text{ and } \bar{R}_{x_i}(0) = 0$$
$$x_1 > 0$$

If we are interested in optimizing system reliability within the total repair resources available ($P = \Sigma p_i$) and, $\bar{R}_{x_i}(t, p_i)$ = Probability of x_i failures in time t is a function of p_i, we therefore can write:

$$f_m(P) = f_m R(t, P)_{System} = \text{Max} \{[\sum_{x_1=0}^{n_1-k_1} \bar{R}_{x_i}(t, p_1)] \cdot f_{m-1}(t, p-p_1)\}$$

Obviously, this solution is only valid if p_i is specialized for use on component i. If we fix the repair rate r_{x_i} and the total number of required operating components K_i, we again can obtain optimum system reliability for a total system allocation of Z where $n_i(Z_i)$ or the degree of redundancy or the quantity of the individual components are functions of the allocation for the state Z_i. Stated in functional equation form, we obtain:

$$f_m(Z) = \text{Max} \{[\sum_{x_1=0}^{n_1-R_1} \bar{R}_{x_i}(t_1, z_1)] \cdot f_{m-1}(Z-Z_1)\}$$

where
$$z = \sum_{i=1}^{m} Z_i$$

and n_i a function of Z_i.

9.1.6 Analysis of Component Failure

Although we only considered catastrophic component failure in the preceding work, it is seldom that a component will fail only as a result of a single failure, even t. In fact, most components will have a multitude of possible catastrophic failure events, each with its own failure rate. The component failure rate is, consequently, the sum of the failure rates of all the possible catastrophic events. If $F_{ij}(t)$ is the probability of failure of the jth event of the

i^{th} component in our system, then assuming constant failure rates of events:

$$F_{ij}(t) = (1-e^{-C_{ij}t}) = R_{ij}(t)$$

where C_{ij} = failure rate of event j of ith component, normally $0 \leq F_{ij}(t) \leq 0.1$, and $\overline{R}(t)_i = R_i(t)$ = the probability of failure of ith component,

$$F_i(t) = \sum_{j=1}^{S} (1-e^{-C_{ij}t}) = \overline{R}(t)_i$$

Breipohl of Sandia Corporation [Ref. 3] has found that component cost and probability of failure can be related by an expression of the form:

$$Z_i = \frac{A_i \exp[-B_i(1-e^{-C_i t})]}{(1-e^{-C_i t})}$$

where A_i and B_i are constants. Similarly, assuming that the probability of failure of the component is the sum of the probabilities of failure of the possible failure events,

$$Z_i = \sum_{j=1}^{S_i} \frac{A_{ij} \exp[-B_{ij}(1-e^{-C_{ij}t})]}{(1-e^{-C_{ij}t})} = \sum_{j=1}^{S_i} Z_{ij}$$

and the cost of the system of m series components, each with redundancy n_i,

$$Z_{System} = \sum_{i=1}^{m} n_i Z_i = \sum_{i=1}^{m} \sum_{j=1}^{S_i} n_i Z_{ij} =$$

$$\sum_{i=1}^{m} \sum_{j=1}^{S_i} [n_i \frac{A_{ij} \exp[-B_{ij}(1-e^{-C_{ij}t})]}{(1-e^{-C_{ij}t})}$$

As formed previously,

$$R_{System}(t) = \prod_{i=1}^{m} R_i(t) = \prod_{i=1}^{m} \{1 - [\sum_{j=1}^{s_i} (1-e^{-C_{ij}t})]^{n_i}\}$$

for series events constituting failure and

$$R_{System}(t) = \prod_{i=1}^{m} R_i(t) = \prod_{i=1}^{m} \{1 - [\prod_{j=1}^{s_i} (1-e^{-C_{ij}t})]^{n_i}\}$$

for parallel events constituting failure.

If we are now interested in finding the minimum cost for a system designed for a set of reliability, dynamic programming will give us an optimum allocation of component and event reliability for minimum system expenditure:

$$f_m(R(t)) = \text{Min}\{[n_i \sum_{j=1}^{s_1} \frac{A_{1j} \exp(-B_{1j}(1-e^{-C_{ij}t}))}{(1-e^{-C_{ij}t})} + f_{m-1}\frac{R(t)}{R(t)_1}\}$$

where $R(t)$ and $R_1(t)$ are obtained from the equations previously developed.

The method can be again generalized for the case of complex systems with nonidentical parallel subsystems and including switching devices. We then obtain Total Cost of System:

$$Z_{System} = \sum_{i=1}^{m} \sum_{j=1}^{s_i} [\frac{A_{ij} \exp[-B_{ij}(1-e^{-C_{ij}t})]}{(1-e^{-C_{ij}t})} + \sum_{i=1}^{m} \sum_{j=1}^{s_i^1} [(n_i-1)]$$

$$\frac{A_{ij}^1 \exp[-B_{ij}(1-C_{1j}^1 t)]}{(1-e^{-C_{ij}t})}$$

where s_i^1 = the number of failure events in switching device i and system reliability.

$$R_{System}(t) = \prod_{i=1}^{m} \{1 - [\sum_{j=1}^{s_i} (1-e^{-C_{1ij}t})] \prod_{k=2}^{n_i} \sum_{j=1}^{s_i^1} (1-e^{-C_{Kij}^1 t})\}$$

$$+ \sum_{j=1}^{s_i} (1-e^{-C_{Kij}t})]\}$$

when component failure events are in series and

$$R_{System}(t) = \prod_{i=1}^{m} \{1-\prod_{j=1}^{s_i}(1-e^{-C_{1ij}t})] \prod_{k=2}^{n_i} \prod_{j=1}^{s_i^1}$$

$$(1-e^{-C_{kij}^1 t}) + \prod_{j=1}^{s_i}(1-e^{-C_{kij}t})]\}$$

when component failure events are parallel. Similarly, again the recurrence equation allocating component and event reliability for minimum cost,

$$f_m(R(t)) = \text{Min } [\{[n_i \sum_{j=1}^{s_{11}} \frac{A_{ij}\exp[-B_{1j}(1-e^{-C_{11j}t})]}{(1-e^{-C_{11j}t})}$$

$$+ [(n_1-1) \frac{A^1_{1j}\exp[-B^1_{1j}(1-e^{-C^1_{11j}t})]}{(1-e^{-C^1_{11j}t})}$$

$$f_{m-1}[R(t)/R_1(t)]\}$$

9.1.7 Conclusions

An approach to the solution of complex system reliability problems under constraints has been presented using the functional equation of dynamic programming for the optimal set of multivariable analysis. The equations so obtained are in a form easily programmed for digital computer application. A next step, directed towards a more realistic model should introduce distribution laws for the failure events and consider out-of-tolerance failure distributions. Design parameters in the form of tolerances must be analyzed as functions of cost and probability of failure for particular components. Finding the intersections or overlap of the failure density distribution of failure events and component failure should also permit a more realistic approach to maintenance scheduling.

9.2 The Use of the Lagrange Multiplier Method

Another form of attack for complex system reliability problems is by the use of the well-known Lagrange Multiplier Method. We again consider a series system made up of n-components in series, each of which may have a redundancy m_i identical components (where i=1, 2,...m)

$$R_{System}(t) = \prod_{i=1}^{n} \{1-(1-e^{-\lambda_i t})^{m_i}\}$$

where

λ_i = failure rate of ith component

Let us now again have an allocation s for the system out of which s_i (i=1,2,...n) is the allocation for component i with a redundancy m_i. The constraint equation is then

$$\psi = \sum_{i=1}^{m} s_i - s = 0$$

which is then applied in the well-known set of Lagrange Multiplier Equations:

$$\frac{\partial R(t)_s}{\partial s_1} - \Lambda \frac{\partial \psi}{\partial s_1} = 0$$

$$\vdots \qquad \vdots \qquad (R_s(t) = \text{System Reliability})$$

$$\frac{\partial R_s(t)}{\partial s_n} - \Lambda \frac{\partial \psi}{\partial s_n} = 0$$

where Λ is the Lagrange Multiplier).

As the components are independent, $R_1(t)$, the component reliability with redundancy m_1 will be a function of s_1 only.

Thus, the equation set becomes:

$$\frac{-\prod_{i=1}^{n} \{1-(1-e^{-\lambda_i t})^{m_i}\} \, m_1(1-e^{-\lambda_1 t})^{m_1-1}}{\{1-(1-e^{-\lambda_1 t})^{m_1}\}} \frac{\partial R_1(t)}{\partial s_1} - \Lambda = 0$$

$$\frac{\prod_{i=1}^{n}\{1-(1-e^{-\lambda_i t})^{m_i}\} \, m_n(1-e^{-\lambda_n t})^{m_n-1}}{\{1-(1-e^{-\lambda_n t})^{m_n}\}} \frac{\partial R_n(t)}{\partial s_n} - \Lambda = 0$$

From this set we obtain the following relationship,

$$\frac{\partial R(t)_n}{\{1-(1-e^{-\lambda_1 t})^{m_1}\} \, \partial s_1} = \ldots = \frac{\partial R_n(t)}{\{1-(1-e^{-\lambda_n t})^{m_n}\}}$$

but as

$$R_i(t) = 1 - \prod_{k_i=1}^{m_i}(1-e^{-\lambda_k t}) = 1-(1-e^{-\lambda_i t})^{m_i}$$

$$\frac{\partial R_i(t)}{\partial s_i} = \frac{\partial R_i(t)}{\partial \lambda_i} \cdot \frac{d\lambda_i}{ds_i} = m_i t e^{-\lambda_i t}(1-e^{-\lambda_i t})^{m_i-1}\frac{\partial \lambda_i}{\partial s_i}$$

the previous relationship becomes:

$$\frac{m_1 e^{-\lambda_1 t}(1-e^{-\lambda_1 t})^{m_1-1}}{\{1-(1-e^{-\lambda_i t})^{m_1}\}} \cdot \frac{\partial \lambda_1}{\partial s_1} = \ldots =$$

$$\frac{m_n e^{-\lambda_n t}(1-e^{-\lambda_n t})^{m_1-1}}{\{1-(1-e^{-\lambda_n t})^{m_n}\}} \cdot \frac{\partial \lambda_n}{\partial s_n}$$

As for most components $\lambda_i t$ is small, then
$$e^{-\lambda_i t} \simeq 1 \text{ and } (1-e^{-\lambda_i t}) \simeq \lambda_i t$$

The equation set can then be simplified using these approximations to

$$\frac{m_1^2(\lambda_1 t)^{2(m_1-1)}}{\{1-(\lambda_1 t)^{m_1}\}} \frac{\partial \lambda_1}{\partial s_1} = \ldots = \frac{m_n^2(\lambda_n t)^{2(m_n-1)}}{\{1-(\lambda_n t)^{m_n}\}} \frac{\partial \lambda_n}{\partial s_n}$$

The above identities show that maximum reliability is attained when allotment is so distributed that the ratio of variation of failure rate with respect to component allotment

$$\frac{1-(\lambda_i t)^{m_i}}{m_i^2 (\lambda_i t)^{2(m_i-1)}} = f(m_i, \lambda_i) \text{ or is equal to a function}$$

of λ_i and m_i and is identical for all subsystems.

If failure rate of all components is equal, then maximum reliability is attained if allotment is distributed equally among subsystems; otherwise the ratios

$$\frac{\partial \lambda_i}{\partial s_i} / f(m_i, \lambda_i)$$

have to be plotted as a set of curves to obtain the optimum point which satisfies the constraint equation

$$\sum_{i=1}^{n} s_i - s = 0$$

It will be found that the method of Lagrange Multipliers becomes rather involved, even for relatively simple systems. Apart from this, the functional equation approach of dynamic programming is straightforward and a simple computer application for most complex problems. The method of Lagrange Multipliers has been presented as a tool that can be effectively used in conjunction with the functional equation of dynamic programming for problems involving more than one set of constraints.

9.2.1 Systems Involving Two Types of Constraints

If two types of resources s and U are the constraining factors on our system, where

$$\sum_{i=1}^{n} s_i \leq S \quad \text{and} \quad \sum_{i=1}^{n} u_i \leq U \quad s_i, u_i \geq 0$$

and if the reliability of the ith component with an allocation s_i and u_i is $h_i(s_i, u_i)$ then the total system reliability is

$$R_n(s_1 \ldots s_n, u_1 \ldots u_n) = \prod_{i=1}^{n} h_i(s_i, u_i)$$

The recurrence equation of dynamic programming then becomes:

$$f_n(s, U) = \text{Max } R_n(s_1 \ldots s_n; u_1 \ldots u_n) = \text{Max}[\prod_{i=1}^{n} h_i(s_i, u_i)]$$

$$\text{Max }\{S, U\} = \text{Max }\{h_n(s_n, u_n) \cdot f_{n-1}(s-s_n, U-u_n)\}$$

$$0 \leq s_n \leq s \qquad 0 \leq u_n \leq U$$

or if

$$T_f(s_1 \ldots s_n, u_1 \ldots u_n) = \sum_{i=1}^{n} g_i(s_i, u_i)$$

$$f_n(s, U) = [\text{Max } g_n(s_n) + f_{n-1}(s-s_n)]$$

Using the Lagrange Multiplier Method as presented by Bellman we would attempt to carry out maximization over s_i independently of the maximization over u_i; or, in other words, we would consider the problem:

$$T_f(s_1 \ldots s_n, u_1 \ldots u_n) = g_1(s_1, u_1) + \ldots + g_n(s_n, u_n)$$

$$- \lambda[u_1 + u_2 + \ldots + u_n]$$

Subject to

$$\sum_{i=1}^{n} s_i = s \qquad s \geq 0 \text{ and } u_i \geq 0$$

Then we can write

$$K_i(s_i, \lambda) = K_i(s_i) = \text{Max }\{g_i(s_i, u_i) - \lambda u_i\} \quad u_i \geq 0$$

For the above expression to be meaningful,

$$\frac{g_i(s_i, u_i)}{u_i}$$

must tend to zero as $u_i \to \infty$, and it only remains now to maximize

$$\sum_{i=1}^{n} K_i(s_i) = \Sigma \, \text{Max} \, \{g_i(s_i, u_i) - \lambda u_i\} \qquad u_i \geq 0$$

It is obvious that this type of problem can easily be solved by means of the functional equation of dynamic programming as expounded by R. Bellman [3] when

$$f_n(s) = \text{Max} \, [\text{Max} \, \{g_n(s_n, u_n) - u_n\} + f_{n-1}(s-s_n)]$$

$$0 \leq s_n \leq s \qquad u_i \geq 0$$

We calculate system reliability then as a function of s_n and thereafter maximize over U_n.

Returning to the problem of optimizing complex system reliability subject to two constraints, let us assume

$$s \geq \sum_{i=1}^{n} s_i \qquad s_i \geq 0$$

$$m_i = 1, 2, \ldots$$

$$U \geq \sum_{i=1}^{n} u_i \qquad u_i \geq 0$$

where s_i and u_i could represent cost and weight of a particular component in the series system with redundancy m_i. As the cost and weight of each component is normally fixed, s_i and u_i will define the degree of redundancy of each component required to achieve optimal system reliability within the cost and weight constraints. Therefore, m_i is a function of s_i and u_i, or $m_i = m_i(s_i, u_i)$.

In order to avoid dealing with sequences of functions of two variables, we will introduce the Lagrange Multiplier.

If the redundant components are identical again, as in our previous example, maximizing the reliability of a system of n components in series each with redundancy m_i implies maximizing

$$R_s(t) = \prod_{i=1}^{n} \{1-(1-e^{-\lambda_i t})^{m_i(s_i, u_i)}\}$$

Let us consider a new problem of maximizing

$$\prod_{i=1}^{n} \{1-(1-e^{-\lambda_i t})^{m_i(s_i)}\} \, e^{-\lambda \sum_{i=1}^{n} u_i}$$

over all the $m_i(s_i)$ with $m_i \geq 1$. Then

$$f_n(s) = \text{Max} \, [\{1-(1-e^{-\lambda_n t})^{m_n(s_n)}\} \, e^{-\lambda u_n} \cdot f_{n-1}(s-s_n)]$$

$$1 \leq m_n(s_n) \leq m_n(s)$$

where

$$f_1(s) = \text{Max} \, [\{1-(1-e^{-\lambda_1 t})^{m_1(s_1)}\} \, e^{-\lambda u_1}]$$

$$1 \leq m_n(s_n \leq m_n(s)$$

The problem consists of evaluating the Lagrange Multiplier λ, so that

$$\sum_{i=1}^{n} u_i = U$$

and then finding the optimum achievable system's reliability by multiplying the value $f_n(s)$ in which the weight $u_i \leq U$ comes closest to $f_n(S,U)$ found by:

$$\exp \lambda \sum_{i}^{u} u_i$$

with the value of λ used. Therefore

$$\text{Max } R_s(t) = f_n(s) \, e^{\lambda \sum_{i=1}^{u} u_i}$$

9.3 Optimum Maintenance Policies by Dynamic Programming

The functional equation can also be used for determining preventive maintenance policies. Let us assume a system has an instant failure rate that is a linear function of time and that it is operated in discrete time intervals of length t_0, defining the length of mission. If the system is required between operating periods, the cost associated with the repair is C_r. If the system fails during an operating interval, the cost is C_f for the repair, and the system cannot be operated again until the next interval.

Define

C_r = cost of scheduled maintenance

C_f = cost of 'in service' maintenance

P_m = probability of failure in interval (m+1), given it has not failed in intervals since last maintenance

$f_n(m)$ = least expected cost of making n more missions, given m missions have been made.

There are two policies that can be followed - either repair the system at m, or do not repair. The following functional equations result:

$$f_n(m) = \text{Min} \begin{cases} C_r + P_o[C_f + f_{n-1}(0)] + (1-P_o)f_{n-1}(1), & \text{Repair} \\ P_m[C_f + f_{n-1}(0)] + (1-P_m)f_{n-1}(1), & \text{Nonrepair} \end{cases}$$

It is logical to assume that $C_f > C_r$.

To show mathematically the effect of the failure repair cost C_f, and the scheduled maintenance cost, C_r, the following problem is postulated:

A system has a failure distribution curve that can be described by the Weibull distribution with the following parameters:

$$p(t) = \frac{\beta t^{\beta-1}}{\alpha} = \frac{2t}{57 \times 10^4}$$

$$f(t) = \frac{\beta t^{\beta-1}}{\alpha} e^{-\frac{t\beta}{a}}$$

$$R(t) = e^{-t\beta/a} = e^{-t^2/57 \times 10^4}$$

$$M(t) = N(t) = \text{interval time} = 100 \text{ hours}$$

β = shape parameter = 2

α = scale parameter = 57×10^4 hours

$\alpha^{1/\beta}$ = characteristic lifetime = η = 755 hours.

The question to be answered is how does the maintenance policy vary as a function of m, n, C_f, and r^2. The results of this investigation for different ratios of C_f/C_r are given in Figure 9.3 and allow an estimate of optimum maintenance scheduling.

The method of determining preventative maintenance scheduling by dynamic programming is more suitable to a marine propulsion plant than a method employing continuous time derivatives. The reason for this is that with operating schedules to maintain, it is probable that the optimum preventative maintenance times found will not coincide with the ship's schedule. This method also provides a continuous picture of what policy to take regarding preventative maintenance over all combinations of time intervals (missions) completed and time intervals still to be completed. From Figure 9.1 in which a system with a Weibull failure density distribution has been investigated, the expected cost of maintenance can be minimized for any operational situation. The graph is only applicable for m, n 0. If m=0, the system has just been repaired; and it does no good to repair it again. If n=0, there are no more trips to make; therefore, the expected cost is zero.

Although the graphs are self-explanatory, one interesting observation can be made, which is not intuitively obvious, nor would it normally be expected if the investigation had not been made. For example, with $C_f/C_r=5$, it is seen that if 8 trips have been made and 2 more are planned (m=8, n=2), the policy should be to repair the system. However, with the same part history and if 3 more trips are required (n=3), the system would not be repaired. The explanation of this is as follows:

1. At large m the probability of failure in n intervals is relatively high; and for small n the repair is the best policy since, in effect, it starts the system off at m=0.

2. As n increases, this effect of starting at m=0 becomes less important; and the cost of preventive maintenance is the dominant factor. The reason is that at

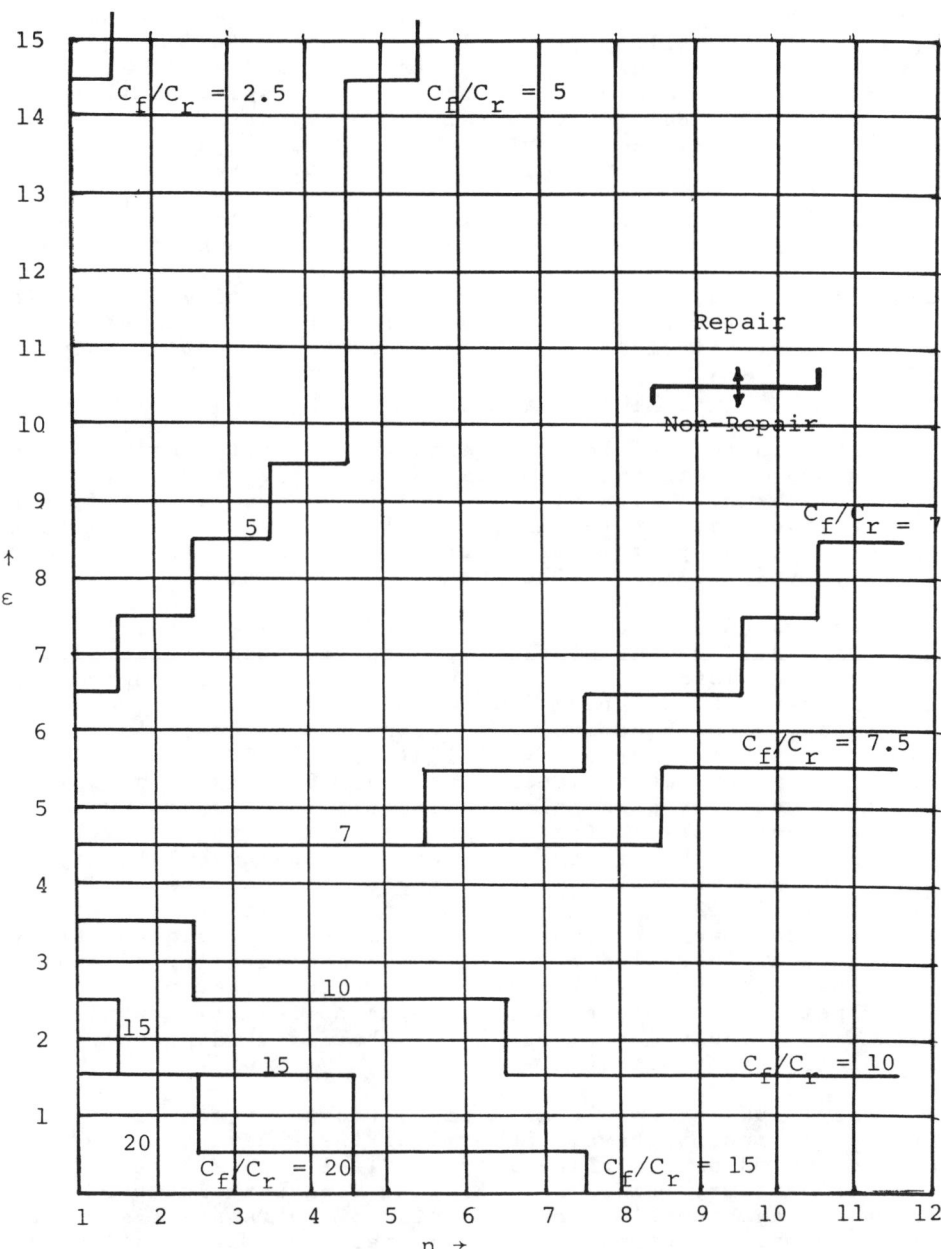

FIGURE 9.3: PREVENTATIVE MAINTENANCE POLICY

large n, even if maintenance is performed, there is still a significant probability of failure in the following (n-1) intervals which causes a greater expected cost than if no repair is made; and a high probability of system failure is accepted in the next n intervals. Policy determination is, therefore, solely dependent upon failure distribution, duration of mission, and ratio C_f / C_r. This fact greatly

lessens the number of calculations that have to be made in a system analysis. As the values of C_f/C_r are bound to change

with time, the use of a single graph representing this ratio allows cross-plotting for optimization of system maintenance where repair, material, and failure cost vary continuously.

9.4 Spare Part Provisioning Models

Provisioning of spare parts has great influence in maintenance decisions, operational effectiveness, and operating costs. Maintenance decisions are affected, for example, by repair or replacement alternatives, which in turn depend on spare part availability and cost versus repair facility availability, effectiveness, and cost.

Various objectives can be applied to spare provisioning, such as system availability, operational reliability, operating efficiency, spare part or equipment operating costs, and weight or space taken up by spares when carried. In reality, the objectives used for the determination of a spare part purchasing policy will always include several, if not all, of the above considerations. Some, such as weight or space, would be more relevant factors in a spare purchase and carriage policy for weight or volume-limited systems. Other factors such as the impact of the availability of spares on reliability and operating efficiency are harder to determine. It is usual to simply set a minimum acceptable level of equipment reliability, availability, and operating efficiency as affected by the ready availability of spares when needed. As a result, the objectives used in the determination of a spares provisioning policy will usually be based on costs of purchase, holding, disposal, and unavailability of spares.

Many models for spare part provisioning have been proposed in recent years. Several of these have been refined and are now used by airlines, truck fleet operators, and, to a lesser extent, railways. Similar work was also performed for spares and supplies provisioning of offshore platforms. More recently methods were adopted for application to shipboard and shore-based spare inventory purchase planning.

An attractive approach can be adopted from the basic work by Black and Proschan in which we maximize the probability of not running out of spares when needed subject to constraints such as maximum spare inventory expenditure, weight or space of spares carried, and other appropriate constraints. In other words the model maximizes spare availability for each level of constraints set and will allocate expenditure and other constraints according to spare essentiality and probability of need for the particular spare.

The following are defined:

A_j = number of units of the jth item in use

N_j = number of spares provided for the jth item

C_j = cost per unit of the jth item

λ_j = average failure rate of the jth item application or use

t = expected number of working hours between spare inventory replacement

μ_j = expected number of failures of the jth item during expected number of working hours between spare replacement where $\mu_j = A_j \lambda_j t$

W_j = total number of failures of jth item during period t

Assuming the failure distribution is exponential and failures among individual items is independent:

$$P_j(N_j) = \text{Probability } (W_j \leq N_j) = \sum_{x=0}^{N_j} e^{-\mu_j} \frac{\mu_j^x}{x!}$$

Considering all the different z items

$$\sum_{j=1}^{z} N_j = n$$

$$P(N) = \prod_{j=1}^{z} P_j(N_j)$$

and the total cost of this spares provision plan is then

$$C(N) = \sum_{j=1}^{z} N_j C_j = C_{Total} = C_T$$

If $C(N) = C_T$ is fixed then the objective is to maximize $P(N)$.

Considering a diesel engine as an example, for the purposes of simplicity, we assume that there are only five critical paths whose unavailability would cause the engine to be either inoperable or reduce its performance, as shown in Table 9.4-1.

TABLE 9.4-1: DIESEL ENGINE PART PERFORMANCE

j	Part	Failure Rate Per Hour of Operation λ_j	t	Number In Use	Cost per Item $	μ_j = Expected Number of Failures in t
1	Injectors	0.00008	4000	6	200	1.92
2	Fuel Pumps	0.00006	4000	6	400	1.44
3	Piston Ring Sets	0.00012	4000	6	60	2.88
4	Pistons	0.00002	4000	6	800	0.24
5	Liners	0.00001	4000	6	1600	0.06

To get a starting point let us choose a value for $N_1 = \mu_1 + 2\sqrt{\mu_1}$ = expected number of failures of part j=1 plus 2 standard deviations $N_1 = 1.92 + 2\sqrt{1.92} = 4.69$, say almost 5.

Next defining:

$$b = \frac{1}{C_1} e^{-\mu_1} \frac{\mu_1^{N_1+1}}{(N_1+1)!} = \frac{1}{200} e^{-1.92} \frac{1.92^6}{6!}$$

To obtain the value of N_2 we compute

$$e^{-\mu_2} \frac{\mu_2^{N_2+1}}{(N_2+1)!} \geq b \cdot C_2$$

from which we obtain N_2. We similarly obtain N_3, N_4, and N_5 from

$$e^{-\mu_j} \frac{\mu_j^{N_j+1}}{(N_j+1)!} \geq b \, C_i \text{ where } j=3, 4, \text{ or } 5 \text{ respectively}$$

This now gives us the vector $N=(N_1, N_2, N_3, N_4, N_5)$ with a corresponding

$$P(N) = \prod_{j=1}^{5} \sum_{x=0}^{N_j} e^{-\mu_j} \frac{\mu_j^x}{x_j}$$

and

$$\text{Total Costs of } C(N) = \sum_{j=1}^{5} C_j N_j = C_T$$

This result in terms of the number of spares of each type represents the best buy of spares for the expenditure of C_T. To obtain another optimal value of $P(N)$ and C_T we simply add one unit to N_1 or make $N_1=6$ and again compute

$$b^1 = \frac{1}{C_1} e^{-\mu} \left(\frac{\mu_1^{N_1+1}}{(N_1+1)!} \right) = \frac{1}{200} e^{-1.92} \frac{1.92^7}{7!}$$

and then determine the new values of N_j for $j=2,\ldots 5$ using the expression

$$b'C_j \leq e^{-\mu_j} \left(\frac{\mu_j^{N_j+1}}{(N_j+1)!} \right)$$

This process is then continued until $P(N)$ reaches a value of 0.995 or other acceptable level of probability of spare part availability when needed. The results are readily plotted as $P(N)$ against $C(N) = C_T$. The graph increases stepwise, but may be approximated by an exponential curve.

A computer program can readily be developed to perform these computations for normal ship-sized problems so that 'essential' machinery or other spares could be provisioned on a more rational basis. The algorithm described provides stepwise solutions in terms of alternative spare part kits supplied. To obtain intermediate solutions, various and different incrementing modes can be derived and tested. An effective method computes for each item the value

$$b_j = \frac{\mu_j^{N_j+1} e^{-\mu_j}}{(N_j+1)! \, C_j}$$

computed with $N_j = \mu_j + 2\sqrt{\mu_j}$ as a starting point.

The largest among the values of b_j from among the items, is then used to determine the value of the numbers of spares for all other items. Various other incrementing methods were tried.

Another development is the inclusion of an essentiality factor, E_j, which indicates the effect of the availability of a spare when needed on ship or equipment operation. E_j varies from 0 to 1 and assumes a value of 1.0 if it is essential for operations, or if the lack of availability of a spare for item j when demanded would cause ship or equipment failure or unavailability.

Essentiality can be included by converting the expected failure rate into expected fatal failure rate

$$\mu_j' = E_j \lambda_j N_j t_j$$

where t is now the time between the replacement of spares type j.

The most efficient incrementing method is one in which the marginal improvement of expected failure risk is maximized. The largest value of

$$b_j' = \frac{\mu_j'^{N_j+1} e^{-\mu_j}}{(N_j+1)!}$$

is used to determine the value N_j which is incremented to (N_j+1) to obtain the next optimal point.

Various extensions of the basic model can be introduced such as the cost of shortage. Instead of, or in addition to, the optimum P(N) for a given total spare part cost the model can be used to compute the minimum total sum of spare part inventory, spare part holding and shortage costs for a given acceptable value of P(N).

Next the model can be expanded to handle different replacement periods t_j for different types (and origins) of spares.

The incrementing method used in the modified Block-Proschan model does not provide all the optimum points, but additional points can readily be computed for intermediate values of C(N) or total spare part inventory costs. [Ref. 2] The model recognizes that unavailability of an inexpensive part can cause a system's performance failure just as well as an expensive part. It will therefore provide a best mix of inexpensive and expensive parts to assure optimal total system reliability and availability in terms of spare part availability by procuring more cheaper parts when warranted to assure a higher level of P(N) for a given value of C(N).

By calculating the 'spare part' reliability of each equipment and system, total 'spare part' reliability can be determined with due regard to systems whose failure, resulting from the unavailability of essential spares would cause only reduced performance. Similarly, equipment or systems which have stand-by would be modeled accordingly.

By determining the relation between P(N) and C(N) for different values of C(N) for each essential piece of equipment or system, we can establish the optimum (minimum cost) mix of ship spares for a desired level of P(N) or probability of not running out of an essential spare part during the period between spare part resupply.

To implement this model, we would usually proceed as shown in Table 9.4-2.

It is often desirable to include an insurance item or provide at least one spare part for each essential component with a failure rate above a minimum value (or a probability of failure during the interspare replacement period of say 0.999). This can readily be introduced into the initial spare part set.

The total 'spare part' reliability can be established by computing the ratio

$$\frac{P_k(N)}{C_k(N)}$$

for each essential equipment k for the different values of $P_k(N)$

$$P_{System}(N) = \prod_{k=1}^{s} P_k(N)$$

assuming all equipment is independent. Equipment dependence can readily be introduced using statistical theory. The corresponding total system spare set costs is

TABLE 9.4-2: PROCEDURE FOR SPARE PART PROVISIONING

1. Select a starting value for the set of number of spares, say $N_j = \mu_j$ to the nearest integer.

2. If we assume a Poisson type failure distribution then the probability of no shortage of part j for period t, given N_j spares are supplied is:

$$P_j(N_j) = \sum_{x=0}^{N_j} \frac{e^{-\mu_j} \mu_j^x}{x!}$$

3. Next we compute for each part

$$P_j(N_j+1) = \sum_{x=0}^{N_j+1} \frac{e^{-\mu_j} \mu_j^x}{x!}$$

   For example if $\mu_1 = 0.9$ and $N_1 = 1$, then $P_1(N_1=1)=0.772$ and $P_1(N_1+1=2) = 0.937$. The change in probability of shortage of this essential part is there $P_1(N_1+1) - P_1(N_1) = 0.937 - 0.772 - 0.165$.

4. Dividing this change in probability of shortage by the cost of the spare (including inventory holding costs, say $C_1 = 300$), we obtain

$$b_1 = \frac{P_1(N_1+1) - P(N_1)}{C_1} = \frac{0.165}{300} = 0.00055$$

   We compute the value of b_j for each part j for the initial spare cost.

5. Next we compute the value of the resulting system spare part reliability

$$P(N) = \prod_{j=1}^{m} \sum_{x=0}^{N_j} \frac{e^{-\mu_j} \mu_j^x}{x!} \quad \text{and the total cost} \quad C(N) = \sum_{j=1}^{m} C_j N_j$$

6. Now we select the component with the largest value of

$$b_j = \frac{P_j(N_j+1) - P_j(N_j)}{C_j}$$

   and add one spare to its stock, and we compute all values of b_j, $P(N)$, and $C(N)$.

7. We continue this process until we achieve the required value of $P(N)$ or reach the budgeted value of $C(N)$.

$$C(N)_{system} = \sum_{k=1}^{s} C(N)_k.$$

To compute the optimum allocation of spare part investment to the various equipment, we again start with an initial allocation of $C_k(N)$ to each of the k equipment and compute the ratios

$$r_k = \frac{P_k(N)}{C_k(N)}$$

for each of the $C(N)_k$. The equipment with the largest r_k will then be incremented by adding an amount of funding to $C_k(N)$ when we recompute $P_{System}(N)$ and all values of the ratios r_k. This procedure is continued until we reach the desired level of $P_{System}(N)$ or exhaust the budgeted amount of $C_k(N)$.

When a part can be repaired or replaced when failed (parts which fail by wear or lack of adjustment), then the model can be designed to consider the effect of the alternative of part repair versus part replacement. Furthermore, the model can include consideration of system performance (efficiency, capacity, consumption, speed, etc.) in addition to spare part kit reliability. This is important when a part failure and lack of space may result in degraded system performance and not system failure. The cost of degraded performance in terms of added fuel cost, lost revenue, etc. is then included.

9.5 System Performance Evaluation

The performance of a system depends on a number of factors in addition and complementary to its attributes regarding effectiveness of operations. Although efficiency, weight, volume, cost, environmental impact, etc. are usually considered the characteristic factors of a system, its performance will be greatly affected by design for reliability and maintainability, operating strategy, maintenance policy, and spare part inventory and procurement policy.

The above considerations must be considered part of the systems development process if the system is to perform to expectation. As a result, we must know more about a system than expected load, required efficiency, etc. In fact, the most common reasons for the lack of performance of a system are usually that:

1. objectives for its operation were not well-defined;
2. operating, maintenance, and spare part policies or strategies were not developed; and,
3. the environment in which the system was to perform was not effectively evaluated and considered in the design and operating policy development for the system.

Objectives for a system's performance often contain a combination of physical, economic, and social factors. A system may be required to perform most efficiently at the least cost in an environmentally acceptable manner. Unfortunately, many times objectives are stated in such imprecise terms that they are less than useful for effective systems design and operation.

An effective approach to system performance evaluation and systems performance policy determination is provided by a Markov approach. Given the transition probabilities of a given system are known, we can usually assume the dynamics of the system to be represented by a Markov chain in the form of a transition probability or stochastic matrix $\underline{P}$. This transition matrix relates the possible states by their respective steady state transition probabilities P_{ij}.

$$\underline{P} = \begin{array}{c} \\ 0 \\ 1 \\ 2 \end{array} \begin{array}{c} \phantom{P_{00}}0\phantom{P_{00}} \phantom{P_{00}}1\phantom{P_{00}} \phantom{P_{00}}2\phantom{P_{00}} \\ \left[\begin{array}{ccc} P_{00} & P_{01} & P_{02} \\ P_{10} & P_{11} & P_{12} \\ P_{20} & P_{21} & P_{22} \end{array} \right] \end{array}$$

Such a stochastic matrix expresses the operating and/or maintenance policy of the system whose perfect state is 0 with states 1 and 2 increasingly degraded states. For example if all $P_{ij}=0$ for $j \leq 1$, in other words if transition probabilities under the diagonal are zero for such a system, then the system can be said to be nonmaintained. If state 0 is the perfect as-new state, state 1 a degraded operating state, and state 2 the failed state, then P_{22} must be 1 if the system once failed cannot return to a higher state, and state 2 is then said to be absorbing. If, on the other hand, the policy is to renew or repair the system whenever it fails, then we would have to enter a one in the location of P_{20}. We could define a number of possible decisions for such a 3-state system.

Decision	Course of Action
1	Do Nothing
2	Repair
3	Replace

These three decisions can be said to be our policy or decision variables which can be used in a decision vector. For example, D=[1,1,3] simply implies that the given system is in state 1 or 2, do nothing, but if system is in state 3, replace it. These decisions can be associated with all the possible states of a system. For example, we could say that whenever the system is failed, replace it with a probability of one and whenever it is degraded, repair it with a probability of one. If we assume that in either case the system is as good as new then given the initial stochastic matrix of the nonmaintained system was:

$$\underline{P} = \begin{array}{c} \\ 0 \\ 1 \\ 2 \end{array} \begin{array}{c} 0 \quad\quad 1 \quad\quad 2 \end{array} \\ \left[\begin{array}{ccc} 1/2 & 1/4 & 1/4 \\ 0 & 1/2 & 1/2 \\ 0 & 0 & 1 \end{array} \right]$$

the new stochastic matrix is:

$$\underline{P} = \left[\begin{array}{ccc} 1/2 & 1/4 & 1/4 \\ 1 & 0 & 0 \\ 1 & 0 & 0 \end{array} \right]$$

If we next consider a larger, say four-state system, where state zero defines a state 'as good as new', state one worn in and operating, state two operable but defective, and state three failed, and assume the stochastic matrix

$$\underline{P} = \begin{array}{c} \\ 0 \\ 1 \\ 2 \\ 4 \end{array} \begin{array}{cccc} 0 & 1 & 2 & 3 \end{array} \\ \left[\begin{array}{cccc} 0 & 1/2 & 1/4 & 1/4 \\ 0 & 1/4 & 1/2 & 1/4 \\ 0 & 0 & 1/2 & 1/2 \\ 0 & 0 & 0 & 1 \end{array} \right]$$

using the decisions we may next assume that we can either replace or repair the system when failed. Given that replacement brings the system back to state 0 and repair only brings it back to state 1, we could then define a number of operating policies, represented by a policy vector D_K such as

$D_K = [1,1,2,3]$ which implies that the policy is to do nothing if the system is in state 0, or 1, repair it if it is in state 2, and renew it if it is in state 3.

The resulting stochastic matrix is

$$\underline{P} = \begin{array}{c} \\ 0 \\ 1 \\ 2 \\ 3 \end{array} \begin{array}{c} \begin{array}{cccc} 0 & 1 & 2 & 3 \end{array} \\ \left[\begin{array}{cccc} 0 & 1/2 & 1/4 & 1/4 \\ 0 & 1/4 & 1/2 & 1/4 \\ 0 & 1 & 0 & 0 \\ 1 & 0 & 0 & 0 \end{array} \right] \end{array}$$

Similarly, other policies can be developed which again are each defined by their respective policy vector.

Each maintenance policy therefore defines a stochastic matrix which in turn can be solved for the steady state by transforming it into a set of simultaneous steady state equations. For example, the four-state policy above can be written (leaving out the argument $P_i(\infty) = \underline{P}_i$) as:

$P_0 = P_3$

$P_1 = 1/2 P_0 + 1/4 P_1 + P_2$

$P_2 = 1/4 P_0 + 1/2 P_1$

$P_3 = 1/4 P_0 + 1/4 P_1$

In addition $P_0 + P_1 + P_2 + P_3 = 1$.

The simultaneous solution is:

$P_0 = 4/27$

$P_1 = 12/27$

$P_2 = 7/27$

$P_3 = 4/27$

These then are the average ratios of time spent by the system in the four states given the above maintenance policy is implemented. It is important to note that these are steady state results.

If we next assume that the expected cost resulting from being in a state i is C_i per unit time, then the long run expected average cost per unit time of the system is given by:

$$E(C) = C_a = \sum_{i=0}^{3} P_i C_i$$

For instance, if in the above example, the expected average cost permit time is

State	Expected Cost/Unit Time
0	0
1	$100
2	$1000
3	$4000

then the expected average cost per unit time of the above system using the stipulated policy will be

$$E(C) = (4 \times 0 + 12 \times 100 + 7 \times 1000 + 4 \times 4000)/27 = 896.3/\text{unit time}$$

To make our analysis more versatile, we may next want to define a whole array of average costs/unit time for different states and related decisions, as costs may not only be associated with the state the system is in during any period of time, but also with the transition, if any, from that state to another state.

Let C_{ik} be the steady state cost of transition from state i given decision K, we could now derive a state-decision cost matrix C_{ik}

State	Decision 1	Decision 2	Decision 3
0	C_{01}	C_{02}	C_{03}
1	C_{11}	C_{12}	C_{13}
2	C_{21}	C_{22}	C_{23}
3	C_{31}	C_{32}	C_{33}

These state-decision costs can be time costs, replacement, or repair costs. They can similarly include the disbenefits or costs of defective performance.

Considering a policy K, $D_K = [d_{iK}, d_{2K} \ldots d_{nK}]$ which defines the decisions d_{iK} taken under this policy given the system is in states 1,2....n, we can define a decision matrix $[D_K]$ which relates the various decisions under policy K to the possible system states.

$$[D_K] = \text{State} \begin{array}{c|cccc} & 1 & 2 & 3 & 4 \\ \hline 0 & d_{01} & d_{02} & d_{03} & d_{om} \\ 1 & d_{11} & d_{12} & d_{13} & d_{1m} \\ \cdot & & & & \\ \cdot & & & & \\ \cdot & & & & \\ n & d_{n1} & d_{n2} & d_{n3} & d_{nm} \end{array}$$

If in each state only one decision applies, then each row has only a single one with the rest of the element equal to zero. If on the other hand, there is a positive probability that one or more decisions would be taken given the system is in a certain state, then we deal with a conditional decision probability distribution where not:

$d_{i\ell}$ = probability that decision is taken given the system is in state i

$0 \leq d_{i\ell} \leq 1$ and $\sum_{\ell=1}^{m} d_{i\ell} = 1$

If P_i is the steady state probability $P_i(\infty)$, that the system is in state i then we can define $F_{i\ell}$ as the probability that the system is in state i and the decision ℓ is made or

$$F_{i\ell} = \underline{P}_i d_{i\ell}$$

also obviously

$$\underline{P}_i = \sum_{\ell=1}^{m} F_{i\ell}$$

and

$$d_{i\ell} = \frac{F_{i\ell}}{\underline{P}_i} = \frac{F_{i\ell}}{\sum_{\ell=1}^{m} F_{i\ell}}$$

If $C_{i\ell}$ is the cost of making decision ℓ in state i, then the long run average cost of the system can be obtained by the cost equation

$$E(C) = \sum_{i=0}^{u} \sum_{\ell=1}^{m} P_i C_{i\ell} d_{i\ell} = \sum_{i=0}^{u} \sum_{\ell=1}^{m} C_{i\ell} F_{i\ell}$$

The problem is to determine the values $F_{i\ell}$ that will minimize this expectancy long run system's cost. A linear program can be defined as:

$$\text{Minimize } \sum_{i=0}^{n} \sum_{\ell=1}^{m} C_{i\ell} F_{i\ell}$$

Subject to:

$$\sum_{i=0}^{n} \sum_{\ell=1}^{m} F_{i\ell} = 1$$

$$\sum_{q=1}^{m} F_{i\ell} - \sum_{i=0}^{n} \sum_{\ell=1}^{m} F_{i\ell} P_{ij}(K) = 0$$

because

$$P_i = \sum_{j=0}^{n} P_j P_{ij} \text{ and } P_i = \sum_{\ell=1}^{m} F_{i\ell}, \quad P_j = \sum_{\ell=1}^{m} F_{j\ell}$$

and $F_{i\ell} \geq 0$

Having found the optimum values of $F_{i\ell}$ we can readily determine the values of $d_{i\ell}$ and therefore the decision matrix which minimizes the cost of running the system.

9.6 Exercises

1. A two component repairable on-line parallel system has a constant age dependent component failure rate of $\lambda = 8 \times 10$ per hour and repair rate of $\mu = 2$ per hour. Both components are identical and there is only one repairman available. What is the percentage utilization of the repairman and the percentage time the system is operable? Assume only one of the two components is required for the operation of the system. If you could obtain two repairmen each with a repair rate of $\mu = 1$ per hour for the price of the one more proficient repairman, what would be the choice if you wanted to maximize the percentage uptime of the system?

2. Assume that you operate a one bay car repair shop. The probability of arrival of a customer is 0.4. If the

customer finds that a car is already in your repair ship there is a probability of 0.5 that he will leave and go elsewhere. If there is another customer already waiting, in addition to the one being served, the probability of a newly arriving customer leaving to go elsewhere is 0.8. If there is no one being served, an arriving customer places his repair with a probability of 1.0. The repair facility performs only one kind of car repair which usually takes 1 hour and costs $20.00. You charge $40.00 for a repair of a car. You could also increase your service rate to 2 cars/hour at a cost of $30.00 per car repair. Assuming the facility is open 24 hours per day, what is your optimum policy?

3. A system has 10 identical independently operating parts. The probability of any of these parts failing during a mission of the system is 2×10^{-2}. The cost of purchase and holding of a part for the duration of the mission is $10.00, while the cost of a breakdown of a part without a spare being available is $1000.00. How many spaces should you carry on the mission so as to minimize the cost of the mission with respect to that part's performance?

4. A truck makes regular cross-country trips. The probability of failure of the truck during such a trip is a function of the number of trips it has made since its last overhaul. If $\underline{P}_i$ is the probability of the truck not failing during a trip, given it was overhauled i trips before and the cost of overhaul is $1,000, while the cost of breakdown during a trip is $10,000, determine the optimum time between overhauls, given the probability of non-failure during the next trip is:

$\underline{P}_0 = 1.00$

$\underline{P}_1 = 0.95$

$\underline{P}_2 = 0.85$

$\underline{P}_3 = 0.65$

$\underline{P}_4 = 0.40$

$\underline{P}_5 = 0.10$

5. In a four-component series system, the first component has a reliability of

$$R_1(t) = e^{-\lambda_1 t}$$

while subsequent components have reliabilities of

$$R_i(t) = e^{-\lambda_i t} \qquad (i = 2, 3, \text{ or } 4)$$

Assume that the failure rate of each component is a function of the cost C_i of the component as shown in the following table.

C_i	$\lambda_1 \times 10^{-4}$	$\lambda_2 \times 10^{-4}$	$\lambda_3 \times 10^{-4}$	$\lambda_4 \times 10^{-4}$
1	1.0	2.0	0.5	1.0
2	1.6	3.0	1.0	2.0
3	1.9	3.5	1.3	2.5
4	2.1	3.8	1.5	2.8
5	2.2	4.0	1.6	3.0

Assuming that the total money available $\Sigma C_i = C_1 + C_2 + C_3 + C_4$ is 10, how much should you spend for each component if you desire to maximize the reliability of the series system under the given total cost constraint?

6. A system has four different parts that are subject to failure. Given their failure rates/hour, the number of each part in use, and the cost per item are:

Part	$\lambda_i \times 10^{-4}$	Number of Use	Cost per Item $
1	0.1	4	1.0
2	0.2	2	5.0
3	0.4	1	20.0
4	0.1	2	2.0

the required time of operation of the system is 1000 hours. Find the best mix of spare parts if you are to maximize system availability (minimize probability of unavailability of spare when needed) given the total amount of money available is $40.00.

9.7 References

1. Hillier, F.S. and Lieberman, G.J., "Operations Research",

Holden-Day, Inc., San Francisco, CA, 1974.

2. Howard, R.A. "Dynamic Programming and Markov Processes", MIT Press, Cambridge, MA, 1960.

3. White, D.J., "Dynamic Programming", Hoden-Day, San Francisco, 1969.

4. Barlow, R.E. and Proschan, F. "Mathematical Theory of Reliability", Wiley, New York, 1965.

5. Sandler, G.H. "Sysstems Reliability Engineering", Prentice-Hall, Englewood Cliffs, New Jersey, 1963.

6. Smith, C.O. "Introduction to Reliability in Design", McGraw-Hill, Nw York, 1976.

7. Howard, A.R. "Dynamic Probabilistic Systems", Vol. I and II, Wiley, New York, 1971.

8. Morse, P.M. "Queues, Inventories, and Maintenance", McGraw-Hill, New York, 1969.

9. Blanchard, B.S. 3rd and Lawery, E.E. "Maintainability", McGraw-Hill, New York, 1969.

10. Grouchko, D., editor, "Operations Research and Reliability", Gordon and Breach, New York, 1969.

11. Kapur, K.C. and Lamberson, L.R. "Reliability in Engineering Design", Wiley, New York, 1977.

12. Jorgenson, D.W., McCall, J.J., and Radner, R. "Optimal Replacement Policy", Rand McNally, Chicago, 1967.

10.0 EFFECTS OF COMPONENT INTERACTION

Ernst G. Frankel

In previous chapters the overall reliability of hypothetical complex systems has been derived on the assumption of complete independence of component failure rates. In situations where the system comprises mechanical, thermal, hydraulic, chemical, etc. components, it is found that although these methods are mathematically correct, they do not yield the actual reliability observed in practice. Intuitively, it might appear that this poor correlation is because the model is not a good functional representation of the real system. Although this may, on occasion, be a possible reason, further analysis may show that the model is correct; but that the assumption of independence of components was unjustified. It can readily be shown that in systems where components are reduced by wear, chemical reaction, environmental attack, etc., or where component subsystems share a component medium, interaction will invariantly exist. This often results in a change of component failure distribution and may also greatly affect optimum maintenance and replacement scheduling. Because of this, the failure distribution observed in the single component life test is not applicable and must be modified.

The problem of finding what type of function this modification factor is will in actual practice be very difficult. However, if reliability investigations are to be accurate, component failure rate interaction must be considered. It can be shown to have a pronounced effect on system availability and repair policies. It is consequently imperative that laboratory life tests of components suspected of interaction and data assimilation of real system operation be analyzed so that interaction functions can be determined. For the purpose of illustration, two possible situations where interaction may play a substantial role are given:

a. The failure distribution of a diesel cylinder liner will be affected by the wear of the piston rings, the piston, the connecting rod bearings, the fuel injectors, and the quality of the fuel. As faulty combustion may be an interfacing factor, fuel pump wear and cam shaft distortion may also have to be included.

b. The failure distribution of a reduction gear will be dependent upon the wear of the shaft bearings, the coupling and the effectiveness of lubrication.

In the above examples, faulty operating conditions, such as improper cooling water or oil supply, have not been mentioned. As these are normally under the control of the human operator, they could be included as a human fault factor.

Since no data is available on interaction, only hypothetical interaction functions are postulated and examined. In a later part, means of arriving at interaction factors and methods of investigating their effects based on actual operating data will be presented.

10.1 Effect of Interaction of Component Reliability

The time dependent reliabilility of a component is described by:

$$R(t) = \exp\left[-\int_0^t f(\tau)d\tau\right] \tag{10.1}$$

which is a function of a single variable, $f(t)$ = the instant or age dependent failure rate. An equation illustrating the main factors that influence component failure rate is characterized by:

$$f(t) = h(\alpha, \beta, \phi, t)$$

α = derating factor – accounts for the fact that the component can operate at different outputs

β = environmental stress factor – a function of the operational environment the component experiences; i.e., temperature, vibration, acceleration stress levels, etc.

ϕ = interaction factor – accounts for the effect of interaction among the component failure distributions

t = operating time

It is normally possible to find the instant failure rate $f(t)$ as a function of derating factor, stress factor, and time frame experimental life testing of the component. An example of such a function can be shown in a diagram for a component that follows the exponential failure distribution (Figure 10.1).

Because reliability engineering is relatively new, little experimental data is available to find such functional relationships; and what is available only applies to

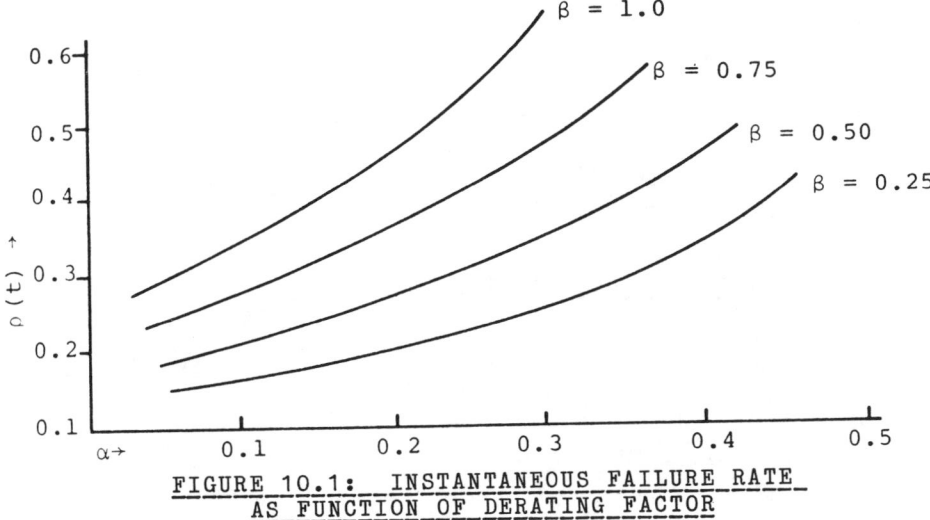

FIGURE 10.1: INSTANTANEOUS FAILURE RATE AS FUNCTION OF DERATING FACTOR

electronic components resulting in exponential failure distributions. If we assume that similar types of relationships can be found for other types of components and failure distributions, the problem of how to handle the interaction factor still presents itself. As a start, only conjectural relationships will be assumed for the purpose of investigation.

Define $\lambda(t)$ = component instant failure rate excluding the interaction factor ϕ. Therefore,

$$f(t) = h(\lambda(t), \phi)$$

Although most non-electrical components will not obey a Poisson failure distribution and consequently have a time dependent $f(t)$, we will consider $f(t) = \lambda$ = constant for ease of computation. The effect of the interaction factor which is determined in this exponential case will apply equally to other failure distributions.

Case 1. ϕ = Additive & Constant

$$f(t) = \lambda + \sum_i c_i$$

$$R(t) = \exp - [(\lambda + \sum_i c_i)t]$$

$$T_f = 1/(\lambda + \sum_i c_i) = MTBF$$

Case 2. ϕ = Additive & Time Dependent

$$f(t) = \lambda + \sum_i c_i \cdot t$$

$$R(t) = \exp[-\lambda t] \exp[-1/2 \sum_i c_i t^2]$$

$$T_f = \left\{ \exp\left[\frac{\lambda^2}{2 \sum_i c_i}\right] \right\} \left\{ \sqrt{\frac{\pi}{2\sum_i c_i}} \left(1 - \mathrm{erf}\left[\frac{\lambda}{\sqrt{2\sum_i c_i}}\right]\right) \right\} = \mathrm{MTBF}$$

Case 3. ϕ = Multiplicative & Time Dependent

$$f(t) = (ct)\lambda$$

$$R(t) = \exp[-c/2 \cdot \lambda t^2]$$

$$t_f = \frac{1.254}{(c\lambda)^{1/2}} = \mathrm{MTBF}$$

Case 4. ϕ = Additive & Function of a Constant Plus a Time Dependent Term

$$f(t) = \lambda + c_1 + c_2 t$$

$$R(t) = \exp[-(\lambda+c_1)t] \exp[-1/2 c_2 t^2]$$

$$T(f) = \left\{ \exp\left[\frac{(\lambda+c_1)}{2c_2}\right] \right\} \left\{ \sqrt{\frac{\pi}{2c_2}} \left(1 - \mathrm{erf}\left[\frac{\lambda+c_1}{2c_2}\right]\right) \right\} = \mathrm{MTBF}$$

It can be seen that these values for $R(t)$ and T_f can differ greatly from the values obtained for the standard exponential failure distribution; i.e.,

$$R(t) = \exp[-\lambda t] \qquad T_f = 1/\lambda = \mathrm{MTBF}$$

10.2 Analysis of "Wear" Rates

As pointed out before, "wear" will in many cases be the major factor in component failure rate interaction; and, therefore, a valid approach for mechanical parts subject to "wear" is to analyze wear interaction. In other words, we require a method of computing the effects of the wear of one component on the wear of all other components in the system. Using the diesel as an illustration again, the wear of the liner, piston, rings, and bearings may all interact or, in other words, affect the respective wear rates.

If $x_i(t)$ $i=1\ldots n$ is the wear of component i in the n component system, we may use regression analysis to find a best fit in a least square or quadratic sense for component wear rate. Let us assume that we obtain data for

$$s_i(t) \quad i=1\ldots n$$

for the piston-liner assembly of a diesel engine at intervals of time and that we ignore component wear interactions. We then try to fit a polynomial of the form:

$$\frac{d x_i}{dt} = a_1 x_i + a_2 x_i^2 + \ldots, \quad x_i(0) = c_i$$

The determination of the constants that yield the best quadratic fit

$$\int_0^T (X_i - x_i)^2 dt$$

over the interval [0, t] (with $X_i(t)$ - measured wear) is then carried out in standard fashion. We can use a linear growth model:

$$\frac{dx_i(t)}{dt} = a_1 x_i(t) \qquad x_i(0) = c_i \quad (i=1 \ldots n)$$

and

$$x_i(t) = c_i e^{a_1 t}$$

or a self-interacting model

$$\frac{dx_i(t)}{dt} = a_2 x_i^2(t)$$

when

$$x_i(t) = \frac{c_i}{1 - a_2 c_i t}$$

or

$$\frac{dx_i(t)}{dt} = a_1 x_1(t) + \text{higher terms}$$

$$x_i(0) = c_i \qquad (i=1\ldots n)$$

Normally, we would start with the linear model and continue to second and higher order polynomials if the fit is unsatisfactory. Assuming that even a fairly high order polynomial does not yield a reasonable fit, interaction of component wear is introduced:

$$\frac{dx_1}{dt} = g(x_1, x_2, \ldots x_n) \qquad x_1(0) = c_1$$

$$\frac{dx_2}{dt} = h(x_1, x_2, \ldots x_n) \qquad x_2(0) = c_2$$

$$\frac{dx_n}{dt} = k(x_1, x_2, \ldots x_n) \qquad x_n(0) = c_n$$

($x_i = x_i(t)$ and $X_i(t) = X_i$ for notational simplicity.)

A criterion of fit of the firm

$$\int_0^T \sum_i [(X_i - x_i)]^2 dt$$

can then be assumed. Obviously, other criteria of success, such as

$$\operatorname*{Max}_i \sum_i (X_1 - x_i)^2$$
$$0 \leq t \leq T$$

can be used in minimizing the error.

Assuming we are concerned with the simple case of two-component interaction, we may start with the linear regression model:

$$\frac{dx_1}{dt} = a_{11}x_1 + a_{12}x_2 \qquad x_1(0) = c_1$$

$$\frac{dx_2}{dt} = a_{21}x_1 + a_{22}x_2 \qquad x_2(0) = c_2$$

If the error is too big, a second-order model is introduced:

$$\frac{dx_1}{dt} = a_{11}x_1 + a_{12}x_2 + a_{13}x_1^2 + a_{14}x_1x_2 + a_{15}x_2^2, x_1(0) = c_1$$

$$\frac{dx_2}{dt} = a_{21}x_1 + a_{22}x_2 + a_{23}x_1^2 + a_{24}x_1x_2 + a_{25}x_2^2, x_2(0) = c_2$$

A similar method is used for interaction among a larger number of components.

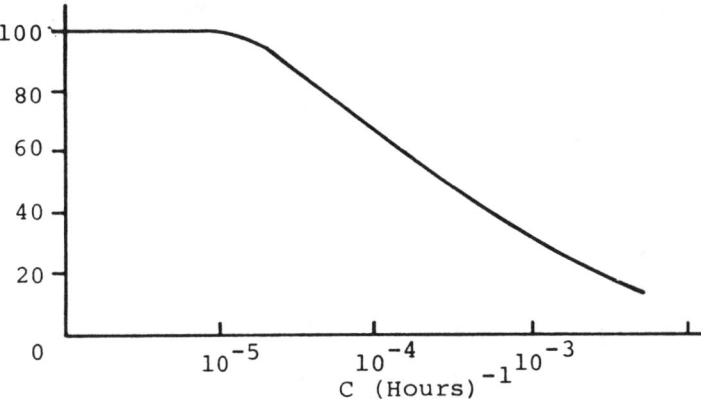

FIGURE 10.2: EFFECT OF INTERACTION ON MTBF $(t) = 0.01 + Ct$

Should the wear curve be unexpectedly obstinate, other steps can be taken to reduce the error to acceptable limits. Under certain conditions, interaction of wear rates may occur with a time lag, whereupon the above expressions take the form:

$$\frac{dx_1(t)}{dt} = g(x_1(t-\tau_1), x_2(t-\tau_2), \ldots, x_n(t-\tau_n)), \quad x_1(0) = c_1$$
$$\vdots$$
$$\frac{dx_n(t)}{dt} = h(x_1(t-\tau_1), \ldots, x_n(t)), \quad x_n(0) = c_n$$

Using a two-component interacting system as an example again, the linear interacting system:

$$\frac{dx_1(t)}{dt} = a_{11}x_1(t) + a_{12}x_2(t-\tau_2)$$

$$\frac{dx_2(t)}{dt} = a_{21}x_1(t-\tau_1) + a_{22}x_2(t)$$

The general approach is similar to that used in introducing hereditary effects in the form of convolution integrals in computing population growth.

A simple example of this method is found in considering the wear rates of diesel engine liners and rings comprising a two-component system. It is practice to renew sets of rings at intervals of about 6,000 working hours while liners are normally renewed every 20-30,000 hours. Yet, as shown in Figure 10.3, the wear rate of rings varies greatly with the wear of the liners; and the rate of wear of liners is dependent on the number and spacing of ring changes.

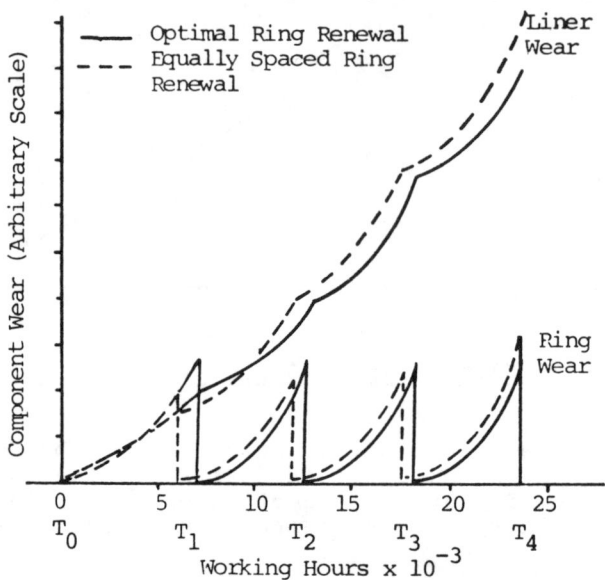

FIGURE 10.3: EFFECTS OF WEAR INTERACTION OF DIESEL CYLINDERS, LINER, AND PISTON RINGS

A large amount of operational data has defined the limited of piston liner and ring wear before the probability of failure becomes large. Consequently, the limiting wear X_w establishes the wear-out time for each component and

$$X_{w_i} = \int_0^{T_{w_i}} \frac{dx_1(t)\,dt}{dt} = \int_0^{T_{w_i}} h_i(x_1 \ldots x_n)\,dt = \text{Limiting wear-out}$$

Value of ith component is used to compute T_i, the wear-out time of the ith component.

Wear-out failure density distribution is normally assumed to be Gaussian and wear failure rate density:

$$f_{w_i}(t) = \rho_{w_i}(t) e^{-\int_0^{T_{w_i}} \rho_{w_i}(t)dt} = \frac{1}{\sigma_t \sqrt{2\pi}} \left[\exp - \frac{(T_{w_i} - M)^2}{2\sigma_t^2} \right]$$

where

$$\text{mean} = M = \frac{\sum^N T_{w_i}}{N} \quad \text{and} \quad \sigma_t = \sqrt{\frac{\Sigma(T_{w_i} - M)^2}{(N-1)}}$$

N = number of samples

$\rho_{w_i}(t)$ = instantaneous wear failure rate

$$= \frac{\exp-(T_{w_i}-M)^2/2\sigma_t^2}{\int_t^\infty [\exp(-(\tau-M)^2/2\sigma_t^2)]d\tau}$$

If X_{w_i} is the value of x_i at which wear-out failure occurs and if the mean is

$$\overline{X}_{w_i} = \frac{\sum_{i=1}^N X_{w_i}}{N}$$

then

$$\sigma_x = \frac{\sum_{i=1}^N (X_{w_i} - \overline{X}_{w_i})^2}{(N-1)}$$

and

$$f_{w_i}(x_i) = \frac{1}{\sigma_x \sqrt{2\pi}} \exp - \left[\frac{(X_i - \overline{X}_{w_i})^2}{2\sigma_x^2} \right]$$

or

$$\rho_{w_i}(x_i) = \frac{\left[B \exp - \frac{(X_i - \overline{X}_{w_i})^2}{2\sigma_x^2} \right]}{\left[\int_t^\infty \exp - \frac{(X_i - \overline{X}_{w_i})^2}{2\sigma_x^2} d\tau \right]}$$

or the wear failure rate density is a function of wear. It is normally easier to use $f_{w_i}(x_i)$, the wear failure rate as a function of wear, particularly as the mean wear to failure $\overline{X}_{w_i}$ of a component is normally given.

Wear failure rate distribution can often be shown to be more closely represented by a Weibull distribution when

$$\rho_{w_i}(t) = r\, G x_i^{r-1}$$

and

$$f_{w_i}(t) = r\, G x_i^{r-1} \exp[-(G x_i^r)]$$

$$G > 0 \qquad r > 2$$

where

$$x_i = \int_0^t h_i(x_1 \ldots x_n)\, dt$$

The importance of using a wear failure rate distribution based on component wear $x_i(t)$ and not implicitly on time stems from the fact that interfacing components are not renewed simultaneously throughout the working life of the system. In fact, as illustrated by the simple diesel example, it is usual practice to renew components at widely varying intervals. The renewal of any component in the system may affect the wear rate and consequent wear failure rate of any other component.

Considering Figure 10.3, it will be noticed that renewing rings at equally spaced intervals will not be the optimum replacement procedure. The cost of the liner is manifold larger than the cost of rings. A decreasing interval should be used for ring renewal. This may decrease the wear rate of the liner as well and result in a reduction of both ring and liner replacements over time. It is usual for a new set of rings to be fitted whenever liners are renewed. Using this premise, an optimum strategy for ring and liner replacement can be developed which attains maximum reliability and availability of the system at minimum cost.

If in Figure 10.3, the ring renewal times are defined by T_j, j=0, 1,...,M while the liner is only renewed at T_M, then the wear rate of the rings will be:

$$\frac{dx_1}{dt} = g(x_1(t-T_j), x_2(t-T_M))$$

and the liner wear rate

$$\frac{dx_2}{dt} = h(x_1(t-T_j), x_2(t-T_M))$$

$$0 \leq t - T_j \leq (T_{j+1} - T_j)$$

$$0 \leq t - T_M \leq 2T_M - T_M = T_M$$

10.3 Component Reliability

The instantaneous failure rate of a component in an interacting system

$$\rho_T(t) = k(\alpha, \beta, \phi, t) = \rho_c(t) + \rho_w(t) + \rho_s(t) + \rho_d(t)$$

is equal to the sum of the instantaneous failure rates.

$\rho_c(t)$ = the rate of chance failure occurrence, independent wear, can be considered essentially constant with time. Chance failures can be shown to result in a Poisson failure distribution; and, therefore, $\rho_c(t) = c_1$.

$\rho_s(t)$ = the rate of failure occurrence as a result of environmental stress is often taken as constant. In mechanical devices it is more closely approximated by a linear function of time and an operational stress factor, s. Therefore, $\rho_s(t) = st$.

$\rho_d(t)$ = the rate of failure occurrence due to overloading (derating factor) is normally assumed to be a constant, the value of which is defined by the expected severity of operation.

$\rho_w(t)$ = the rate of failure occurrence due to wear may be expressed as a Weibull distribution or an explicit function of time, or in case of non-simultaneous renewal of interacting components as an implicit function of time.

$$\rho_{w_i} = r\, G x_i^{r-1}(t)$$

$$f_w(t) = r\, G x_i^{r-1}(t)\, \exp[-G x_i^r(t)]$$

$$x_i(t) = \int_0^t f(x_1 \ldots x_n)\, dt$$

If a normal distribution is assumed for wear-out failure rate,

$$f_{w_i}(t) = \frac{A}{\sigma_x \sqrt{2\pi}} \exp\left[-\frac{(X_{w_i} - \bar{X}_{w_i})^2}{2\sigma_x^2}\right] = \frac{1}{\sigma_2 \sqrt{2\pi}} \exp\left[-\frac{(T_1 - M)^2}{2\sigma_t^2}\right]$$

The total instantaneous component failure rate for component i,

$$\rho T_i(t) = (c_1 + c_2 + st + rG x_i^{r-1}(t)$$

and component reliability with Weibull wear rate of failure

$$R_i(t) = \exp\left[-\int_0^t \rho_T(t)\, dt\right] = \exp\left[-\{[c_1+c_2+st/2]t + rG \int_0^t x_1^{r-1}(t)\, dt\}\right]$$

$$x_i(t) = \int_0^t f(x_1 \ldots x_n)\, dt$$

With normal wear rate of failure

$$R_i(t) = \{\exp-[c_1+c_2+st/2]t\} \int_t^{-\infty} \left[\frac{1}{\sigma_x \sqrt{2\pi}} \exp\left(-\frac{(X_i - \bar{X}_{w_i})^2}{2\sigma_x}\right)\right] dt$$

$$= \{\exp[-(c_1+c_2+st/2)t]\} \int_t^\infty [\frac{1}{\sigma_t \sqrt{2\pi}} \exp\left(-\frac{(T_1-M)^2}{2\sigma_t^2}\right)]dt$$

10.4 System Reliability

From a reliability point of view, a system consisting of n parts, which will fail whenever one of the component parts fails, is a series system. We thereby imply that if any part fails, the system fails to perform to specification or stops. We can restore performance by replacing one or several parts.

When the system is new, system reliability with Weibull wear-rate failure distribution is:

$$R_s(t) = \prod_1^n R_i(t) = \exp-\Sigma]c_{1i}+c_{2i}+ \frac{s_i t}{2})t+rG\int_o^t x_i^{r-1}(t)dt$$

With Gaussian wear-rate failure distribution

$$R_s(t) = \{\exp-\Sigma_i]c_{1i}+c_{2i}+ \frac{s_i t}{2}]t\} \prod_i^n \int_t^\infty \frac{A}{\sigma_{x_i}\sqrt{2\pi}} \exp\left(-\frac{(X_{w_i}-\bar{X}_{w_i})^2}{2\sigma_{x_i}^2}\right)dt$$

where $t<T_1$ = first replacement time.

The system mean time before failure is given by:

$$T_f = \int_0^\infty R_s(t)dt = \exp(A) \cdot B \cdot (1-\mathrm{erf}(C))$$

for a Weibull wear-rate density distribution. A similar expression can be derived for the normal wear-rate distribution.

If $t>T$, and some parts of the system are worn out and replaced at T_i and so on, then

$$R_s(t) = \prod_{i=1}^n \frac{R_i(t-T_i)}{R_i(T-T_i)} = \prod_{i=1}^n R_i(t,T-T_i)$$

where

T = time from zero, operating age of system

T_i = operating age of system at last renewal of component i

$T-T_i$ = operating age of component i

$R_i(t, T-T_i)$ = reliability of ith component given it survived time $(T-T_i)$

This expression is used for the overall system reliability with component renewal and interaction by introducing the appropriate formulas for component reliability for the discrete intervals between replacements as found before.

It should be noted that the above expressions for system reliability assume independence of the system components. It is proposed to use the coupling terms in the wear failure density expression $f_{w_i}(X_i(t))$ to represent interaction. This actually means an assumption of quasi-independence of components in operation.

As noted, wear can be shown to be the major interacting factor for mechanical, structural, and even social or economic systems.

Interaction of mechanical component performance due to flow media as, for instance, failure of pressure regulating devices resulting in pipe failures, etc., is not considered. It is found that these failures are, in general, chance occurrences. We are only just beginning to understand how and why interaction affects reliability.

A cautious start is presently being made to analyze operational data. Long-life mechanical systems with continuous part replacements, such as diesel engines, etc., may attain a higher operational reliability at a lower maintenance and space parts cost if component wear interaction is taken into account.

10.5 Use of Networks in the Analysis of Interactive Systems Reliability, Maintainability, and Availability

Systems performance in terms of reliability and availability depends not only on system design as defined by systems structure, component quality, capacity and performance, but also on its operability, maintainability, accessibility, observability, component interaction, interaction with the environment and external factors, operating and maintenance policy, operating and maintenance skills, and other factors. For realistic analysis these factors must be included.

While Markov and flow graph models provide effective approaches to the analysis of systems structure, conditional probabilistic and stochastic network simulation techniques have been found to provide effective tools for analysis of interactive reliability, maintainability, and availability of complex systems.

Research in the use of such models indicates their effective use not only for the structural design of systems but also for the analysis of alternative operating and maintenance strategies to achieve the desired performance in terms of output, reliability and availability at minimum cost or other parameter. Such network simulation allows systems to be analyzed, changed, or put into different environments without disturbing them. Reliability analysis must not be concerned solely with component and resulting systems performance in terms of failure, but must include cause and effect relationships. Causes affecting states of the system may be originating within or without the system. The most important cause is often externally induced human error. The range of possible human errors should be included in the evaluation of transitions among states or conditions of a system in terms of their probable occurrence and effect on the system. While effective cause and effect analysis is usually performed in systems risk assessment, we often consider reliability analysis a design exercise with the system operating in isolation from its environment.

Returning to consideration of simple degrading systems such as mechanical systems, there are a number of basic considerations which should be remembered:

1. In most physical systems component failure and consequential systems failure is caused by component wear (including stress, etc.).

2. Component wear is usually affected by the wear of other interacting components.

3. Wear failure rate distributions must be based on component wear and not implicitly on time because interfacing components are not usually all renewed simultaneously.

4. Wear rate estimation must include consideration of the effect of wear on all interacting components, on component wear rate, and resulting wear failure rate.

5. Wear failure rate must be superimposed on the initial or start up as well as the random failure rates unless the start up and random failure periods are very small, or the component is subject to a limited number of short operations as part of a standby system.

In the consideration of component interaction we usually confront a number of basic problems such as:

a. poor correlation of the component wear and failure data because of the dependence of failure causes and events,

b. component failure functions can therefore usually not be expressed as functions of single variations,

c. in most real systems the failure rate of components and systems is time variant and dependent on the failure rate of interacting components and events within as well as without the system, and

d. as a result, component failure rates should be time variant, dependent, and conditional to represent reality.

The major factors affecting mechanical component failure rates are shown diagramatically in Figure 10.4

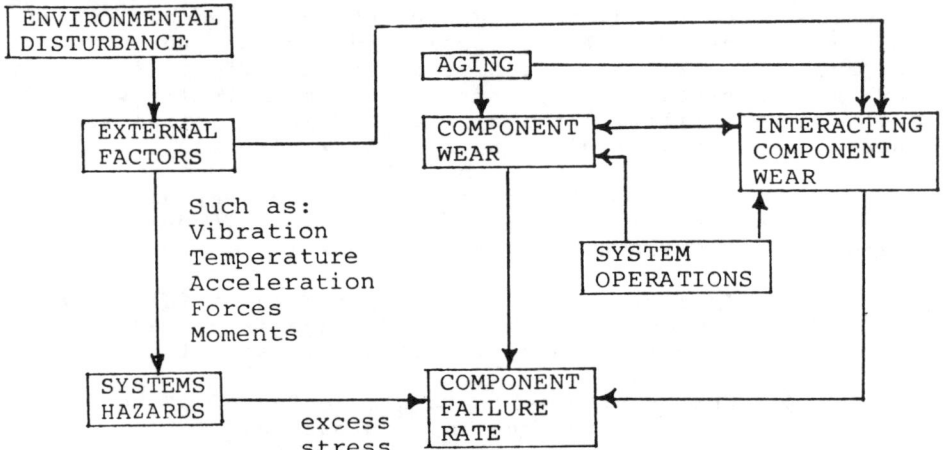

FIGURE 10.4: FACTORS AFFECTING MECHANICAL COMPONENT FAILURE RATE

All of these factors must be included for effective representation of component failure rate and consequent reliability and availability. Unfortunately data is not always available to permit a formal evaluation of all the interacting factors affecting mechanical component failure rates. As a result simplifying assumptions are usually made which emphasize the factors which provide the dominating effects causing component failure, such as component wear and interacting component wear. Yet other, largely random, factors such as vibrations, temperative excursions, and accelerations may be dominating factors in their effects on component failure rate. The probability of a random catastrophic component failure occurring as a result of environmental or external factors, for example, is usually dominant during the life of the component before wear-out.

Only when wear exceeds a certain permissible limit will the probability of wear-caused catastrophic component failure usually exceed the probability of environmental or external factor caused random catastrophic component failure.

The most effective method for the analysis of multiple interacting factor component or systems failure analysis is probably the use of conditional probabilistic networks such as GERT, which permits the effects of monotonically increasing factors such as wear and random factors caused by environmental or external events, on component and system failure rate, to be effectively modeled and analyzed.

10.6 Exercises

1. Consider a passenger automobile to be composed of major subsystems of components such as engine, gear, transmission or clutch, drive shaft, differential, wheels, brakes, etc. from the point of view of subsystem or component interaction. Develop a logic diagram showing component interaction and indicate the measurements you believe would indicate the interactions affecting automobile operating reliability.

2. In a diesel engine of a certain type, the wear rate of the rings can be expressed as a linear function of the time since replacement of the liner in which the rings operate. Assume that the wear rate of the rings is

$$2 \times T \times 10^{-6} \text{ mm/hr}$$

where T is the operating time since liner replacement. If rings are considered worn out when worn 1 mm and must be replaced then, and if the cost of spare rings and liners is very small compared to the cost of downtime to change parts, what is the optimum time to replace rings, assuming that liner wear rate is not affected by the ring replacement policy as long as rings are always replaced before being worn out. The liner will be worn out when T = 10,000. For simplicity, assume a constant ring wear between ring replacement equal to the average rate of wear.

10.7 References

1. Frankel, E. and Roberts, R. "The Application of the Functional Equation to Complex Reliability Problems", Journal of American Society of Naval Engineers, 1964.

2. Frankel, E. and Pollack, G. "On the Effects of Component Wear and Interaction on Systems Reliability", International Shipbuilding Progress, March 1965.

3. Barlow, R.E. and Proschan, F. "Planned Replacement", Sylvania Electronic Defense Laboratories, Technical Memorandum No. EDL-M296, 1960.

4. Welker, E.L. "Relationship between Equipment Reliability, Preventative Maintenance Policy, and Operating Costs", Fifth National Symposium on Reliability and Quality Control, 1959.

5. Bender, D.M., "The Statistical Dynamics of Preventative Replacement", Wescon Convention Record, 1959.

6. Barlow, R.E. and Hunter, L.C. "Mathematical Models for Systems Reliability", The Sylvania Technologist, Vol. XIII, No. 1 and 2, 1960.

11.0 APPLICATION OF FAULT TREE AND OTHER NETWORK TECHNIQUES

Ernst G. Frankel

As discussed before, fault tree analysis is a technique by which many events affecting a system which interact to provide other events, and ultimately system failure, can be related using simple logical relationships as part of a tree network structure. The logical relationships define the interaction of the events and allow the methodical construction of the fault tree structure. As noted, a fault tree usually starts with a top event, which is incurred as a result of the occurrence of primary events which in turn are caused by secondary, lower order, and command events. A simple example of an on-line redundant system is shown in Figure 11.1.

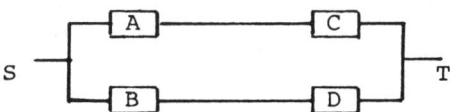

FIGURE 11.1: SIMPLE ON-LINE REDUNDANT SYSTEM

If we assume that events A, B, C, and D imply that the respective components fail and the event F is the event that neither the series of components A - C or B - D are operable, then a fault tree can be constructed as shown in Figure 11.2.

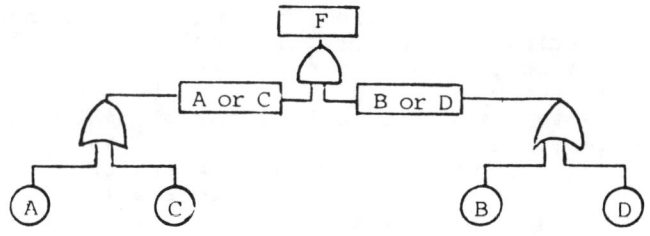

FIGURE 11.2: FAULT TREE OF SIMPLE ON-LINE REDUNDANT SYSTEM

The top or system failure event occurs when

$$F = AD + AB + CB + CD$$

These events usually have a failure probability, but the failure probability is, in most cases, uncertain. Similarly the degree of dependence (or independence) among events is usually uncertain.

Finally, there is the issue of inclusion of all failure events. It is very difficult to assure that all failure events or failure modes of significance are identified. A typical example of uncertainty in the estimated failure probability is the case where two systems or components with different failure probability distributions but equal expected failure probabilities are considered. Obviously if only the expected probability of failure is of concern, then a component with a distribution which has a large deviation from the mean will be considered equal to that with the same expected probability of failure but with little or no deviation in its probability distribution. In other words if we only care about the expected probability of system failure, then a system with a failure distribution I will be considered equal to a system with a failure distribution II as shown in Figure 11.3, or in the language of probability theory we are indifferent between Systems I and II, where $P(F)$ = Probability of Failure.

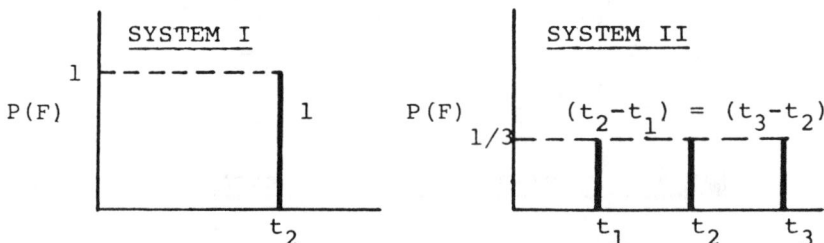

FIGURE 11.3: SYSTEM DISTRIBUTION

On the other hand, if we are interested in the expectation of the number of independent trials until first failure, the probability of two or more consecutive failures within a certain time period, or the probability of no failure during a fixed time period, then we prefer System I. The main issue is really the purpose of the fault tree analysis. Its purpose may be to perform a reliability or risk analysis of a specific system. In reliability analysis we usually assume the system's structure (fault event hierarchy) to be known. In risk analysis, on the other hand, we start with a sketchy knowledge of the system's structure and must use the knowledge of engineers, operators, or other experts to develop a reasonable structure and structural alternatives. In reliability analysis we usually require extensive fault tree analysis codes to handle the complex structure, while in risk analysis simple fault tree methods such as cause and effect tables, failure logic models, and other basic approaches are usually adequate. The approach taken in fault tree analysis as well as in the use of failure probability (point or distribution) estimates therefore depends on the purpose of the analysis.

Furthermore we must often include more subjective issues than simple aspects relating to uncertainty of systems or component failure probability and failure events structure. Such subjective issues may include various preferences relating to different systems failure events. In turn preferences may be affected by personal or group utility, environmental, or economic factors, as well as a multitude of other issues.

There are many issues involved in the selection of distributions which effectively represent failure probabilities. The most common statistical distributions used are the Erlang, Weibull, and Log-normal distributions. Distributions are usually chosen on the basis of fit of experimental or operational data, but there are many problems which are usually difficult to consider, such as:

1. Realism and replicability of environment in which data was obtained.

2. Modeling of dependence among component failure probabilities.

3. Modeling of overt "common mode" causes.

4. Breadth of distribution and its higher order statistics characteristics.

5. Uncertainty intrinsic to each component in a class, which affects the determination of its dependence on other component performance.

Finally there is the issue of uncertainty in the state of our knowledge of the components, systems structure, and system versus their inherent uncertainty. In the performance of data analysis, we usually try to obtain specific results such as:

- Statistical results which can be used for component and system's failure probability determination

    1. failure rates for predominant failure modes
    2. confidence bounds
    3. common cause failure probability
    4. trends and abnormal behaviors

- Engineering or structural knowledge which can help structure the fault tree network

    1. predominant failure mechanisms
    2. failure mode breakdown
    3. common cause mechanism
    4. environmental effects
    5. system effects

In turn, fault tree evaluation is designed to provide qualitative results such as:

- minimum cut sets - combination of component failures causing system failures

- qualitative ranking of contributions of failure events

- common cause potentials - susceptability to single cause failure

and quantitative results such as:

- numerical possibilities - failure probability of systems, etc.

- quantitative rankings of contributions to systems failures by various component and basic failure events.

11.1 Implementation of Fault Tree Analysis

As discussed before, a fault tree analysis consists of logically-structured trees of fault sequences. Fault tree analysis is a technique by which many interacting events, which in turn produce other events, are related in simple logical relationships, which in turn permits a methodical building of a structure which represents the system. Fault trees consist of 'top events' which are broken down into 'primary', 'secondary', etc. and 'command' faults (events). It is usual to have only one top event which is often defined as the catastrophic failure event.

The use of fault trees is usually designed to provide quantitative and qualitative results. It can be designed to determine probabilities of failure events as well as help design a process or system. In other words, fault tree analysis is a systematic procedure used to examine systems in order to determine component failure modes and other events. Fault tree analysis was introduced by Bell Telephone Laboratories (1961) to perform safety evaluations of launch control systems for the Minuteman program. Fault tree analysis has been found to be a general tool for modeling system failures and system failure contributions. As large fault trees require the handling of large amounts of data and structural information, it has been found useful to develop and use computer codes for fault tree evaluation. There are different computer codes for qualitative and quantitative analysis, as well as for direct, dual purpose, and common cause failure analysis.

11.1.1 Representing Fault Trees by Networks

A fault tree 'AND' gate can be represented by parallel branches of a flow graph (or s-t graph), while an 'OR' gate can be represented by series branches. We can now build up a flow or s-t network to represent the fault tree. Starting from the bottom we move towards the events triggered by the basic inputs or events. The top event is realized when and only when all paths from the initiating events to the top event are cut off. To determine what non-primary events must be accomplished we proceed as in the method of Minimal Cut Sets, by identifying the branches of the network, which are most difficult to cut. A branch is cut if an event, depending on the performance of that branch, is realized.

The gates of a fault network can also be represented by the input node characteristics of a GERT type network as described before. The 'Exclusive-OR' type of input node then represents an 'OR' gate while the 'AND' type of input node represents an 'AND' node. An added advantage of the use of a GERT-type network for modeling fault trees is that in addition to providing an effective technique for the representation of the structure of the fault tree, and its probabilities of event realization, it permits introduction of the dynamics of the system under study. In other words while a fault tree and a deterministic flow network representing the fault tree only allow consideration of the probabilistic relationship of the various levels of events leading to the final event, a GERT representation permits consideration of the time variance of the probabilities of the different events.

In many systems event probabilities are time varying and conditional. The proposed approach allows not only determination of the probability of event occurrence at a specific time, but of event occurrence over the time period of interest. Various statistical distributions of the time to event occurrence can be used.

Let us consider first a network representation of the fault tree of the simple on-line redundant system shown in Figure 11.2. If $P(A)$, $P(B)$, $P(C)$, and $P(D)$ are the probabilities of failure of components A, B, C, and D respectively in time t, and $P(F)$ is the probability of the event F, total systems failure, then the fault tree could be represented by an s-t network as shown in Figure 11.4.

The s-t graph representation is obtained by transferring AND gates to parallel connections and OR gates to connections

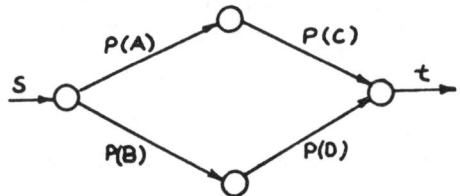

FIGURE 11.4: S-T GRAPH OF SIMPLE FAULT TREE

in series. When a subevent is realized the corresponding link must be cut off. The final or top event is realized when and only when all paths from s to t are cut off. An s-t graph, while instructive, does not permit the introduction of time variance.

In many cases the concept of time dependence should be introduced to make a fault tree a useful device. Therefore there are often two types of complementary solutions to a fault tree. The first type, as discussed before, is a time independent solution which contains information on all the various events and their realtionships independent of time, while the second is a time dependent solution which identifies the required relations of events which, during different times, have a certain relationship. The dependence among events may be affected by time delays required for the establishment of the relationship or dependence. Similarly the degree of dependence may be a function of time measured from some t=0. For example, two events may be complementary only during a given period (t_2-t_1) and not complementary at

any other time. Similarly the degree of their complementarity may be a function of time.

If we are interested in the failure of different interacting components leading to the failure of a system composed of these components, then the probability of failure or failure rate of various components may be functions of time. In that case we would be interested in representing the probability of failure of the total system, the top event, as a time dependent sequence of time-dependent events. Such systems are most effectively analyzed using a semi-Markov technique.

Use of semi-Markov techniques obviously requires extensive knowledge about the system structure but, in addition, we must have information on the time distribution for each transition between events.

An analytical solution of a semi-Markov model is quite complex for systems subject to multiple modes of failure. As such systems constitute the bulk of problems of practical importance, a numerical approximation is often resorted to to solve the semi-Markov model. There are other options such as formal solution of a discrete semi-Markov process and use of the Graphical Evaluation and Review Technique, which we discussed before in Chapter 8. The discrete semi-Markov process differs from the continuous semi-Markov model in that delay, holding, and other time distributions are discrete. Such an approach facilitates the task of translating our analytical method into a numerical approximation. GERT, a technique for the analysis of stochastic networks, is based on semi-Markovian principles.

The ability of translating the semi-Markov process into a numerical approximation offers a wide degree of applicability. A few of its must successful applications to date include the "Semi-Markov Model of a Flow System"; GERT's analysis of a space vehicle countdown, and zone refining of semiconductor material.

11.2 Uncertainty in Reliability Analysis

A number of different approaches have been proposed to handle uncertainty in component and systems failure rates, in reliability as well as structural relationships such as inter- and intra-dependence of components and systems. A basic question is often whether uncertainty is intrinsic to each component in a class or whether the uncertainty is basically a function of the state of our knowledge of that class. For example, if P_i is the failure probability of component i, is it proper to consider P_i as independent of i or as dependent in i, which in turn would imply that if we know one or some of the P_i's, then we would essentially know the failure probabilities of all the components i. One measure of probabilistic importance or PI(i) of component i developed by Lambert, is expressed as:

$$PI(s) = \frac{P_i}{S} \frac{\partial S}{\partial P_i}$$ where S = system failure probability

Another measure of "Uncertainty Importance = UI(i)" proposed by Bier assumes the P_i's to be random variables. In this case, the "Uncertainty Importance" of a component i is expressed as:

$$UI(i) = \frac{Var\ P_i}{Var\ S} \frac{\partial Var\ S}{\partial Var\ P_i}$$

It should be noted though that the numerical value of UI(i) would not change if the standard deviations of P_i and S had been chosen instead of the variances. In case of small variations in the variances of P_i, say by b percent, the variance of S would be reduced by b·UI(i) percent, given that P_i is independent of all other component failure probabilities. In other words UI(i) serves as a constant of proportionality.

When the component failure probabilities are dependent then the variance of S is much more difficult to obtain as is obviously S, in terms of the P_i. This type of problem is usually modeled using computer algorithms. To obtain the "Uncertainty Importance" for a set of components which make up a system S we simply compute

$$UI(S) = \sum_{i \varepsilon S} UI(i)$$

if and only if the component failure probabilities are independent. Similarly if P_i's are perfectly correlated for all i then changing the variance of P_i changes variances of all other i in the set, such that

$$UI(i) \simeq \sum_{j: i \varepsilon I_j} \frac{Var(S_j)}{Var(S)}$$

where S_j's are the failure probabilities of the cut sets containing component i.

Various approximations have been proposed to deal with the problem of component dependence in determining uncertainty importance. Many of these use a log-normal distribution for analysis. This is an area where substantial research is underway to develop methods for effective analysis of systems with inter- and intra-dependent failure rates in a formal manner.

11.2.1 GERT Reliability Networks with Uncertainty

Networks provide a convenient method for the representation of reliability problems. Using GERT networks we can model not only the failure distribution of various components and resulting failure events but we can also introduce the conditional probability of the different failures occurring. We use the equivalent function as the product of the conditional probability that a failure event will occur and the distribution of the time to occurrence of the failure. Consider a system such as shown in Figure 11.5.

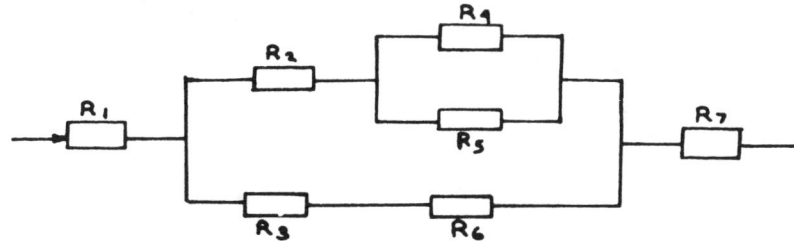

FIGURE 11.5: RELIABILITY BLOCK DIAGRAM FOR PARALLEL SYSTEM WITH PARALLEL SUBSYSTEM

The GERT equivalent of this system is shown in Figure 11.6.

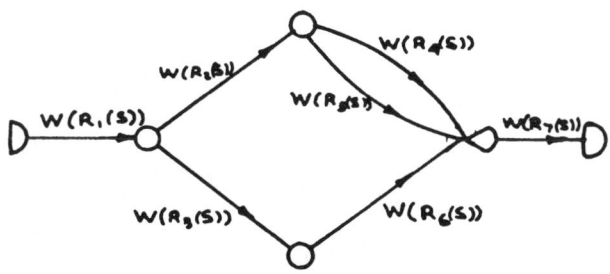

FIGURE 11.6: GERT NETWORK FOR PARALLEL SYSTEM WITH PARALLEL SUBSYSTEM

The network includes AND and Inclusive-OR nodes and can readily be reduced to a single transform from input to 1 to output from 6, using the methods discussed in Chapter 8.

Returning to basics, the most fundamental problem in reliability analysis consists of reducing simple series and parallel systems. A system of two series components with component reliabilities $R_A(t)$ and $R_B(t)$ respectively and a systems reliability of $R_T(t) = R_A(t)R_B(t)$ can be expressed as a series Exclusive-OR GERT network as shown in Figure 11.7. In order to represent a parallel on-line stand-by system of two components with reliabilities $R_A(t)$ and $R_B(t)$ as an Exclusive-OR GERT network, it is usually convenient to compute $Q_T(t) = 1-R_T(t)$ or the unreliability of the system

TYPE I

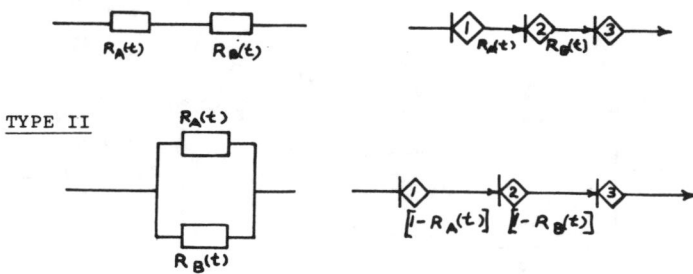

TYPE II

FIGURE 11.7: SIMPLE EXCLUSIVE OR RELIABILITY NETWORK

when

$$Q_T(t) = 1-R_T(t) = (1-R_A(t))(1-R_B(t))$$
$$= 1-R_A(t)-R_B(t)+R_A(t)R_B(t)$$

and

$$R_T(t) = R_A(t)+R_B(t)-R_A(t)R_B(t)$$

Instead of component reliabilities (or unreliabilities) we can obviously use component equivalent functions $W_A(s)$ and $W_B(s)$ which represent the product of the conditional probability that the component is subject to failure and the distribution of the time to failure. Use of Exclusive-OR GERT networks makes the analysis quite simple as reliability networks expressed as Exclusive-OR networks can be readily reduced by the use of Mason's reduction techniques.

Considering next a series of two on-line redundant or parallel component systems, in other words a series of two systems of Type II, we obtain

$$R_T(t) = [1-(1-R_A(t))(1-R_B(t))]^2 = R_{T1}(t)R_{T2}(t)$$

or

$$P_T M_T(t) = P_{T1} M_{t1}(t) P_{T2} M_{T2}(t)$$

from which the total reliability of the system $R_T(t)$ is

obtained. As mentioned before, the use of GERT network methods is most relevant to problems with uncertainty in failure occurrence and failure time distribution and where therefore use of equivalent functions is advisable. As discussed, GERT also permits, in addition to probabilistic branching, other features such as network looping, multiple sink nodes, multiple node or event realizations, and multiple probability distributions.

The network looping feature is useful in modeling sequential failures or the behavior of maintained systems subject to periodic or otherwise scheduled maintenance. Transition probabilities and transition interval time distributions can readily be determined using such network techniques for complex problems. Instead of the equivalent function, made up of the product of the conditional probability of a failure event and the moment generating function or transform of the time distribution of the failure event, as discussed before, the Mellin integral transform is often used as a powerful tool for analyzing the probability density and distribution functions of products and quotients of random variables. This is particularly important when the reliability of components or subsystems in series are assumed to be random variables. The Mellin integral transform of the probability function f(t) of a random variable t is defined as

$$M(f(t)) = \int_0^\infty t^{s-1} f(t) dt$$

where s is a complex variable. Also $M(f(t)) = E(t^{s-1})$ gives the moments of f(t) for real values of s. If $T = t_1 t_2$ and t_1 as well as t_2 are independent positive random variables, then

$$M(f(t)) = E(T^{s-1}) = E(t_1^{s-1}) E(t_2^{s-1}) = M(f(t_1)) M(f(t_2))$$

The density function of the variable T can be obtained from its Mellin transform by evaluating the inversion integral

$$f(t) = \frac{1}{2\pi i} \int_{C-i\infty}^{C+i\infty} T^{s-1} M(f(t)) ds$$

GERT network techniques are also useful in fault tree analysis, particularly when transitions among events are subject to time-varying conditional probabilities and when the time of transition is random or subject to some statistical distribution.

For example, AND or OR gates may be time varying. In other words, the dependence of an event on lower level or basic events may vary over time, be valid only during certain time periods or intervals, or be otherwise time dependent. It may also be conditioned by the occurrence of lower level events at specific times or during specific time intervals. Each branch in a fault tree GERT network gives the probability that the branch is realized, given that the previous event is realized and the time distribution of the effect of that event on the next higher event.

Sometimes more than one parameter is of interest such as time and cost of moving from one event to another. This is quite readily considered in a GERT nerwork model, by introducing two complex variables s_1 and s_2 into the Moment Generating Function or Transform representing the time (and cost) distributuon of the transition.

11.3 References

1. Dhillon, B.S. and Singh, C. "Engineering Reliability", Wiley Interscience Publications, New York, 1981.

2. Barlow, R.E. and Proschan, F. "Statistical Theory of Reliability and Life Testing - Probability Models", Rinehart and Winston, New York, 1975.

3. Fussell, J.B. "Fault Tree Analysis - Concepts and Techniques", Proceedings of the NATO Advanced Study Institute on Generic Techniques of Systems Reliability Assessment, Nordoff, Leidess, The Netherlands, 1975.

4. Bazovsky, I., "Fault Trees, Block Diagrams, and Markov Graphs", Proceedings of the Annual Reliability Maintainability Symposium, IEEE, New York, 1977.

5. Barlow, R.E., Fussel, J.B., and Singpurwalla, N.D. "Reliability and Fault Tree Analysis", SIAM, Philadelphia, 1975.

6. Eagle, K.H. "Fault Tree and Reliability Comparison", Proceedings of the Synmposium on Reliability", IEEE, New York, 1969.

7. Haasl, D.F., "Advanced Concepts in Fault Tree Analysis", Systems Safety Symposium, University of Washington, 1965.

8. Young, J. "Reliability and Fault Tree Analysis", SIAM, Philadelphia, 1975.

9. Wolfe, W.A. "Fault Trees Revisited", Microelectron-Reliability, 17, 1978.

10. Aitchison, J. and Brown, J.A.C. "The Lognormal Distribution", Cambridge University Press, New York, 1963.

11. Bhattacharya, G.K. and Johnson, R.A. "Stress Strength Modules for Systems Reliability, Reliability and Fault Tree Analysis", SIAM, Philadelphia, 1975.

12.0 RELIABILITY AND RISK IN PERSPECTIVE

Ernst G. Frankel

12.1 General Considerations

All systems, natural or man-made, are subject to failure. We have discussed methods of analysis of failure phenomena and approaches which may help in the design of more reliable systems. Man plays an important role both in the design and use of systems, and contributes to the failure of both natural and man-made systems. On the other hand, nature may also cause failure through natural events or interactive factors. The nature of failures varies widely as do the causes of failure events and the events leading to failures. Man's contributions to the failure of systems involves

1. conception or misconception of system requirements, capability, or environment,
2. design deficiencies and erroneous assumptions,
3. faulty construction, manufacture, erection and installation,
4. mistakes in the methods or procedures of operation, and
5. control and management of the system.

For many years attempts have been made to reduce failures of man-made systems through introduction of safety measures, quality control, stress factors and other approaches designed to reduce failure probability. Use of measures such as factors of safety, sometimes called factors of ignorance, is usually resorted to to improve the design for reliability. Yet this approach often works against improvements in performance associated with output, operability, and efficiency.

There is also a basic difference in society's perception of risk versus designers'/operators'/investors' perception of risk associated with man-made systems, as societal impacts are different from risks and their impacts on designers, operators, investors, and users. The concepts of systems reliability and risks resulting from system failure represent a cause and effect relationship. While systems users/owners are primarily concerned with system reliability and performance, society is primarily concerned with the risks posed by a system. Acceptability of risk of major systems though is often determined by the publics' attitude as interpreted by the political system. Availability of information greatly affects acceptability of risk. In a well informed society and under conditions of rational decision making, acceptability of risk should usually be measured against the perception of alternate risk. This is not always the case in reality though as there is usually a tremendous gap between perceived and actual risk. In considering risk we must distinguish between:

- user or owner/investor risk
- public or societal risk, and
- catastrophic risk.

Each risk is different from a perception and consequence point of view. The more we know about a risk, or the better we think we can estimate the probability of occurrence of the undesired event, our willingness of acceptance of the risk is usually greater.

For example, incident probabilities of an airplane crash are 1×10^{-3} /airplane year while those of nuclear reactor failures or core damage are 4×10^{-4} /reactor year. Although an airplane only costs about 1% of the cost of a nuclear reactor, financial losses caused by airplane disasters are usually much larger and average $200 million/crash or 4.27 times airplane replacement costs. Similarly the probability of human loss resulting from an airplane crash are proportionally much larger than those of a comparable reactor failure. Yet we are willing to accept this exposure to risk because it is assumed to be a much better known risk and a risk taken primarily by providers and users but not the public or society at large. This appears to be the general perception and approach to risk taking.

12.1.1 Risk Attitudes

In considering the approach taken by individuals to risk taking, we must recognize that the preference for risky alternatives by an individual is affected by two factors:

1. preference or aversion of consequences of possible outcome, and
2. attitude towards risk taking.

Risk preference or aversion by an individual must therefore be expressed as a multidimensional function which judges an individual's preference or aversion in terms of both the relative utility of the outcome or consequence and the attitudinal factors influencing an individual's behavior. Risk preference is therefore a relative and conditional measure and the perceptions of consequences vary from individual to individual. Similarly risk evaluation must consider risk aversion and risk attitude or acceptance. A major problem in risk analysis is that current methods of probabilistic risk assessment assume that:

1. All important failure modes have been identified.

2. Uncertainties in estimates of component failure probability are well documented and supportable.

3. Multiple failures induced by a single obscure initiating event are recognized.

4. Consequential failures or effects on performance which may ultimately lead to failure are effectively identified.

5. Reduced performance or partial failure induced risks have been included.

6. Estimates used have all been validated.

7. Extremely large accidents of infinitesimally small probability of occurrence have been included.

It must be recognized though that when judgements on risk are made by the public, they focus on consequences and ignore risk - often including the risk of the user. Risk, as defined before, is the potential for the realization of unwanted consequences from impending events. Risk as such cannot be regulated and can only be estimated as a consequence of such impending events. A major problem in risk estimation is usually differentiation between events caused by inherent risk and those caused as a result of inherent problems. We are usually interested in the magnitude of both risk and risk exposure.

12.2 Analysis of Risk

Risk assessment is often broken down into risk determination which consists of risk identification and risk estimation, and risk evaluation which in turn consists of risk aversion or consequence analysis and risk acceptance or attitude analysis.

In risk determination we identify new risks, changes in risks, and risk parameters and determine the occurrence and magnitude of consequences of risks. In risk evaluation we determine degrees of possible risk reduction and avoidance, establish risk aversion and acceptance references and evaluate the impacts of risks. Of particular importance in public risk evaluation is usually the inclusion of catastrophic risk - however rare - because the cost of such occurrences may be extremely high. At the other extreme it is important to assure also that risk of small effects be included as these may have high probability and low cost per event. In other words it is suggested not to assume a threshold hypothesis, which is often done, but which hides important consequences of risk.

12.2.1 Reliability and Risk Assessment

Risk of man-made systems failures has become a major public issue in recent years as system size has increased and public awareness has been enhanced. As a result, reliability of systems is of increasing concern, as costs of failures as well as system unavailability become more and more unacceptable. Therefore it is increasingly required now that formal reliability analysis and risk assessment form an integral part of the system design process where design includes determination of system structure and operation. In general two types of risk are distinguished

1. risk of failure, and
2. risk inherent in the system which functions as designed.

From a societal impact point of view, systems failures are usually differentiated by

1. the magnitude of the problem, and
2. the basic failure mechanisms.

An increasingly important issue in reliability and risk analysis is the human influence on systems design and operation and therefore systems reliability and risk. Humans cannot be counted on to act in a predictable manner. They can act in both a helpful or a destructive manner. As a result, human influence must be included in both reliability and risk analysis.

We must also recognize that there is an interaction between design, construction, or manufacture and operation of a system. Errors in design are usually carried into construction and operation of the system. Important considerations in the design of systems now include:

- producibility or operability of the design
- redundancy and fail safe features
- logical safety and inspection procedures
- self-inspection and ready indicator measures
- maintainability which includes accessibility.

All of these considerations are attempts at improving risk in the system design procedure. Similarly, alternative approaches to the design or selection of a system can be evaluated using a procedure such as shown in Figure 12.1, in which we evaluate alternative systems or system designs in the dynamic environment in which they can be expected to operate. The most difficult part of risk analysis, as noted before, is the identification of all possible failure events and causes, particularly when the environment in which the system is expected to operate is not perfectly known. Notwithstanding the difficulty it is important to consider systems reliability and risk in all phases of conception, design, manufacture, and operation of systems because it not only raises our consciousness but usually allows real improvements in reliability and risk reduction to be introduced.

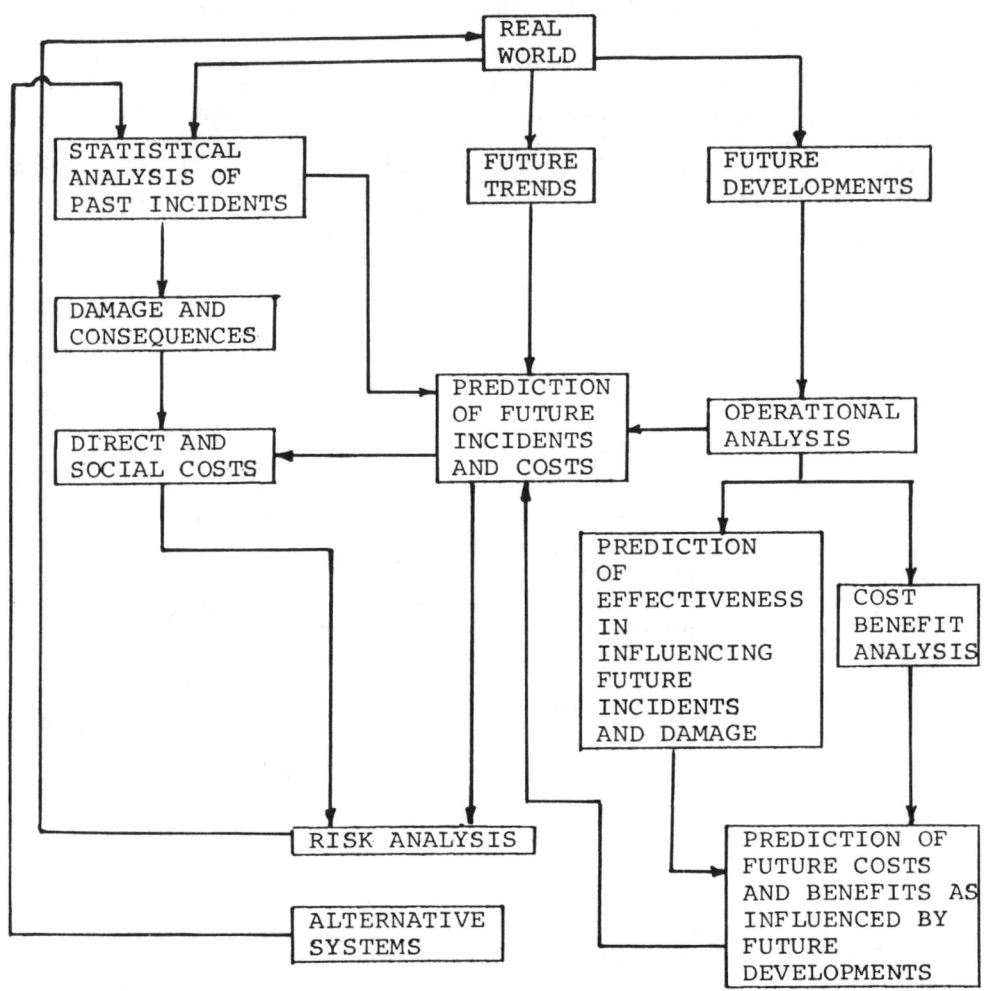

FIGURE 12.1: OPERATIONAL ANALYSIS OF SYSTEMS SUBJECT TO RISK

12.3 Issues and Concerns

We are increasingly dependent on man-made systems for

most of the things we need in our daily life. We are surrounded by man-made systems, at least in urban life, until man-made systems largely supplant nature in providing our environment. We rely on man-made systems to process our food, provide shelter, heat, power, transportation, communication, and most of the things associated with our way of life or, as some say, quality of life. The advances of technology have left the average person often in awe and unable to understand how man-made systems are built or operated. He simply accepts them as part of the environment and assumes that man-made systems are constructed to operate without risk of failure. This applies to rigid structures such as buildings, bridges, roads, tunnels, dams, breakwaters, and runways, as well as operating fixed systems such as power stations, pipelines, and tanks, or mobile systems such as airplanes, road vehicles, and ships.

Many were recently surprised to hear that a large number of our bridges are near collapse. They would be equally surprised to hear that many of our dams are not very safe. While natural factors such as earthquakes, wind, rain, erosion, and corrosion among them may have contributed to the degradation of many of these engineering structures, the major contributions of their risk of failure are generally man-made. Man's contribution to the risk of failure may be in the perception, engineering, site selection, design, construction, operation, or use of the structure or system. It may also be caused by the lack of effective maintenance of the system. While most natural systems are self-maintaining, man-made systems - no matter how advanced - usually require regular maintenance to sustain their performance and low propensity for failure.

Operation of modern facilities poses problems in training and procedures. While 50 years ago an operator of any man-made system was usually intimately familiar with its construction and method of operation, operators today are, by and large, only drivers, often highly skilled drivers, who can manipulate and assure the proper functioning of a facility or system, but only as long as the system, or at least back-up components of the system, are operative. In fact, most modern facilities and systems are designed to discourage any intervention by the operator.

Operators work largely by procedures and rely on monitors to assure fail-safe and effective performance of the system, often without being able to interpret the implications of monitor reading excursions except their knowledge that something abnormal has happened. They then follow, if possible, another set of procedures which may or may not correct the abnormality.

Most system failures are consequential and the result of not one break or malfunction, but a sequence of many failures. As a result, reliability and risk analysis must not be conceived as a static exercise performed once, and which provides a unique answer. Reliability and risk analysis is a dynamic process, which must be updated as additional or improved information becomes available. In other words, effective reliability and risk analysis must be performed itteratively using new and additional information on systems and fault tree structure and component performance. Failure events, their inter and intra dependence, causal factors and more can usually only be estimated during the initial design phase of a system, using knowledge of the system's structure, performance of components under 'similar' conditions, information on expected methods of manufacture and installation of system, expected operating procedures, and finally knowledge of the projected operating environment. Yet usually things change between the design and operating stage. Furthermore, some information only becomes available after start-up of the system.

In conclusion it is therefore emphasized that reliability analysis and risk assessment are but preliminary estimates of expected outcomes of a system, which must be periodically updated, both as to obtain improved estimates of system reliability and risk as well as to permit determination of strategies designed to improve system performance.

APPENDIX A - BASIC CONCEPTS OF PROBABILITY AND STATISTICS

Ernst G. Frankel

Introduction

An operation can be described as a pattern of activities to achieve a certain goal. In many areas, the pattern of activity is governed by a series of events, most of which are haphazard or uncontrollable. We are, therefore, usually confronted with the task of evaluating an operation and predicting trends on the basis of scant and erratic data. Forecasting and judging the complexity of factors influencing operations requires the analysis of data by the use of probability theory and statistics. In the following notes the basic theory will be presented, and it is hoped that the reader will recognize the effectiveness of these methods and invest time for further study of the subject.

A.1 Elementary Considerations

Permutations, Combinations, and the Binomial Theories

a. **Permutations of n things, all together n!**

   or $_nP_n = n! = n(n-1)(n-2)\ldots\ldots(2)(1)$.

b. **Permutations of n things, r at a time (no repetitions)**

   $_nP_r = \dfrac{n!}{(n-r)!} = n(n-1)(n-2)\ldots\ldots(n-r+1)$

c. **Combinations**

   This is a selection of objects considered without regard to their order. A subset of r objects is selected without regard to their order, from a set of n different objects.

   $_nC_r = \binom{n}{r}$ where $r \leq n$

   $= \dfrac{n!}{r!(n-r)!}$ or $_nC_r \times r! = \,_nP_r = \dfrac{n!}{(n-r)!}$

Example: The number of combinations of n things taken n-r at a time is the same as the number taken r at a time.

$$\binom{n}{n-r} = \frac{n!}{(n-r)!\,r!} = \binom{n}{r} = \frac{n!}{(n-(n-r))!\,(n-r)!}$$

This result leads us to Pascal's Rule:

$$\binom{n+1}{r} = \binom{n}{r-1} + \binom{n}{r} \text{ for } 1 \le r \le n$$

$$= \frac{n!}{((n+1)-r)!\,r!} = \frac{n!}{(n-(r-1))!\,(r-1)!} - \frac{n!}{(n-r)!\,r!}$$

Pascal's Rule is a simple way of building up a table of $\binom{n}{r}$; i.e., $\binom{3}{2} = \binom{2}{2} + \binom{2}{1}$.

Permutations of objects not all different:

$$\frac{n!}{n_1!\,n_2!\,\ldots\,n_k!} \text{ where } n_1 + n_2 + \ldots + n_k = n$$

For example, if we have two kinds of objects, r elements of one and (n-r) elements of the second, then the number of permutations of n objects taken all together:

$$\frac{n!}{r!\,(n-r)!} = \binom{n}{r} = \binom{n}{n-r}$$

d. <u>The Binomial Theorem</u>

$$(a+x)^n = \binom{n}{0} a^n + \binom{n}{1} a^{n-1} x + \binom{n}{2} a^{n-2} x^2 + \ldots + \binom{n}{r} a^{n-r}$$

$$+ \ldots + \binom{n}{n} x^n = \sum_{r=0}^{n} \binom{n}{r} a^{n-r} x^r$$

$$= a^n + \frac{n}{1!} a^{n-1} x - \frac{n(n-1)}{2!} a^{n-2} x^2 + \ldots x^n$$

If $x \ll 1$ then $(1+x)^n \simeq 1 + nx$

Considering probability as a concept, we generally accept that the

Probability of Favorable Outcome = $\frac{\text{Number of Favorable Outcomes}}{\text{Total Number of Possible Outcomes}}$

$0 \leq P \text{ (Favorable Outcome)} \leq 1.$

This is the classical definition. Actually it represents the relative frequency of the favorable outcome in relation to the possible outcomes. The possible outcomes may obviously be limited, non-representative, biased, or otherwise not truly random or too small in number. Only if the total number of possible outcomes of the experiment tends to infinity (or all possible outcomes) is the ratio of the number of outcomes representing the event of interest to the total number of outcomes, also called the sample size, truly equal to the probability of the favorable outcome. If A is the event of interest and N(A) is the number of times the event of interest occurred in a test involving N(T) total number of outcomes, then the probability of the favorable outcome or event of interest P(A) is only equal to the ratio of N(A) and N(T) if N(T) tends to infinity or

$$P(A) = \lim_{N(T) \to \infty} \frac{N(A)}{N(T)}$$

and

$0 \leq P(A) \leq 1$

Example

Assume for example that there are three manufacturers producing the same part in proportion of 50%, 30%, and 20%, respectively. A test of 100 parts selected randomly from deliveries of the three manufacturers in proportion to their production results is shown in the following

Test Outcome	Manufacturer			Totals
	A	B	C	
Pass P	40	24	16	80
Fail F	10	6	4	20
Totals	50	30	20	100

Analyzing this data we may be interested in the:

Probability that a test part fails $P(F) = \frac{20}{100} = 0.2$

Probability that a part tested is manufactured by C and fails in test = $P(CF) = \frac{4}{100} = 0.04$

Probability that a part that failed is manufactured by C, given that only parts manufactured by C are tested $P(F/C) = \frac{4}{20} = 0.2$

The example above considered only discrete outcomes. It is often desirable to use a continuous mathematical function to describe the results of a test either because outcomes are not discrete or because they are so numerous that a continuous representation of the results is more effective. In that case the relative frequency of an outcome which in the limit converges onto the probability of the outcome, is represented by the probability density function or pdf. Assuming, for example, that a random variable, say x, can take any value between some lower and upper limits, then $f(x) = dx =$ probability that the random variable x will assume a value between $x-dx/2$ and $x+dx/2$ and $x+dx$ where dx tends to zero or

$$f(x)dx = Pr(x-dx/2 \leq x \leq x+dx/2)$$

Similarly

$$\int_{x_a}^{x_b} f(x)dx = Pr(x_a \leq x \leq x_b)$$

Such a pdf must satisfy the following conditions:

1. it must be larger or equal to zero

   $$f(x) \geq 0$$

2. the area under the pdf curve must be one

   $$\int_{-\infty}^{+\infty} f(x)dx = 1$$

Any function satisfying the above two conditions can represent a pdf. We often are also interested in the cumulative probability or probability distribution function which is the probability that our discrete or random variable representing the outcome has a value equal to or less than some defined number.

For discrete distribution the probability distribution function:

$$P(\text{outcome less or equal to M}) = \sum_{i=-\infty}^{M} \underline{P}(i)$$

Similarly for continuous distributions the probability distribution function is:

$$P_r(x \leq x_a) = P(x_a) = \int_{-\infty}^{x_a} f(x)\,dx$$

A.2 Binomial Probability Function

In a binomial experiment of n trials where in each trial only two outcomes are possible, say success and failure, if the number of successes registered after n trials is x then the number of failures must be (n-x). Let p be the probability of success in an independent trial and q = (1-p) be the probability of failure in an independent trial. Assuming the outcomes for n trials to be independent, then the probability of x successes and (n-x) failures in any given order is

$$p^x q^{(n-x)}$$

But there are $\binom{n}{x} = \dfrac{n!}{x!\,(n-x)!}$ permutation of the order in which x successes and (n-x) failures can occur in n independent trials. Therefore the total probability of x successes and (n-x) failures in n trials is:

$$\binom{n}{x} p^x q^{n-x} = \dfrac{n!}{x!\,(n-x)!}\, p^x p^{n-x}*$$

Let us now consider two simple reliability applications in the binomial probability function.

Examples of Binomial Probability Function

1. Suppose the probability of an electric circuit failing during a mission is 25%. What is the probability of the circuit failing in four sequential missions? On first sight we may want to assume that if its failure probability is 25% during a mission, the probability of failure during four sequential missions should be certainty or 100%. If we assume that the circuit is repaired or replaced on failure then what we are interested in computing is not the average number of failures of circuits in four missions or trials, but the probability of at least one failure in four trials which is:

---

* It is noted that this is the (x+1)st term in the binomial expansion $(q+p)^n = q^n + \binom{n}{1} p q^{n-1} + \ldots + \binom{n}{x} p^x q^{n-x} + \ldots + p^n = 1$.

$$p(x \geq 1) = 1 - p(x=0) = 1 - p^0 q^4 = 1 - (\tfrac{1}{4})^0 (\tfrac{3}{4})^4$$

$$= \frac{175}{276}$$

2. Another interesting binomial probability function problem arises in the determination of redundancy of engines, pumps, etc. Let us assume that we have to install pumps in a nuclear power plant. We have a choice of installing two large pumps (one on-line and one standby) or four smaller pumps (two on-line and two standby). Given the probability of failure p of the small and large pumps is the same, which of the two approaches offers more reliable configuration? The two pump plant has a probability of success of:

$$p(x \geq 1) = 1 - p^0 q^2 = 1 - q^2$$

while the four pump plant has a probability of success of:

$$p(x \geq 2) = 1 - q^4 - 4pq^3 = 1 - 4q^3 + 3q^3$$

For the two pump plant to have a higher reliability therefore $(1-q^2)$ must be larger than $1 - 4q^3 + 3q^4$. It can readily be shown that the two plants have identical reliability for $q^2 (1-q)(1-3q) \geq 0$ when q=1 or q=1/3. Between q=0 and q=1/3 the four pump plant has a higher reliability, while for q>1/3 the two pump plant has a higher reliability. While it is highly unlikely that q≤1/3 and therefore the two pump design proves superior, it should be noted that for larger number of pumps (units) or independent trials it may well be found that a smaller degree of redundancy may offer advantages.

A.3 Elementary Algebra of Events

The Venn diagram, which shows the relations between events in graphical form, is a convenient method for discussing definitions, concepts, and operations of the algebra of events. Considering the major relations among events and the total event or sample space we can define in Figure A.1.

Formally, the following set of axioms provide all the major relations of interest in the algebra of events.

1. $A + B = B + A$ Commutative Law

2. $A + (B+C) = (A+B) + C$ Associative Law

3. $A(B+C) = AB + AC$ Distributive Law

4. $(A')' = A$

5. $(AB)' = A' + B'$

6. $AA' = \phi$

7. $AU = A$

<u>Events (or sets)</u> as collections of points or areas in a space. The collection of all points in the entire space is called the universal set (U).

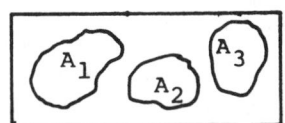

<u>The Complement of Event A</u>, often written A' or Ā, is the collection of all points in the universal set not included in event A. The null set, , is the complement of the universal set and contains no points.

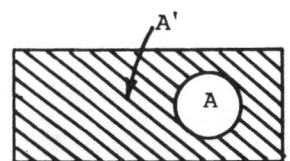

<u>The Union of Two Events A and B</u> is the collection of all points either in A or in B or in both. The union of events A and B is written $A + B$ or $A \cup B$ and represents the "inclusive or" in logic.

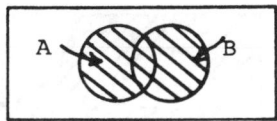

<u>The Intersection of Two Events A and B</u> is the collection of all points common to events A and B. The intersection of events A and B may be written as AB or A B and represents the "and" in logic.

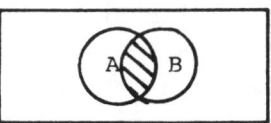

If all the points of U which are contained in B are also in A, then event B is said to be included in Event A.

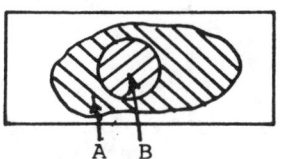

Two events A and B are said to be equal if and only if all the points of U contained in event A are also included in event B, and further that all the points of U in A' are also contained in B'.

FIGURE A.1: VENN DIAGRAM OF MAJOR RELATIONS AMONG EVENTS

The above axioms are provided to insure completeness, and the reader may consider visualizing them on a Venn diagram. Other relations which are valid in the algebra of events, and perhaps intuitively obvious, may be proven by the use of these axioms. Some representative relations, which may be interpreted by use of a Venn diagram are also

$$A + A = A$$

$$A + AB = A$$

$$A\phi = \phi$$

$$A + A' = U$$

$$A(BC) = (AB)C$$

Furthermore we should define Mutually Exclusive Events (see Figure A.2) as a list of events composed of exclusive events if there is no point in the universal set, U, which is included in more than one event in the list. The Venn diagram may be as shown but in no case may there be any overlap of events.

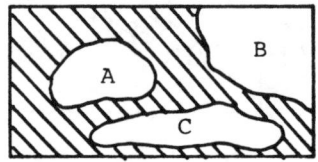 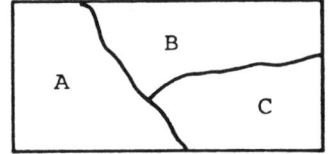

FIGURE A.2: MUTUALLY EXCLUSIVE EVENTS

Employing the algebra of events, a list of events $A_1, A_2, \ldots, A_n$ is said to be mutually exclusive if and only if

$$A_i A_j = \begin{cases} A_i & i = j \quad i = 1,2,3,\ldots,n \\ \phi & i \neq j \quad j = 1,2,3,\ldots,n \end{cases}$$

Collectively Exhaustive Events (see Figure A.3) are defined as a list of events where each point in the universal set is included in at least one event in the list. The Venn diagram above demonstrates collectively exhaustive universal sets. Mathematically, a list of events $A_1, A_2, A_3, \ldots, A_n$ is said to be collectively exhaustive if and only if $A_1 + A_2 + A_3 \ldots A_n = U$ or

$$\sum_{i=1}^{n} A_i = U$$

It should be noted here that a list may be mutually exclusive or collectively exhaustive, or both, or neither.

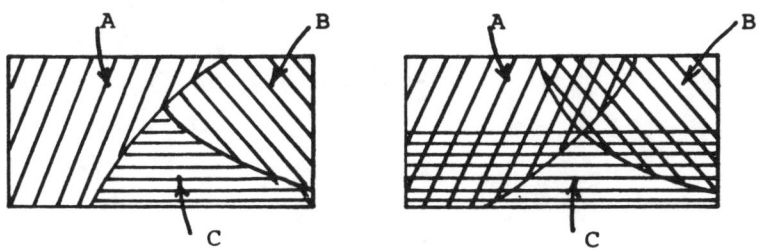

FIGURE A.3: COLLECTIVELY EXHAUSTIVE EVENTS

A.4 Sample Spaces

A sample space can be defined as the mutually exclusive, collectively exhaustive listing of all possible outcomes of an experiment.

Of course, in modeling an experiment, we will usually choose those outcomes of interest to the particular experiment. For example, if a gambler were interested in the outcome of the roll of dice, his universal set of outcomes would include only the possible sums of the number of dots on the up faces. Other possible outcomes such as the height of the bounce would hardly be considered in the calculation of his odds. Sample spaces may look like almost anything from a simple listing to a multidimensional array of all possible experimental outcomes.

Included among the common types of sample spaces is the sequential sample space, which is a convenient and thorough method of listing all possible outcomes of a particular experiment, in terms of some convenient parameters. Consider the experiment where a fair coin is flipped three times, and where we denote the outcome heads on the nth row by H_n and the outcome tails by T_n. The sequential sample space describing this experiment can then be represented by a decision or outcome tree as shown in Figure A.4. Of course the listing of the eight possible outcomes is also a sample space, but in listing a sequential sample space, one normally pictures the entire tree and the outcomes associated with the various branch paths.

At times it may be more convenient to use an event space in the modeling of an experiment. Such a space has all the attributes of a sample space, except that it might not separately list all distinguishable outcomes. If a list of events A_1, A_2, ..., A_n forms an event space, then each possible, finest grain outcome is included in exactly one event in the list. However, more than one distinguishable outcome may be included in any event of the list.

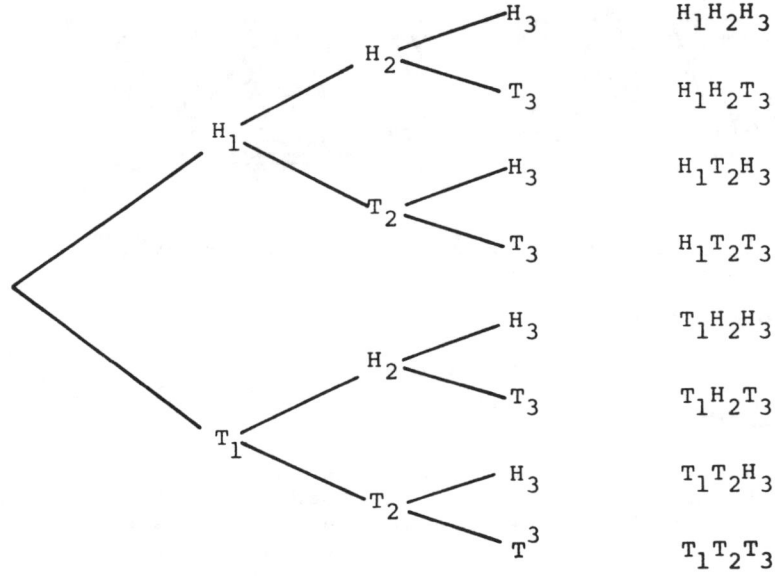

FIGURE A.4: OUTCOME TREE

A.5 Probability Measure

Considering probability as a concept, we generally accept that

$$\frac{\text{Probability of a}}{\text{Favorable Outcome}} = \frac{\text{Number of Favorable Outcomes}}{\text{Total Number of Possible Outcomes}}$$

This expression corresponds to the chance or relative likelihood that a particular event will occur when an experiment takes place. If we use the notation $P(A)$ for the probability associated with event A, then the three axioms of probability measure are:

1. For any event A, $P(A) \geq 0$
2. $P(U) = 1$
3. If $AB = \phi$, then $P(A+B) = P(A) + P(B)$

These axioms are implied by our concept of probability measure. The first axiom states that all probabilities are to be non-negative. The second axiom states that the probability of the occurrence of the universal set is one. This provides a normalization for probability measure, thus leading to the relative likelihood of events. The third axiom states that if two events A and B have no points in common, then the probability of the union of the events is the sum of their individual probabilities.

These three axioms, along with the seven axioms of the algebra of events, can be used to prove various relations such as the following:

$P(A') = 1 - P(A)$

$P(\phi) = 0$

$P(A+B+C) = 1 - P(A'B'C') = P(A)+P(B)+P(C)-(P(AB)-P(AC)-P(BC)+P(ABC))$

$P(A+B) = P(A) + P(B) - P(AB)$

This last relationship is not a contradiction of the axioms of probability measure, but instead it is a more general case of axiom three.

It might be worthwhile to note at this point that:

1. Events are combined and operated upon only in accordance with the seven axioms of the algebra of events.

2. The probabilities of events are numbers and can be computed only in accordance with the three axioms of probability measure and the valid relations which can be derived therefrom.

3. Arithmetic is something else.

As an illustration using our coin toss experiment, a probability of 1/8 is assigned to each sample point. If we define events $A = (H_1 H_2 H_3)$, $B = (T_1 T_2 T_3)$

$P(A+B) = P(A) + P(B) = 1/4$

$P(AB) = P(\phi) = 0$

$P(A+B') = P(A) + P(B') - P(AB') = 7/8$

A.5.1 Conditional Probability

Conditional probabilities are obtained when one determines the probability of some event, subject to the outcome of another event. As an example consider the experiment which consists of tossing a single die twice, noting the number of dots on the top up face on each toss. Let x_1 identify the value of the up face on the first toss, and x_2 the value of the up face on the second toss.

Each sample point defined by the pair (x_1, x_2) of this experiment has a probability of occurrence of 1/36. Assuming the first toss resulted in a five, called event B, we may want to determine the probability that the sum of both tosses resulted in either a 7 or an 11. (Define event A to be $x_1 + x_1 = 7$ or $x_1 + x_2 = 11$.) If we had no knowledge regarding the outcome of the experiment P(A), the probability of event A, is determined by summing the probabilities of all the sample points contained in event A, or P(A) = 8/36. However, given that event B has occurred we are only interested in that portion of event A which is included in event B. In other words, to determine P(A/B), the conditional probability of event A given event B has occurred, we must consider event B to be the universal event. In order to meet this requirement the probabilities of all points in event B must sum to one; therefore, it is necessary to multiply all the sample points contained in event B by the factor 1/P(B). The desired measure of probability, P(A/B) is now obtained by summing all the conditional probabilities of the points in event AB, the intersection of events A and B. Working with the sample space as P(B) = 6/36 we obtain 1/P(B) = 6 so that each point in the conditional sample space has a probability of 1/6. Summing the probability of all points in event AB leads to P(A/B) = 12/36.

The definition of the conditional probability of event a given event B has occurred is then

$$P(A/B) = \frac{P(AB)}{P(B)} \qquad P(B) \neq 0$$

A.5.2 Interdependence of Events

Intuitively one might consider two events to be independent if knowledge as to whether or not the experimental outcome had attribute B would not affect our measure of the likelihood of an event with attribute A. Formally, two events A and B are defined to be independent if and only if

$$P(A/B) = P(A) \qquad P(B) \neq 0$$

or alternatively

$$P(AB) = P(A)P(B)$$

Extending this definition to n events, $A_1, A_2, A_3, \ldots, A_n$ n events $A_1, A_2, \ldots A_n$ are defined to be mutually independent if and only if

$$P(A_i/A_j, A_k, \ldots, A_n) = P(A_i) \quad i \neq j, k, \ldots, n$$
$$1 \leq i \leq i, j, k, \ldots, n$$

which is equivalent to requiring that the probabilities of all possible intersections of these different events be equal to the products of the individual event probabilities.

A.5.3 Bayes Theorem

The relation known as Bayes Theorem results from a particular application of the definition of conditional probability. The theorem is used extensively where many mutually exclusive events (A_i) may result in the occurrence of some event B to happen. Then given event B has occurred, one is able to determine in terms of probabilistic measure the likelihood that a particular A_i was the cause.

Let A_1, A_2, ..., A_n be a set of n mutually exclusive and collectively exhaustive events which comprise the sample space. Further, let's define another event B over a portion of the same sample space, such as shown in the Venn diagram (Figure A.5).

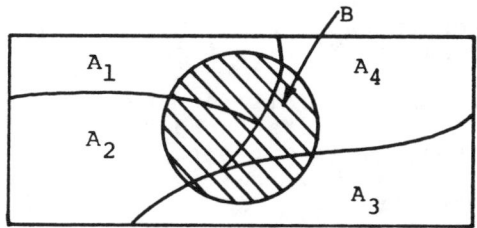

FIGURE A.5: VENN DIAGRAM OF BAYES THEOREM

As long as the conditioning events are not of probability zero, we have

$$P(A_i B) = P(B/A_i)P(A_i) = P(A_i/B)P(B)$$

Suppose now that we know all $P(A_i)$ and $P(B)$ so that the $P(B/A_i)$ can be obtained for $1 \leq i \leq n$. It is now desired to obtain $P(A_i/B)$'s. By application of the definition of conditional probability one obtains:

$$P(A_i/B) = \frac{P(A_iB)}{P(B)} \text{ and } P(B/A_i) = \frac{P(A_iB)}{P(A_i)}$$

since

$$P(B) = P(AB) = P[(A_1+A_2+\ldots,A_n)B]$$

$$= \sum_{i=1}^{n} P(A_iB) = \sum_{i=1}^{n} P(A_i)P(B/A_i)$$

and substituting this expression for $P(B)$ into the equation for $P(A_i/B)$ there results

Bayes Theorem:

$$P(A_i/B) = \frac{P(A_i)P(B/A_i)}{\sum_{i=1}^{n} P(A_i)P(B/A_i)} \qquad \begin{array}{l} P(B) \neq 0 \\ \\ A_1+A_2+\ldots,A_n = u \end{array}$$

An example of a situation which makes use of this theorem follows: let event A_i represent that a particular component was manufactured by firm i. Let B represent the event that the component fails during the first year of operation. The quantity $P(B/A_i)$ then represents the event that a component will fail during the first year of operation, given it was manufactured by firm i. The question of interest is then, given a component which did fail during the first year of operation, what is the probability that it was manufactured by firm i?

Returning to Bayes Theorem:

$$P(A/B) = \frac{P(A)P(B/A)}{P(B)}$$

where

$$P(B) = P(A)P(B/A) + P(A')P(B/A')$$

an A and A' are the events A and not A. In this application $P(A)$ and $PA')$ are usually a priori or prior known probabilities of the events A and A'. It is then necessary to modify $P(A)$ or $P(A')$ on the basis that event B has occurred and that the outcome of the experiment is known to be influenced by the occurrence of events A or A'.

Considering an example of parts manufactured by A, B,

and C respectively, discussed before:

Test Outcome	Manufacturer			Total
	A	B	C	
Pass P	50	16	14	80
Fail F	10	4	6	20
Totals	60	20	20	100

Assuming that a part is selected at random from among 100 parts and tested, we may be interested to know the probability that the part was manufactured by A if it is found to pass the test, $\underline{P}(A/P)$.

$\underline{P}(A)$, the a priori probability of choosing a part manufactured by A, is $60/100 = 6/10$. Similarly the conditional probability that a part manufactured by A will pass $\underline{P}(P/A) = 50/60 = 5/6$ and the conditional probability that a part manufactured by A will not pass $\underline{P}(F/A) = 10/60 = 1/6$. These are the a priori conditional probabilities. We can now compute

$$\underline{P}(A/P) = \frac{\underline{P}(A)\underline{P}(P/A)}{\underline{P}(A)\underline{P}(P/A) + \underline{P}(A')\underline{P}(P/A')}$$

$$= \frac{6/10 \times 5/6}{(6/10 \times 5/6) + (4/10 \times 3/4)} = \frac{5/10}{5/10 \times 3/10} = 5/8$$

there the a priori probability that a part manufactured by A was chosen is modified to the a posteriori probability $\underline{P}(A/P) = 5/8$ on the basis of the information that the part chosen had passed the test. In other words the a posteriori probability 5/8 is found to be different from the a priori probability of choosing a part from the sample.

A.6 Random Variables

Often we have reasons to associate one or more numbers (in addition to probabilities) with each possible outcome of an experiment. In order to study these instances where outcomes may be specified numerically, the following definition is introduced:

> A random variable is defined by a function which assigns a value of the random variable to each sample point in the sample space.

As an example, consider again the coin toss experiment. Of the many random variables we may wish to define on this sample space, let's choose the following:

x = total number of heads resulting from the three flips

y = length of the longest run resulting from the three flips (a run is a set of successive flips which all have the same outcome)

The sample space for this experiment along with the values of random variables is listed below.

Sample Points	Probability	x	y
H_1 H_2 H_3	1/8	3	3
H_1 H_2 T_e	1/8	2	2
H_1 T_2 H_3	1/8	2	1
H_1 T_2 T_3	1/8	1	2
T_1 H_2 H_3	1/8	2	2
T_1 H_2 T_3	1/8	1	1
T_1 T_2 H_3	1/8	1	2
T_1 T_2 T_3	1/8	0	3

One may now ask, "What is the probability of the random variable x taking on a value of 2?" This is equivalent to asking for the probability of obtaining exactly two heads. The result 4/8 is obtained by summing the probabilities of the sample points for which $x = 2$.

The set of values assigned to the random variable is termed the probability function of that particular variable.

$f_x(x_0)$ = probability (that an experimental value of the random variable x obtained on the performance of the experiment is equal to x)

The probability functions for x and y are listed below:

$f_x(0) = P(x=0) = 1/8$

$f_x(1) = P(x=1) = 3/8$

$f_x(2) = P(x=s) = 3/8$

$f_x(3) = P(x=3) = 1/8$

$f_x(1) = P(y=1) = 2/8$

$f_x(2) = P(y=s) = 4/8$

$$f_y(3) = P(y=3) = 2/8$$

It is noted here that for the discrete case (where the random variable may only take on discrete values) the axioms of probability require

$$\sum_{x_o} f_x(x_o) = 1$$

and

$$0 \leq f_x(x_o) \leq 1$$

These conditions are met by the probability functions $f_x(x_o)$ and $f_y(y_o)$.

A.7 Compound Probability Functions

Considering the discussion above, one might consider situations in which values of more than one random variable are assigned to each point. We will again utilize the random variables x_o and y_o in our discussion, but the extension to more than two variables is apparent.

The compound (or joint) probability function is defined:

$f_{x,y}(x_o, y_o)$ = probability that the experimental values taken on by random variables x and y on the performance of an experiment are equal to x_o and y_o respectively

A graphical representation of this function could have possible points marked on an x_o, y_o axis, with each value of $f_{x,y}(x_o, y_o)$ represented as a point above the x_o, y_o plane and perpendicular to it. Frequently a two-dimensional plot is made, whereon each point representing an outcome is labeled with its corresponding value of probability. The individual probabilities are again obtained by summing the probabilities of all points with that particular attribute. Since this joint probability function is still a probability function, the probabilities for all the possible outcomes must sum to one, namely

$$\sum_{x_o} \sum_{y_o} f_{x,y}(x_o, y_o) = 1$$

It is also seen that if the summation is performed over just one random variable, the result is the probability function for the remaining variable (in this case called the marginal probability function). Thus

$$\sum_{y_o} f_{x,y}(x_o, y_o) = f_x(x_o), \quad \sum_{x_o} f_{x,y}(x_o, y_o) = f_y(y_o)$$

It is important to note that in general there is no way to go back from the marginal probability function to the joint probability function.

The probability of any event described in terms of the values of random variables x and y may be easily found, now that the joint probability function $f_{x,y}(x_o, y_o)$ has been determined. The reader may verify the following results by referring to the joint sample space.

$$P \text{ (run of 3)} = f_{x,y}(x_o, 3) = 2/8$$

$$P \text{ (run of 2 and 2 heads)} = f_{x,y}(2,2) = 2/8$$

$$P \text{ (3 heads)} = f_{x,y}(3, y_o) = 1/8$$

A.8 Expectation

The expectation or expected value of a function $g(x)$ is defined as

$$E[g(x)] = \sum_{x_o} g(x_o) f_x(x_o) = \overline{g(x)}$$

By definition then, the expected value of a function for the discrete case is just the sum of the products of the value of the function multiplied by the probability of occurrence of the function at that same value. Considering any discrete probability function, $f_x(x_i)$ of a random variable x_i, and assuming it is desired to obtain the expected value of the function, one has the notion that the expected value is somehow representative of all the values that the function can assume. If we wish to make the analogy at this point that the values of the probability function are masses, then the expected value could be interpreted as the center of mass of the system. Further, if $E(x)$ is the center of mass it would be reasoned that the system must be in equilibrium with itself about that point. This implies then, that upon summing the moments of all the points in the system about its expected value, one will obtain a result of zero. Applying this reasoning to the graph one obtains:

$$f_x(x_1) \cdot (x_1 - \overline{x}) + f_x(x_2) \cdot (x_2 - \overline{x}) + \ldots + f_x(x_n) \cdot (x_n - \overline{x}) = 0$$

where in each product the first term is the probability function evaluated at x_i, and the second term is the distance from x_i to the expected value. Writing this sum in a more convenient form:

$$\sum_{x_i} f_x(x_i) \cdot (x_i - \bar{x}) = \sum_{x_i} f_x(x_i) x_i - \sum_{x_i} f_x(x_i) \bar{x} = 0$$

since $\bar{x}$ is a constant

$$\sum_{x_i} f_x(x_i) x_i - \bar{x} \sum_{x_i} f_x(x_i) = 0$$

and recalling that

$$\sum_{x_i} f_x(x_i) = 1$$

$$\bar{x} = E(x) = \sum_{x_i} x_i f(x_i)$$

It now becomes obvious why the E(x) is often referred to as the first moment of x.

The expected values of other functions may also be obtained. A few of these, along with some additional definitions, are listed.

$E(x^n)$ = n^{th} moment of x

$E[(x-\bar{x})^n]$ = n^{th} central moment of x

σ_x^2 = $E[(x-\bar{x})^2] = E(x^2) - E^2(x)$ = variance of x

σ_x = $\sqrt{\sigma_x^2}$ = standard deviation of x = $S.D._x$

A.9 Continuous Random Variables

Our discussions of the probability functions and the associated relationships have thus far considered the discrete case in which we were limited to random variables which could only assume values at individual points, i.e., p, 1, 2, 3,...etc. It is appropriate now to extend our introduction to probability to the continuous case, in which the random variable may assume values over continuous intervals, i.e.,

$-\infty \leq x_o \leq \infty$, $0 \leq x_o \leq 10$, $x_o > 12.5$

Rather than develop again the various probability

relationships, it is noted here that a similarity exists between the discrete and the continuous probability functions, namely, that wherever summation signs appear for the discrete case, they are normally replaced by the appropriate integral signs in the continuous case. The analogy will become apparent in the following paragraphs.

First let's look at a graph of a continuous probability function (often called a probability density function or PDF, (Figure A.6) since the values of probability are obtained by computing the area under the curve.

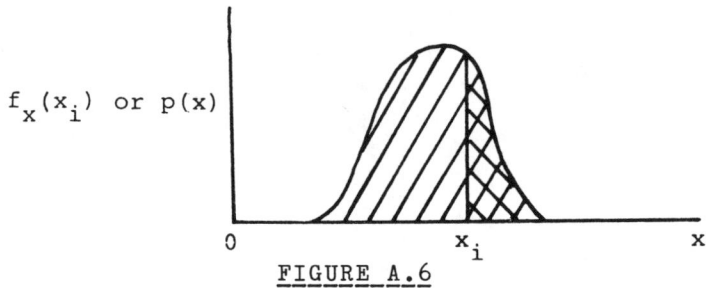

FIGURE A.6

The probability density function for the continuous random variable x is defined by

$$\text{Prob}(a \le x \le b) = \int_a^b f_X(x_o) dx_o$$

Fulfilling the requirements of the axioms of probability we obtain

$$\int_{-\infty}^{\infty} f_X(x_0) dx_o = 1 \text{ and } 0 \le f_X(x_o) \le \infty$$

Note that the probability that x takes on a specific value x_o is equal to zero because although $f_X(x_o)$ may have a non-zero value, the integral of $f_X(x_o)$ over an interval of zero width is zero.

Next we define the cumulative density function, CDF, for continuous random variable x by

$$P(x \le x_o) = F_X(x_o) = \int_{-\infty}^{x_o} f_X(x_o) dx_o$$

The probability represented by the CDF is the probability that x will take on a value less than or equal to x_o. The properties of the CDF are:

$$P(x \leq \infty) = 1$$

$$P(a < x \leq b) = P(x \leq b) - P(x \leq a)$$

$$\frac{d}{dx} P(x \leq x_o) = f_x(x_o)$$

The relationship between a probability density function and its associated cumulative function is illustrated in Figure A.7.

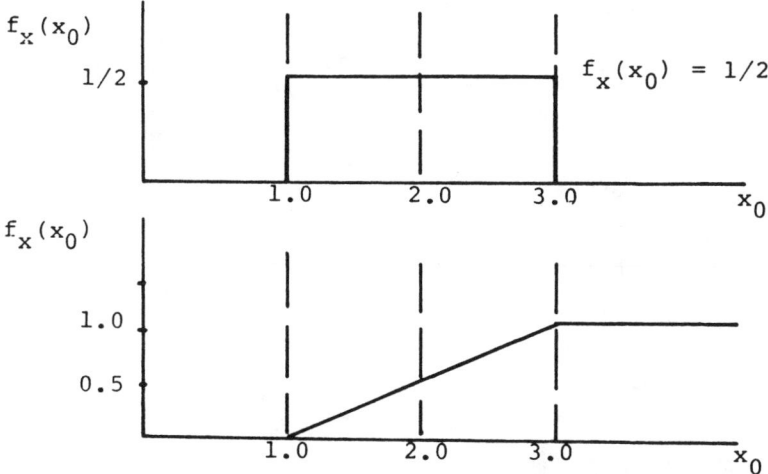

FIGURE A.7: PROBABILITY DENSITY AND CUMULATIVE FUNCTIONS

The joint probability density function is defined as:

$$\int_{x_o} \int_{y_o} f_{x,y}(x_o, y_o) \, dx_o \, dy_o$$

and if the marginal probability density function is desired it is only necessary to integrate over one of the random variables. For example,

$$f_x(x_o) = \int_{y_o} f_{x,y}(x_o, y_o) \, dy_o, \quad f_y(y_o) = \int_{x_o} f_{x,y}(x_o, y_o) \, dx_o$$

The expected value of a continuous function is given as

$$E[g(x)] = \int_{-\infty}^{\infty} g(x) f_x(x_o) \, dx_o$$

A.10 Discrete Probability Distribution of a Single Variable

Considering a discrete random variable x_i which assumes n different values with frequencies f_i (i=1,...,n), where

$$\sum_{i=1}^{n} f_i = N$$

is the total number of trials or data points, then the average value of the variable is:

$$\mu_1^1 = \bar{x} = \frac{1}{N}(f_1 x_1 + f_2 x_2 + \ldots + f_n x_n) = \frac{1}{N}\left(\sum_{i=1}^{n} f_i x_i\right)$$

To find the moments of the variable about the average or mean, we expand the expression for the rth moment about the mean by the binomial theorem:

$$\mu_r = \frac{1}{N}\Sigma(x_i - \bar{x})^r f_i = \mu_r^1 - \binom{r}{1}\mu_{r-1}^1 \bar{x} + \binom{r}{2}\mu_{r-2}^1 \bar{x}^2 \ldots$$

$$= E[(x_i - \bar{x})^r] = \text{expected value of } (x_i - \bar{x})^r$$

where

$$\mu_r^1 = \frac{1}{N}\Sigma x_i^r f_i = r^{th} \text{ moment of the variable}$$

$$= E[x_i^r]$$

about the origin of the variable x_i. Using the above expression, we find that the variance

$$\mu_2 = \frac{1}{N}\Sigma(x_i - \bar{x})^2 f_i = \frac{1}{N}\Sigma x_i^2 f_i - \bar{x}^2 = \mu_2^1 - \bar{x}^2$$

In general, we will use the relative frequency

$$f(x_i) = \frac{f_i}{N}$$

instead of the frequency when we can write

$$\sum_{i=1}^{n} f(x_i) = 1, \quad f(x_o) \simeq P_i$$

where

$$\lim n = \infty, \text{ or } n \to \infty$$

where

$p_i = P(\xi = x_i)$ = probability that the variable assumes the value x_i

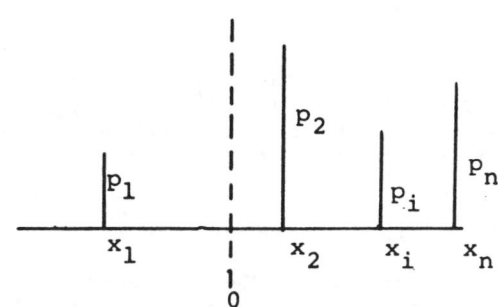

FIGURE A.8: FREQUENCY DISTRIBUTION

Another important concept is the distribution function:

$$F(x_k) = P(\xi \le x_k) = \sum_{i=1}^{k} p_i = \sum_{i=1}^{k} P(\xi = x_i)$$

$$P(a < \xi \le b) = F(b) - F(a)$$

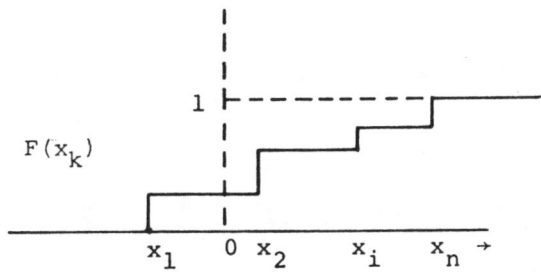

FIGURE A.9: CUMULATIVE OR COMPOUND DISTRIBUTION

The moments of the distribution of a variable are the statistical measures. Depending on the distribution which may or may not follow one of the large number of standard or defined distributions, one, two, or more of these moments are required to describe the distribution, the dispersion of the variable, and the closeness of the distribution to the normal or Gaussian curve with the same moment.

A measure of closeness to the normal curve is the ratio of

$$\frac{\mu_4}{\mu_2^2} = \beta = \text{Kurtosis}$$

This follows from the fact that β is always equal to three for a normal distribution, no matter how large its variance. An important (though conservative) measure of dispersion or closeness to the mean for any type of distribution is Chebycheff's Theorem:

> "If x_i is a discrete random variable with mean $\bar{x}$ and a standard deviation σ, then for any positive constant k, the probability that x_i assumes a value less than $(\bar{x}-k\sigma)$ or greater than $(\bar{x}+k\sigma)$ is less than $1/k^2$."

Or, $P(|X_i - \bar{x}| > K\sigma) < 1/k^2$

This is easily shown by considering that:

$$\sigma^2 = \sum_i (x_i - \bar{x})^2 f(x_i)$$

$$= \sum_{-\infty}^{\bar{x}-k\sigma} (x_i - \bar{x})^2 f(x_i) + \sum_{\bar{x}-k\sigma}^{\bar{x}+k\sigma} (x_i - \bar{x})^2 f(x_i) + \sum_{\bar{x}+k\sigma}^{\infty} (x_i - \bar{x})^2 f(x_i),$$

or

$$\sigma^2 > \sum_{-\infty}^{\bar{x}-k\sigma} k^2 \sigma^2 f(x_i) + \sum_{\bar{x}+k\sigma}^{\infty} k^2 \sigma^2 f(x_i);$$

and finally,

$$\frac{1}{k^2} > \sum_{-\infty}^{\bar{x}-k\sigma} f(x_i) + \sum_{\bar{x}+k\sigma}^{+\infty} f(x_i)$$

Two of the most important discrete distributions are the binomial and Poisson Distributions.

A.10.1 Standard Discrete Distributions

1. Binomial Distribution

The binomial distribution is based on repeated independent trials in which the probability of success of each trial remains constant. If we consider the number of successes x in an experiment of n trials as our random variable and assume a given probability of success p (or failure $q=1-p$), then the probability density function of the binomial distribution is:

$$f(x; n, p) = f(x) = \binom{n}{x} p^x q^{n-x} = \binom{n}{x} p^x (1-p)^{n-x}$$

If the probability distribution function is desired:

$$F(x; n, p) = \sum_{0}^{x} f(\xi) = \sum_{\xi=0}^{x} \binom{n}{\xi} p^\xi q^{n-\xi}$$

= probability of at least x successes in n trials.

To find the different statistical measures of the binomial distribution,

$$\mu_1^1 = E(x) = \sum_{0}^{n} x\, f(x) = np(q+p)^{n-1} = np$$

Similarly, we obtain:

$$\mu_2^1 = E(x^2) = \sum_{0}^{n} x^2\, f(x) = np[(n-1)p+1]$$

from which we find the variance

$$\mu_2 = E[(x-\bar{x})^2] = \sum_{0}^{n} (x-\bar{x})^2 f(x) = \mu_2^1 - [E(x)]^2$$

$$= np[1+(n-1)p] - (np)^2 = npq$$

To find higher moments, the use of the moment generating function

$$M_0(t) = \sum_{0}^{n} e^{tx} f(x), \text{ cumulant } \log M_0(t),$$

or a transform method is recommended.

Using the M.G. function and expanding,

$$M_0(t) = \sum_0^n e^{tx}\binom{n}{x} p^x q^{n-x} = [pe^{-t} + q]^n$$

$$= 1 + npt + [np + n(n-1)p]\frac{t^2}{2} + (np + 3n(n-1)p^2 + n(n-1)(n-2)p^3]\frac{t^3}{3!}..$$

from which we deduce that

$$\mu_2 = npq \quad \mu_3 = npq(q-p) \quad \mu_4 = npq[1 + 3(n-2)pq]$$

by finding all the μ_r^1 and taking the terms whose coefficients are $\frac{t^r}{r!}$.

2. Hypergeometric Distribution

If, unlike in the binomial distribution, our trials take place without replacement with a consequent change in probabilities of success, we first require knowledge of the distribution of successful (good) and failure (bad) events in the total number of elements. Assuming that we intend to draw n items from a total set of N=g+b above where g is the total number of good and b is the total number of bad elements contained in the set, then the probability of drawing x good samples in a total draw of n is:

$$f(x; n, g, b) - f(x) = \frac{\binom{g}{x}\binom{b}{n-x}}{\binom{g+b}{n}} \quad x = 0, 1, 2, \ldots, n.$$

Similarly, the probability distribution function becomes:

$$F(x; n, g, b) = \sum_{\xi=0}^x f(\xi) = \sum_{\xi=0}^x \frac{\binom{g}{\xi}\binom{b}{n-\xi}}{\binom{g+b}{n}}$$

3. Poisson Distribution

This distribution is the limiting form of the binomial distribution when $n \to \infty$, $p \to 0$, and np becomes a constant. If we write $\frac{m}{n} = p$, then in the limit:

$$\lim \binom{n}{x} p^x q^{n-x} = f(x_i m) = f(x) = \frac{m^x e^{-m}}{x!}$$

for
$$m > 0$$

$$\sum_{x=0}^{\infty} f(x) = e^{-m} \sum_{x=0}^{\infty} \frac{m^x}{x!} = e^{-m} e^m = 1$$

The probability distribution function is:

$$F(x) = \sum_{\xi=0}^{x} f(\xi) = \sum_{\xi=0}^{x} \frac{m^\xi e^{-m}}{\xi!}$$

and the moments are:

$$\bar{x} = \mu_1^1 = m e^{-m} (1 + m + \frac{m^2}{2!} + \ldots) = m e^{-m} e^m = m$$

$$\mu_2^1 = e^{-m} (0^2 + 1^2 m + 2^2 \frac{m^2}{2!} + \ldots) = m(m+1)$$

from which we obtain:

$$\mu_2 = \mu_2^1 - \bar{x}^2 = m(m+1) - m^2 = m$$

These results could have been obtained by using the M.G. function:

$$M_0(t) = e^{-m}(e^{0t} + e^t m + e^{2t} m^2/2! + \ldots) = \exp[m(e^t - 1)],$$

or cumulant function

$$K(t) = \log M_0(t) = m(e^t - 1) = m(t + t^2/2! + t^3/3! + \ldots)$$

from which

$$\mu_1^1 = \mu_2 = \mu_3 = m \qquad \mu_4 = m + 3 m^2, \text{ etc.}$$

4. Negative Binomial Distribution

An important probability distribution in reliability theory is the negative binomial distribution which is similar to the binomial, except for the fact that the total number of trials n=x is the variable. In other words, we require the probability that we obtain the rth success on the xth trial (r is a constant).

$$f(x; r, p) = \binom{x-1}{r-1} p^r (1-p)^{x-r} = \binom{x-1}{r-1} p^r q^{x-r}$$

for $x = r, r+1, r+2, \ldots$

$$F(x; r, p) = \sum_{\xi=r}^{x} \binom{\xi-1}{r-1} p^r (1-p)^{\xi-r}$$

= probability of obtaining r successes in at most x trials

Examples of Discrete Distributions

<u>Hypergeometric</u> distributions are used when there are only two possible outcomes: there is a finite population size; and, sampling is performed without replacement, i.e. dependent size. Let us assume that we have lots of 40 parts, of which on the average 5 are defective. If 5 parts are sampled from a lot, what is the probability of (1) exactly two defectives, (2) two or fewer defectives, or (3) more than 2 defectives.

(1) $f(2) = \dfrac{\binom{5}{2}\binom{35}{3}}{\binom{40}{5}} = \dfrac{\frac{5!}{2!3!} \cdot \frac{35!}{3!32!}}{\frac{40!}{5!35!}} = \dfrac{10.6545}{658008} = .09946..$

(2) $F(2) = \sum\limits_{i=0}^{2} \dfrac{\binom{5}{i}\binom{35}{5-i}}{\binom{40}{5}} = \dfrac{\binom{35}{5}}{\binom{40}{5}} \dfrac{\binom{5}{1}\binom{35}{4}}{\binom{40}{5}} + \dfrac{\binom{5}{2}\binom{35}{3}}{\binom{40}{5}}$

$= .493 + .398 + .099 = .990$

(3) Probability of more than two defectives

$F(x>2) = 1 - F(2) = 1 - .99 = .01$

Binomial distribution requires also that there are two possible outcomes, that the probability of each outcome is constant for all trials, and that the trials are independent, which implies that there is no replacement during sampling. Using the example of 40 parts in a lot with an average of 5 defective parts in a lot, then

Probability of 2 or less defectives found in such a lot of 40 chosen randomly from among the lots is

$$F(x \leq 2) = \sum_{i=0}^{2} \binom{40}{1} q^i p^{40-i} = .108$$

where

q = probability of defective = $\frac{5}{40} = \frac{1}{8}$

$1-q = p$ = probability of non-defective = $\frac{7}{8}$

Poisson distribution is often used as an approximation of the binomial distribution or as the distribution of numbers of independent discrete outcomes. Assuming that we select 25 parts from a large lot in which an average of 10% are defects, the probability of two defective parts being found in the sample is then

$$F(x=2) = F(2) = \frac{e^{-(25)(0.1)}[25 \times 0.1]^2}{2!}$$

$$= 0.2565$$

The corresponding binomial probability would be

$$F(2) = \binom{25}{2}(0.1)^2 (0.9)^{23} = 0.2659$$

A.10.2 Continuous Probability Distributions

There are a number of continuous statistical or probability distributions of interest to reliability analysis.

1. Exponential Distribution

The exponential distribution is a special case of the gamma distribution in which the random variable is the time to the first event or outcome. It is primarily used to compute waiting times or times to first failure and assumes that the time to the event (first failure, etc.) is exponentially distributed. The pdf of the exponential distribution is

$$\phi(t) = \lambda e^{-\lambda t}$$

where

$$t \geq 0$$

and

$$\frac{1}{\lambda} > 0$$

The mean and variance of the exponential distribution is

$$\bar{t} = \frac{1}{\lambda}$$

$$\sigma_t^2 = \frac{1}{\lambda^2}$$

where λ is the characteristic coefficient of the exponential distribution.

The exponential function is often taken to represent the pdf of failure of a system. The probability distribution function of failure is then

$$F(t) = \int_0^t \phi(\tau) d\tau = \text{probability of failure in the time interval o to t}$$

and

$$1-F(t) = R(t) = \int_t^\infty \phi(\tau) d\tau = e^{-\lambda t}$$

is then the probability of non-failure in the time interval o to t or the reliability of the system. The mean time to failure is obviously then the mean of the exponential distribution or $1/\lambda$ and the probability of non-failure to the mean life ($=1/\lambda$) is then

$$R(t = \frac{1}{\lambda}) = e^{-\lambda/\lambda} = e^{-1} = 0.368$$

If the reliability over a time interval t is given as $R(t)$, then the mean or expected life before failure can be found from the equation

$$\frac{1}{\lambda} = \frac{-t}{\log R(t)}$$

When the exponential distribution is used as described above, with time as the random variable, then the process is often termed a 'Poisson process'. The exponential distribution is usually assumed to represent failure rate distribution functions, but this assumption must be validated by analysis of failure statistics for soundness of fit using Chi Square and Kolmogorov and other statistical tests.

2. Weibull Distribution

The Weibull distribution is often used to represent failure rate distributions. It is a two parameter distribution with a high degree of flexibility. It uses a scale parameter, a, and a shape parameter, b. When b=1 the Weibull distribution reduces to an exponential distribution. The pdf of the two parameter Weibull distribution is

$$f(t) = \frac{b}{a} t^{b-1} e^{-\frac{1}{a}t^b}$$

$$\text{Mean} = a^{1/b} \Gamma(\frac{1}{b} + 1)$$

$$\text{Variance} = a^{2/b} \cdot \{\Gamma(1+2/b) + \Gamma^2(1+1/b)\}$$

3. Gamma Distribution

The Gamma distribution is also extremely flexible, and the estimated mean life of an exponential distribution has a Gamma density, which allows probability statements for estimates and tests of mean or expected life.

The coefficient a is again a scale parameter, while b is a critical parameter controlling the shape of the distribution. The pdf of the Gamma distribution is

$$f(t) = \frac{e^{-t/a} \cdot t^{b-1}}{(b-1)! a^b} \qquad \begin{array}{l} t \geq 0 \\ a \geq 0 \\ b > 0 \end{array}$$

$$\text{Mean} = ab$$

$$\text{Variance} = b/a^2$$

4. Normal Distribution

This is one of the most widely used distributions with a symmetrically shaped pdf curve. It is usually used as a standard of comparison.

The pdf of the normal distribution is

$$f(t) = \frac{1}{\sigma\sqrt{2\pi}} e^{-(t-\bar{t})^2/2\sigma^2}$$

with
$$-\infty < t < +\infty$$

and*

Mean $= \bar{t}$

Variance $= \sigma^2$

The probability distribution function of the normal distribution is not directly integrable and therefore tables of the normal cumulative distribution function with a mean of zero and a variance of one called the standard normal distribution are usually used. These tables can be used for any normal distribution by transforming the original variable t to a new variable y where

$$y = \frac{t - \bar{t}}{\sigma}$$

The variable y is also normally distributed with a mean of zero and a variance of one. For example, assuming that when $t = 100$ and $\sigma = 5$, we want to find the probability that $95 \leq t \leq 110$ on a single trial

$$y_1 = \frac{95 - 100}{5} = -1$$

$$y_2 = \frac{110 - 100}{5} = 2$$

$$P(95 \leq t \leq 110) = P(-1 \leq y \leq 2)$$
$$= F(2) - F(-1) = 0.977 - 0.159$$
$$= 0.818$$

(Refer to the tables at the end of the book for verification.)

An important characteristic of the normal distribution in reliability analysis is that the mean life and median life are each equal to the time interval for which the reliability of the system is equal to $R(t) = 0.5$.

5. **Log-Normal Distribution**

The log normal distribution applies to variables whose log is normally distributed. The pdf of the log-normal distribution is positively skewed. The pdf of the log-normal distribution is,

$$f(t) = \frac{1}{at\sqrt{2\pi}} e^{-\frac{(\log_e t - b)^2}{2a^2}}$$

---

* Formally speaking, mean and variance cannot be defined for normal distribution.

where

$t > 0$ and $a, b > 0$

Mean $= e^{b+a^2/2}$

Variance $= e^{2b+a^2}(e^{a^2}-1)$

There are many other important probability distributions such as the Chi Square and t distributions which are often used for validation or test of statistical or probability distributions.

A.11 Statistical Sampling

In order to draw statistical inference about a population, a sample is usually drawn from the population which is significantly smaller than the population. This introduces the problem of a best sampling procedure. In general, we are interested in choosing a sampling procedure which yields an unbiased estimate with a minimum variance. The cost of sampling is also important and adds a conflicting constraint on the size of the required sample. In the first case, a sample may be counted more than once. If we draw a simple random sample x out of a population of N members where $x = (x_1, x_2 \ldots x_n)$, then

$$\bar{x} = \text{expected value of } x_i = \frac{1}{n} \sum_{i=1}^{n} x_i$$

If different samples of $x = (x_1, x_2 \ldots x_n)$ are drawn where $0 < n < N$, then the expected value of $\bar{x}$ overall possible samples is:

$$E(\bar{x}) = \mu$$

The sample variance is defined by

$$\sigma^2 = \frac{1}{n} \sum_{i=1}^{n} (x_i - \bar{x})^2$$

but this is not an unbiased estimate of the Var (x). The unbiased estimate of Variance x is:

$$\sigma^2 = \frac{1}{(n-1)} \sum_{i=1}^{n} (x_i - \bar{x})^2 = \text{Var}(x)$$

The sample mean $\bar{x}$ is also a random variable, whose mean is and whose variance is:

$$\text{Var}(\bar{x}) = (1 - \frac{n}{N}) \frac{\text{Var}(x)}{n}$$

If $N \to \infty$, $\text{Var}(\bar{x}) = \frac{\text{Var}(x)}{n}$

If our experiment consists of statistical sampling from a population N divided into k strata when

$$\sum_{j=1}^{k} N_j = N$$

and if a simple random sample n_j is taken from the jth strata of population N_j when

$$\sum_{j=1}^{k} n_j = n$$

where n is the total sample, then μ_k is the true mean of the kth strata and $\bar{x}_k$ is the observed mean of the sample n_k drawn from N_k, the kth strata.

The unbiased estimate of the true mean is:

$$\bar{\mu} = \frac{1}{N} = \sum_{j=1}^{k} N_j \bar{x}_j$$

and

$$\text{Var}(\mu) = \sum_{j=1}^{k} (\frac{N_j}{N})^2 (\frac{N_j - n_j}{N_j}) \frac{\text{Var}(x_j)}{n_j}$$

If population consists of one strata or if k=1, $N_j = N$, and $n_j = n$, then

$$\text{Var}(\bar{\mu}) = \frac{N-n}{N} \frac{\text{Var}(x)}{n} = (1 - n/N) \frac{\text{Var}(x)}{n}$$

as before.

If we draw a proportional stratified sample,

$$\frac{N_j}{N} = \frac{n_j}{n}$$

then

$$\bar{\mu} = \sum_j \frac{n_j \bar{x}_j}{n}$$

and

$$\text{Var}(\bar{\mu}) = \left(1 - \frac{n}{N}\right) \frac{1}{nN} \sum_{j=1}^{k} N_j \text{Var}(x_j)$$

If we want to minimize the variance for the estimate μ in a stratified sampling experiment, then we minimize

$$\text{Var}(\bar{\mu}) = \frac{1}{N^2} \sum_{j=1}^{k} \left(\frac{N_j^2}{n_j} - N_j\right) \text{Var}(x_j)$$

subject to the constraint

$$\sum_{j=1}^{k} n_j = n$$

Using the Lagrange equations

$$\frac{\partial \text{Var}(\bar{\mu})}{\partial n_j} + \lambda \frac{\partial \phi}{\partial n_j} = 0$$

where

$$\phi = \sum_{j=1}^{k} n_j - n = 0$$

we get

$$\frac{-N_j^2}{n_j^2} \text{Var}(x_j) + \lambda = 0 \text{ for all } j$$

or

$$\frac{N_j}{n_j} \sqrt{\text{Var}(x_j)} = \frac{N_{j+1}}{n_{j+1}} \sqrt{\text{Var}(x_{j+1})} \quad \ldots$$

and

$$n_j = \frac{N_j \sqrt{Var(x_j)}}{\sqrt{\lambda}}$$

but

$$\sum_{j=1}^{k} n_j = n = \frac{\sum_{j=1}^{k}[N_j \sqrt{Var(x_j)}]}{\sqrt{\lambda}}$$

or

$$\sqrt{\lambda} = \sum_{j=1}^{n}[N_j(\sqrt{Var(x_j)}]\frac{1}{n}$$

and

$$\frac{n_j}{n} = \frac{N_j \sqrt{Var(x_j)}}{\sum_{j=1}^{k}[N_j(\sqrt{Var(x_j)}]}$$

APPENDIX B - MATRIX ALGEBRA AND TRANSFORMATIONS

Ernst G. Frankel

B.1 Introduction to Matrix Algebra

B.1.1 Basic Concepts and Definitions

When a problem has many states, we obtain many state variables. In that case it is convenient to analyze a problem by determining the results by manipulating vectors and matrices instead of individual variables. We can define an n column vector as the set of n numbers, possibly complex, written in the form

$$\underline{x} = \begin{bmatrix} x_1 \\ x_2 \\ \cdot \\ \cdot \\ \cdot \\ x_n \end{bmatrix} \tag{B.1}$$

where the variables x_i, for $i = 1,2....n$ will be called the components of the vector $\underline{x}$.

The number n is called the dimension of the vector $\underline{x}$. If the dimension of $\underline{x}$ is unity, i.e. n=1, the vector is scalar and the vector simply consists of a variable. If all the variables in a vector are real, then we can say that the vector $\underline{x}$ is real.

A row vector $\underline{x}'$ is defined as the set of n numbers, possibly complex, written in the form:

$$x' = (x_1, x_2 ... x_n), \tag{B.2}$$

A row vector again consists of the variables x_i for $i=1,2...n$ where n is the dimension of the vector $\underline{x}'$. A row vector is the transpose of a column vector.

If $\underline{x}$ is a column vector and $\underline{y}$ is a column vector, then these vectors are equal

$$\underline{x} = \underline{y}$$

if and only if

$$x_i = y_i \quad \text{for all } i = 1, 2 \ldots n.$$

Similarly, if $\underline{z}$ is a n column vector defined by:

$$\underline{z} = \underline{x} + \underline{y}$$

then for this equality to hold

$$z_i = z_i + y_i \quad \text{for all } i = 1, 2, \ldots n$$

Column vectors obey the commutative law and therefore

$$\underline{x} + \underline{y} = \underline{y} + \underline{x}$$

and

$$\underline{x} + (\underline{y} + \underline{z}) = (\underline{x} + \underline{y}) + \underline{z}$$

which is called associativity.

If we multiply a column vector $\underline{x}$ by a scalar a then this is equal to multiplying each component variable of the vector $\underline{x}$ by a.

$$a\underline{x} = \begin{bmatrix} ax_1 \\ ax_2 \\ . \\ . \\ . \\ ax_n \end{bmatrix}$$

A scalar product of two n column vectors $\underline{x}$ and $\underline{y}$ is described by $\langle x, y \rangle$ and defined by

$$\langle x, y \rangle = \sum_{i=1}^{n} x_i y_i. \qquad (B.3)$$

Scalar products obey the laws of vector inversion, commutation, association, etc. and therefore

$$\langle x, y \rangle = \langle y, x \rangle$$
$$\langle \underline{x+y}, \underline{x+z} \rangle = \langle x, w \rangle + \langle y, w \rangle + \langle x, z \rangle + \langle y, z \rangle$$

If a column vector $\underline{x}$ in a scalar product of two vectors is multiplied by a constant a, then this is equivalent to multiplying the scalar product of the vectors by a or

$$<a\underline{x},\underline{y}> = a<\underline{x},\underline{y}>$$

A scalar product of two vectors $\underline{x}$ and $\underline{y}$ can also be written as the product of a row and a column vector or

$$<\underline{x},\underline{y}> = x'y = [x_1 x_2 \ldots x_n] \begin{bmatrix} x_1 \\ x_2 \\ \cdot \\ \cdot \\ \cdot \\ x_n \end{bmatrix}$$

If the real vectors $\underline{x}$ and $\underline{y}$ are such that their scalar product, or the product of the row vector $\underline{x}'$ multiplied by the column vector $\underline{y}$ is zero, then the two vectors are said to be orthogonal.

B.1.2 Matrices

A matrix is an array of variables or numbers, real or complex, which can be written in the form:

$$\underline{x} = \begin{bmatrix} x_{11} & x_{12} & \cdots & x_{1m} \\ x_{21} & x_{22} & \cdots & x_{2m} \\ \cdot & \cdot & & \cdot \\ x_{n1} & x_{n2} & \cdots & x_{nm} \end{bmatrix} \qquad (B.4)$$

The variables x_{ij} for $i = 1,2\ldots n$ and $j=1,2\ldots m$ are called the elements of the matrix $\underline{x}$. A matrix with n rows and m columns is called an nxm matric. If n=m then the matrix is called a square matrix. A column vector $\underline{x}$ is therefore simply a nx1 matrix.

Matrices are usually denoted by a capital underlined letter, such as $\underline{X}$ above or any other appropriate underlined capital letter.

As with vectors, matrices follow the laws of commutation etc. Therefore, for example if $\underline{X}$ and $\underline{Y}$ are matrices with components or elements x_{ij} and y_{ij} respectively, then we can write $\underline{X} + \underline{Y} = [x_{ij} + y_{ij}]$. If a matrix is multiplied by a constant or scalar then this is equivalent to every component or element of the matrix being multiplied by a constant or scalar

$$a\underline{X} = [a\, x_{ij}]$$

Matrix addition is a commutative and associative as was the manipulation of vectors, we can therefore write that

$$\underline{X} + \underline{Y} = \underline{Y} + \underline{X}$$

and

$$X + (\underline{Y} + \underline{Z}) = (\underline{X} + \underline{Y}) + \underline{Z}$$

A square matrix n x n with zeros everywhere except on the diagonal which contains only ones, or a matrix with $x_{ij}=0$ for all $j \neq 1$ and $x_{ij} \neq 1$ for all $j=1$, is called an Identity Matrix:

$$\underline{I} = \begin{bmatrix} 1 & 0 & . & . & . & 0 \\ 0 & 1 & . & . & . & 0 \\ . & . & . & . & . & . \\ 0 & 0 & . & . & . & 1 \end{bmatrix}$$

If a matrix is the denominator of a unit numerator or $1/\underline{X}$ then this can be expressed as the inverse of the matrix $\underline{X}$, or $\underline{X}^{-1}$. In this case $\underline{X}^{-1}\underline{X} = \underline{X}\,\underline{X}^{-1} = \underline{I} =$ Identity Matrix; X^{-1} exists provided that det $|\bar{A}| \neq 0$.

Assuming we have a column vector $\underline{x}$ of order n and a matrix $\underline{X}$ of dimension mxn, the m column vector $\underline{z}$ which is the resulting vector of the product of matrix $\underline{X}$ and vector $\underline{x}$ is then equal to

$$\underline{z} = \underline{X}\,\underline{x}$$

and the individual ements of the vector and matrix or

$$z_i = \sum_{j=1}^{n} x_{ij} x_j \quad ; \quad i = 1, 2, \ldots m$$

Therefore, an n column vector multiplied by an mxn matrix

gives an m column vector.

B.1.3 Multiplication of a Square Matrix by a Square Matrix

If $\underline{X}$ is a nxn matrix and $\underline{Y}$ an nxn matrix, then the nxn matrix $\underline{Z}$ is equal to $\underline{Z} = \underline{X}\,\underline{Y}$ and elements x_{ij} are equal to

$$z_{ij} = \sum_{k=1}^{n} x_{ik} x_{kj}$$

In general, commutativity does not hold in matrix multiplication:

$$\underline{X}\,\underline{Y} \neq \underline{Y}\,\underline{X}$$

But associativity is always true

$$(\underline{X}\underline{Y})\underline{Z} = \underline{X}(\underline{Y}\underline{Z})$$

B.1.4 The Transpose of a Matrix

If (x_{ij}) are the elements of $\underline{X}$, then the transpose of the matrix $\underline{X}$, similar to the transpose of a vector, can be written $\underline{X}'$ with elements (x_{ij}). In other words, the columns of $\underline{X}'$ are the rows of $\underline{X}$ and the rows of $\underline{X}'$ are the columns of $\underline{X}$.

If the matrix is equal to its transpose, or $\underline{X}=\underline{X}'$, then the matrix $\underline{X}$ or its transpose $\underline{X}'$ are called symmetric.

B.1.5 Determinants of Square Matrices

Define M_{ij} as a minor determinant of element a_{ij}, with the ith row and j^{th} column cancelled. Define the cofactor of a_{ij} as the signed minor determinant of a_{ij}

$$\text{cofactor of } a_{ij} = |A_{ij}| = (-1)^{i+j} |M_{ij}|$$

For example

$$|A_{11}| = \begin{bmatrix} a_{22} & \cdots & a_{2n} \\ \vdots & & \vdots \\ a_{n2} & & a_{nn} \end{bmatrix} = \text{cofactor } a_{11}$$

We can now evaluate the determinant by expanding it in terms of cofactors by the i row

$$|A| = \sum_{j=1}^{n} a_{ij} |A_{ij}|$$

or by the j column

$$|A| = \sum_{j=1}^{m} a_{ij} |A_{ij}|$$

The determinant of a square matrix is a unique value and is the same whether the determinant is expanded by row or by column, no matter which row or column.

The cofactors can be further expanded in terms of their own cofactors until the minor determinants are of order two.

The determinant of a 2x2 matrix is a scalar equal to $a_{11}a_{22} - a_{12}a_{21}$.

B.1.6 Matrix Inversion

A set of n linear equations, $y_1 = a_{ij} x_j$; i, j = 1...n, where a_{ij} are constants (or transitional probabilities), defines a linear transformation of the set of n variables x_i into a new set y when

$$x_i = \frac{|A_{ji}|}{|A|}$$

where $|A_{ji}|$ = adjoint of a_{ij} in the determinant $|A| = a_{ij}$.
The set of equations where

$$\underline{x} = \begin{bmatrix} x_1 \\ \cdot \\ \cdot \\ \cdot \\ x_n \end{bmatrix} \quad \text{and} \quad \underline{y} = \begin{bmatrix} y_1 \\ \cdot \\ \cdot \\ y_n \end{bmatrix}$$

is then $\underline{X} = \underline{A}^{-1} \underline{Y}$

$$\overline{A}^{-1} = \text{Inverse Matrix } \overline{A} = \begin{bmatrix} \frac{|A_{ij}|}{|A|} \cdot & & \\ & \cdot & \\ & & \cdot \end{bmatrix} = \begin{bmatrix} \frac{A_{11}}{|A|} & \cdots & \frac{A_{n1}}{|A|} \\ & & \\ \frac{A_{1n}}{|A|} & & \frac{A_{nn}}{|A|} \end{bmatrix}$$

The inverse matrix is obtained by dividing the adjoint of matrix $\underline{A}_{ij}$ by the determinant $|A|$. The adjoint of a matrix is defined as the transposed matrix of the cofactors of $\underline{A}$ or

$$\underline{A}_{ji} = [(-1)^{i+j} |M_{ij}|]^T$$

The inverse matrix of $\bar{A}$ can be constructed whenever A is nonsingular (i.e., A does not vanish)

$$\bar{A}\,\bar{A}^{-1} = (\delta_{ij}) = I = \text{Identity Matrix}$$

B.1.7 Eigenvalues of a Matrix

An nxn matrix has n eigenvalues, denoted by $\lambda_1, \lambda_2, \ldots, \lambda_n$. The eigenvalues of a matrix $\underline{A}$ can be found from

$$\det(\underline{A} - \lambda \underline{I}) = 0$$

where I is the identify matrix which yields an nth order polynomial in λ, whose n roots, defined by $\lambda_1, \ldots, \lambda_n$, are the eigenvalues of the matrix. For example, considering the matrix

$$\underline{A} = \begin{bmatrix} a_{11} & a_{12} \\ a_{21} & a_{22} \end{bmatrix}$$

$$A - \lambda I = \begin{bmatrix} a_{11} - \lambda & a_{12} \\ a_{21} & a_{22} \end{bmatrix} - \lambda$$

$$\det(\underline{A} - \underline{\lambda}I) = (a_{11} - \lambda)(a_{22} - \lambda) - a_{12}a_{21} = 0$$

or

$$\lambda^2 - (a_{11} + a_{22})\lambda + a_{11}A_{22} - a_{12}a_{21} = 0$$

and λ_1 and λ_2 are the roots of the above second-order equation.

The eigenvalues of $\underline{A}$ and $\underline{A}'$ are the same. The eigenvalues, if complex, come in conjugate pairs.

Right Eigenvectors of a Matrix

If $\underline{A}$ is matrix with eigenvalues, $\lambda_1 \ldots \lambda_n$, and we let $\underline{\phi}_i$, $i=1,\ldots n$, be a vector such that $\underline{A}\, \underline{\phi}_i = \lambda_i \phi_i$ $i=1,\ldots n$
then the vectors $\underline{\phi}_1, \underline{\phi}_2, \ldots \underline{\phi}_n$ are called eigenvectors of $\underline{A}$ corresponding to the eigenvalues $\lambda_1, \lambda_2, \ldots \lambda_n$, respectively.

Left Eigenvectors of a Matrix

If we let $\underline{\phi}_i$, $i=1,2,\ldots n$ be the right eigenvectors of a matrix $\underline{A}$, and $\underline{\phi}_j$, $j=11,2,\ldots n$ be vectors such that

$$\langle \underline{\phi}_i, \underline{\phi}_j \rangle = \delta_{ij} = \begin{cases} 1 \text{ if } i=j \\ 0 \text{ if } i \neq j \end{cases}$$

Then the vectors $\underline{\phi}_j$ $j=1,2,\ldots n$, are called the left eigenvectors of $\underline{A}$.

B.1.8 Orthogonal Matrices

A matrix $\underline{A}$ is an orthogonal matrix if and only if

$$\underline{A}'\underline{A} = \underline{I}$$

and

$$\underline{A}' = \underline{A}^{-1}$$

It is easy to see that if $\underline{A}$ is orthogonal, then

$$\langle \underline{A}\,\underline{x}, \underline{A}\,\underline{y} \rangle = \langle \underline{x}, \underline{y} \rangle \text{ for any } \underline{x}, \underline{y}$$

We measure the length of a vector $\underline{x}$ by its norm $||\underline{x}||$, such that

$$||\underline{x}||^2 = \langle \underline{x}, \underline{x} \rangle = x_i^2 + \ldots + x_n^2$$

We use the vector $\underline{A}\,\underline{x}$ where $\underline{A}$ is orthogonal to determine the length of $\underline{A}\,\underline{x}_j$

$$||\underline{A}\,\underline{x}||^2 = \langle \underline{A}\,\underline{x}, \underline{A}\,\underline{x} \rangle = \langle \underline{x}, \underline{x} \rangle$$

and

$$\|\underline{x}\| = \|\underline{A}\,\underline{x}\|$$

which indicates that an orthogonal matrix acts like a linear transformation on a vector with the property that the length of the vector remains unchanged.

B.2 Z-Transform Methods for Markov Chains

A very effective way of dealing with n step Markov chains is by use of the z-transform which will be described as follows. First, consider a time function f(n) that takes on arbitrary values f(0), f(1), f(2), and so on, at non-negative, discrete, integrally-spaced points in time. Such a time function is shown in Figure B.1.

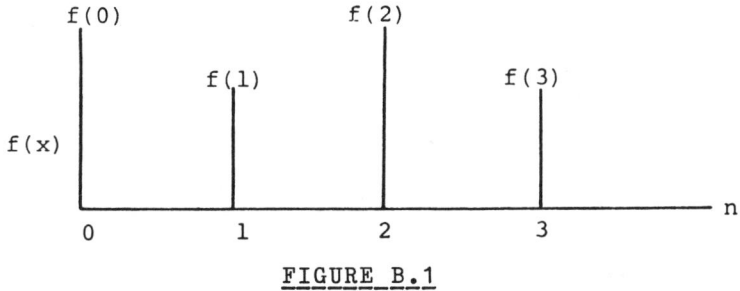

FIGURE B.1

For time functions that do not increase in magnitude with n faster than a geometric sequence, it is possible to define a z-transform f(z) such that

$$f(z) = \sum_{n=0}^{\infty} f(n) z^n$$

The relationship between f(n) and its transform f(z) is unique; each time function has only one transform, and the inverse transformation of the transform will produce once more the original time function. The z-transform is useful in Markov processes because the probability transients in Markov processes are geometric sequences. The z-transform provides a closed-form expression for such sequences.

To see the power of the z-transform, consider the z-transform of some typical time functions that will be often encountered. First, consider the step function

$$f(n) = \begin{cases} 1 & n=0, 1, 2, 3\ldots \\ 0 & n<0 \end{cases}$$

The z-transform is

$$f(z) = \sum_{n=0}^{\infty} f(n) z^n = 1 + z + z^2 + \ldots = \frac{1}{1-z}, \quad 0 < z < 1$$

For the geometric sequence $f(n) = \alpha^n$, $n \geq 0$

$$f(z) = \sum_{n=0}^{\infty} f(n) z^n = \sum_{n=0}^{\infty} (\alpha z)^n = \frac{1}{1-\alpha z}, \quad 0 < z < 1/\alpha$$

Note that if

$$f(z) = \sum_{n=0}^{\infty} \alpha^n z^n \quad \text{then} \quad \frac{d}{dz} f(z) = \sum_{n=0}^{\infty} n \alpha^n z^{n-1}$$

and

$$\sum_{n=0}^{\infty} n \alpha^n z^n = z \frac{d}{dz} f(z) = z \frac{d}{dz} \left(\frac{1}{1-\alpha z} \right) = \frac{\alpha z}{(1-\alpha z)^2}$$

Thus, we have obtained as a derived result that, if the time function is $f(n) = n \alpha^n$, its z-transform is

$$f(z) = \frac{\alpha z}{(1-\alpha z)^2}$$

Also, if a time function $f(n)$ with transform $f(z)$ is shifted to the right so as to become $f(n+1)$, then the transform of the shifted function is:

$$\sum_{n=0}^{\infty} f(n+1) z^n = \sum_{m=1}^{\infty} f(m) z^{m-1} = z^{-1}(f(z) - f(0))$$

The following table includes the most useful transform pairs (Table B.1):

TABLE B.1:

Time function $n \geq 0$	z-transform
$f(n)$	$f(z)$
$f(n) + f(n)$	$f(z) + f(z)$
$kf(n)$ (k is a constant)	$kf(z)$
$f(n-1)$	$zf(z)$

$f(n+1)$	$z^{-1}(f(z) - f(0))$
α^n	$\dfrac{1}{1-\alpha z}$
1 unit step at $n=0$	$\dfrac{1}{1-z}$
$n\alpha^n$	$\dfrac{\alpha z}{(1-\alpha z)^2}$
n unit ramp	$\dfrac{\alpha z}{(1-z)^2}$
$\alpha^n f(n)$	$f(\alpha z)$
$\delta(n) = \begin{cases} f(0) = 1 \\ f(n) = 0 \; n > 0 \end{cases}$	$\dfrac{1}{1-2}$, $0 < z < 1$

Now, to show how these z-transforms are useful to analyze Markov chains, it is possible to take the z-transforms of vectors and matrices by taking the z-transform of each component of the array.

To demonstrate, take the z-transform of the identity

$$P(n+1) = P(n) \cdot P \qquad (B.2)$$

which was derived from (6.9) by noticing that

$$P(n) = P(0)P^n$$

and

$$P(n+1) = P(0)P^{n+1} = (P(0)P^n)P = P(n) \cdot P$$

So, if the transform of the vector $P(n)$ is given the symbol $\Pi(z)$, then the transform of 6B.1 becomes

$$z^{-1}(\Pi(z) - P(0)) = \Pi(z) \cdot P$$

Rearranging:

$$\Pi(z) - z\Pi(z)P = P(0)$$

$$\Pi(z)(I - zP) = P(0)$$

$$\Pi(z) = P(0)(I - zP)^{-1} \quad \text{where } I = \begin{bmatrix} 1 & 0 & 0 \\ 0 & 1 & 0 \\ 0 & 0 & 1 \end{bmatrix} = \text{Identity Matrix}$$

Thus, the transform of the state probability vector, $\Pi(z)$, is equal to the initial state probability vector multiplied by the inverse of the matrix $I-zP$ which will always exist. If the inverse transform of 6B.1 is taken, then we have:

$$P(n) = P(0) \text{ (inverse transform of } (I-zP)^{-1})$$

However, from (6.9) it is known that:

$$P(n) = P(0)P^n$$

so that the inverse transform of $(I-zP)^{-1}$ is equal to P^n.

To demonstrate the use of the z-transform method, consider the two examples discussed previously. First, consider the general two state chain which had the transition matrix

$$P = \begin{bmatrix} P_{0,0} & P_{0,1} \\ P_{1,0} & P_{1,1} \end{bmatrix}$$

In this case

$$(I - zP)^{-1} = \left\{ \begin{bmatrix} 1 & 0 \\ 0 & 1 \end{bmatrix} - \begin{bmatrix} zP_{0,0} & zP_{0,1} \\ zP_{1,0} & zP_{1,1} \end{bmatrix} \right\}^{-1}$$

$$= \begin{bmatrix} 1-zP_{0,0} & -zP_{0,1} \\ -zP_{1,0} & 1-zP_{1,1} \end{bmatrix}^{-1}$$

$$= \left(\frac{1}{1-z(P_{0,0}+P_{1,1}) + z^2(P_{0,0}P_{1,1}-P_{0,1}P_{1,0})} \right)$$

$$\begin{bmatrix} 1-zP_{1,1} & zP_{0,1} \\ zP_{1,0} & 1-zP_{0,0} \end{bmatrix}$$

$$= \left(\frac{1}{1-z(P_{0,0}+P_{1,1}) + z^2(P_{0,0}+P_{1,1}-1)} \right)$$

$$\begin{bmatrix} 1-zP_{1,1} & z(1-P_{0,0}) \\ z(1-P_{1,1}) & 1-zP_{0,0} \end{bmatrix}$$

$$= \begin{bmatrix} \dfrac{1-zP_{1,1}}{(1-z)(1-z[P_{0,0}+P_{1,1}-1])} & \dfrac{z(1-P_{0,0})}{(1-z)(1-z[P_{0,0}+P_{1,1}-1])} \\ \\ \dfrac{z(1-P_{1,1})}{(1-z)(1-z[P_{0,0}+P_{1,1}-1])} & \dfrac{1-zP_{0,0}}{(1-z)(1-z[P_{0,0}+P_{1,1}-1])} \end{bmatrix}$$

Inverse transform of

$$(I-zP)^{-1} = P^n = \frac{1}{2-P_{0,0}-P_{1,1}}$$

$$\begin{bmatrix} (1-P_{1,1})+(1-P_{0,0})(P_{0,0}+P_{1,1}-1)^n & (1-P_{0,0})-(1-P_{0,0})(P_{0,0}+P_{1,1}-1)^n \\ (1-P_{1,1}(-(1-P_{1,1})(P_{0,0}+P_{1,1}-1)^n & (1-P_{0,0})+(1-P_{1,1})(P_{0,0}+P_{1,1}-1)^n \end{bmatrix}$$

$$P^n = \frac{1}{2-P_{0,0}-P_{1,1}} \begin{bmatrix} 1-P_{1,1} & 1-P_{0,0} \\ \\ 1-P_{1,1} & 1-P_{0,0} \end{bmatrix} + \frac{(P_{0,0}+P_{1,1}-1)^n}{2-P_{0,0}-P_{1,1}} \cdot \begin{bmatrix} 1-P_{0,0} & -(1-P_{0,0}) \\ \\ -(1-P_{1,1}) & 1-P_{1,1} \end{bmatrix}$$

The above result agrees with that obtained previously.

Now consider the three state Markov chain numerical example worked previously in this case:

$$(I-zP)^{-1} = \left[\begin{bmatrix} 1 & 0 & 0 \\ 0 & 1 & 0 \\ 0 & 0 & 1 \end{bmatrix} - z \begin{bmatrix} 1/2 & 1/4 & 1/4 \\ 1/2 & 1/2 & 0 \\ 1/2 & 0 & 1/2 \end{bmatrix}\right]^{-1}$$

$$= \begin{bmatrix} 1-1/2z & -1/4z & -1/4z \\ -1/2z & 1-1/2z & 0 \\ -1/2z & 0 & 1-1/2z \end{bmatrix}^{-1}$$

$$= \frac{1}{(1-1/2z)^3 - 1/8z^2(1-1/2z) - 1/8z^2(1-1/2z)}$$

$$\cdot \begin{bmatrix} (1-1/2z)^2 & 1/4z(1-1/2z) & 1/4z(1-1/2z) \\ 1/2z(1-1/2z) & (1-1/2z)^2 - 1/8z^2 & 1/8z^2 \\ 1/2z(1-1/2z) & 1/8z^2 & (1-1/2z)^2 - 1/8z^2 \end{bmatrix}$$

$$= \begin{bmatrix} \dfrac{(1-1/2z)}{1-z} & \dfrac{1/4z}{1-z} & \dfrac{1/4z}{1-z} \\[6pt] \dfrac{1/2z}{1-z} & \dfrac{1/8z^2 - z + 1}{(1-z)(1-1/2z)} & \dfrac{1/8z^2}{(1-z)(1-1/2z)} \\[6pt] \dfrac{1/2z}{1-z} & \dfrac{1/8z^2}{(1-z)(1-1/2z)} & \dfrac{1/8z^2 - z + 1}{(1-z)(1-1/2z)} \end{bmatrix}$$

and by partial fraction expansion,

$$= \begin{bmatrix} \dfrac{1}{2} + \dfrac{1/2}{1-z} & -\dfrac{1}{4} + \dfrac{1/4}{1-z} & -\dfrac{1}{4} + \dfrac{1/4}{1-z} \\[6pt] -\dfrac{1}{2} + \dfrac{1/2}{1-z} & \dfrac{1}{4} + \dfrac{1/4}{1-z} + \dfrac{1/2}{1-1/2z} & \dfrac{1}{4} + \dfrac{1/4}{1-z} - \dfrac{1/2}{1-1/2z} \\[6pt] -\dfrac{1}{2} + \dfrac{1/2}{1-z} & \dfrac{1}{4} + \dfrac{1/4}{1-z} - \dfrac{1/2}{1-1/2z} & \dfrac{1}{4} + \dfrac{1/4}{1-z} + \dfrac{1/2}{1-1/2z} \end{bmatrix}$$

So,

$$(I-zP)^{-1} = \frac{1}{1-z}\begin{bmatrix} 1/2 & 1/4 & 1/4 \\ 1/2 & 1/4 & 1/4 \\ 1/2 & 1/4 & 1/4 \end{bmatrix} + \frac{1}{1-1/2z}\begin{bmatrix} 0 & 0 & 0 \\ 0 & 1/2 & -1/2 \\ 0 & -1/2 & 1/2 \end{bmatrix}$$

$$= \begin{bmatrix} 1/2 & -1/4 & -1/4 \\ -1/2 & 1/4 & 1/4 \\ -1/2 & 1/4 & 1/4 \end{bmatrix}$$

The inverse transform of $(I-zP)^{-1}$ will be:

$$P^n = \begin{bmatrix} 1/2 & 1/4 & 1/4 \\ 1/2 & 1/4 & 1/4 \\ 1/2 & 1/4 & 1/4 \end{bmatrix} + \frac{1}{2}^{n+1}\begin{bmatrix} 0 & 0 & 0 \\ 0 & 1 & -1 \\ 0 & -1 & 1 \end{bmatrix} + \delta(n)\begin{bmatrix} 1/2 & -1/4 & -1/4 \\ -1/2 & 1/4 & 1/4 \\ -1/2 & 1/4 & 1/4 \end{bmatrix}$$

where the last term is zero for $n>0$.

Thus the steady state matrix Q will be:

$$\lim_{n \to \infty} P^n = Q = \begin{bmatrix} 1/2 & 1/4 & 1/4 \\ 1/2 & 1/4 & 1/4 \\ 1/2 & 1/4 & 1/4 \end{bmatrix} \text{ since } \lim_{n \to \infty}(1/2)^{n+1} = 0$$

This result agrees with that previously obtained.

It should be noted that the steady state matrix may also be obtained by taking the limit

$$\lim_{z \to 1^-}(I-zP)^{-1}(1-z) = \text{steady state matrix}$$

If this limit is applied to matrix (A) above:

$$\text{steady state matrix} = \lim_{z \to 1^-}(I-zP)^{-1}(1-z) = \lim_{z \to 1^-}$$

$$\begin{bmatrix} 1-1/2z & 1/4z & 1/4z \\ 1/2z & \dfrac{1/8z^2-z+1}{1-1/2z} & \dfrac{1/8z^2}{1-1/2z} \\ 1/2z & \dfrac{1/8z^2}{1-1/2z} & \dfrac{1/8z^2-z+1}{1-1/2z} \end{bmatrix}$$

Steady state matrix = $\begin{bmatrix} 1/2 & 1/4 & 1/4 \\ 1/2 & 1/4 & 1/4 \\ 1/2 & 1/4 & 1/4 \end{bmatrix}$

in agreement with the result above.

B.3 Laplace Transformation

Transformation methods are often used to facilitate the solving of linear constant coefficient integrodifferential equations and integral equations involving convolutions. The Laplace transformation serves a similar purpose to the Fourier, Heaviside, and z transformation. It is a convenient tool for the analysis of continuous systems. In reliability theory it is primarily used in the determination of transient conditions of complex systems and in particular to the derivation of the time dependent probabilities that a system will be in any of its possible states.

Let t be a real variable, say time, and s a complex variable. Then if we assume that f(t) is a function of t which is zero for t smaller than zero or

$$f(t) = 0 \text{ if } t < 0$$

and F(s) is a function of s then

$$F(s) = \int_0^\infty e^{-st} f(t) dt = L[f(t)]$$

is called the direct Laplace transform of f(t). To obtain the Laplace transform of the nth derivative of the function f(t)

$$L[\frac{d^n f(t)}{dt^n}] = sF(s) - \sum_{k=0}^{n-1} \frac{d^k f}{dt^k} s^{n-1-k}$$

where the second term is evaluated for $t > 0$ and

$$\frac{d^0 f}{dt^0}$$

means $f(0^+)$.

Similarly the Laplace transform of the nth ingegral

$$L[\int\int...\int f(t)...dt] = s^{-n} F(s) + \sum_{k=-1}^{-n} [\int...\int f(t)...dt] s^{n-1-k}$$

To obtain the solution in the time domain we inverse the Laplace transformation, which is possibly only under certain conditions. The most common representation of the inversion process is

$$f(t) = L^{-1}[F(s)] = \frac{1}{2\pi i} \int_{C-i\infty}^{C+i\infty} e^{ts} F(s) ds$$

where C is a real constant.

To obtain the inverse Laplace transformation of a product of transforms $F_1(s)$ and $F_2(s)$ where

$$L^{-1}[F_1(s)] = f_1(t)$$

and

$$L^{-1}[F_2(s)] = f_2(t)$$

then

$$L^{-1}[F_1(s)F_2(s)] = \int_0^t f_1(t-T) f_2(t) dt$$

Similarly linear transformation of transforms and inverse transforms can be expressed as follows

$$L(A_1 f_1(t) + A_2 f_2(t)) = A_1 L[f_1(t)] + A_2 L[f_2(t)]$$

and

$$L^{-1}[A_1 F_1(s) + A_2 F_2(s)] = A_1 L^{-1}[F_1(s)] + A_2 L^{-1}[F_2(s)]$$

where A_1 and A_2 are real constants.

Using the above principles we can obtain the following transform pairs (Table B.2):

TABLE B.2

$f(t)$	$F(s)$
$a\, f(t)$	$a\, F(s)$
$f_1(t) + f_2(t)$	$F_1(s) + F_2(t)$
$\dfrac{df(t)}{dt}$	$s\, F(s) - f(0+)$
$\int f(t)\, dt$	$\dfrac{F(s)}{s} + \dfrac{f^{(-1)}(0+)}{s}$
$f\left(\dfrac{t}{a}\right)$	$a\, F(as)$
$\int_0^t f_1(t-\tau)\, f_2(\tau)\, d\tau$	$F_1(s)\, F_2(s)$
$f(t-a)$	$e^{-as} F(s)$
$f(t+a)$	$e^{as} F(s)$
e^{-at}	$\dfrac{1}{s+a}$
1	$\dfrac{1}{s}$
$e^{-at} f(t)$	$F(s+a)$
$e^{at} f(t)$	$F(s-a)$
$\dfrac{r}{(n-1)}\, t^{n-1}\, e^{-at}$	$\dfrac{1}{(s+a)^n}$

t	$\dfrac{1}{s^2}$
$\dfrac{1}{(n-1)!} t^{n-1}$	$\dfrac{1}{s^n}$
$\dfrac{e^{-at} + at - 1}{a^2}$	$\dfrac{1}{(s+a)^{n+1}}$
$e^{-at} \sum_{K=0}^{n} \dfrac{n!(-a)^k}{(n-K)!(K!)^2}$	$\dfrac{s^n}{(s+a)^{n+1}}$
$\dfrac{1-(1+at)e^{-at}}{a^2}$	$\dfrac{1}{s(s+a)^2}$

To apply the transformation principles given we want to obtain the solution to an equation

$$A_1 \frac{dx(t)}{dt} + A_2 x(t) = u(t),$$

we first obtain the transforms

$$L[A_1 \frac{dx(t)}{dt}] = A_1 L(\frac{dx(t)}{dt}) = A_1[sX(s)-x(0)]$$

where $x(0)$ is the value of the function $x(t)$ at $t=0$ if $x(0) = 1$ and

$$\delta(A_2 x(t)) = A_2 X(s), \quad L(u(t)) = \upsilon(s),$$

then

$$\upsilon(s) = A_1[sX(s)-1] + A_2 X(s) = \upsilon(s)$$

and

$$X(s) = \frac{\upsilon(s)+1}{A_1 s + A_2} \quad \text{and} \quad x(t) = L^{-1}[\frac{\upsilon(s)+1}{A_1 s + A_2}]$$

APPENDIX C - TESTING FOR MARKOV PROPERTIES

Ernst G. Frankel

Most of the reliability models discussed in this book assume that the systems under consideration exhibit stationary Markov properties. To test whether this assumption is valid or not requires the use of various Chi Square and maximum likelihood statistical techniques. In this section much of the theory for testing Markov properties will be developed for Markov chains. Methods for extending the theory for continuous time parameter Markov processes will be mentioned. In particular, methods will be provided to estimate the transition probabilities from data, to test whether the transition probabilities indicate that the system is stationary, Markovian, or statistically independent, to test whether two processes are identical and to test whether the data is from a system with specific transition probabilities.

C.1 Estimation of Transition Probabilities

Suppose that an m state (1,2,...m) stationary Markov chain with no known transition probabilities

$$P = \begin{bmatrix} p_{11} p_{12} \cdots \cdots p_{1m} \\ p_{21} p_{22} \cdots \cdots p_{2m} \\ \\ p_{m1} p_{m2} \cdots \cdots p_{mm} \end{bmatrix}$$

is observed until n transactions take place. If n_{ij} represents the number of transactions from state i to state j (i,j=1,2...m) and

$$\sum_{j=1}^{m} n_{ij} = n_i$$

then the transition frequencies may be represented by the following table.

States	1	2	m	Row Sum
1	n_{11}	n_{12}	n_{1m}	n_1
2	n_{21}	n_{22}	n_{2m}	n_2
.	.	.	.	.
.	.	.	.	.
.	.	.	.	.
m	n_m	n_{m2}	n_{mm}	n_m
				n

(Note: $\sum_{i=1}^{m} n_i = n$)

The estimates of P_{ij}, P_{ij}, will be determined from the entries of this table by maximum likelihood techniques discussed below.

First, it is observed that for a given initial state i and a number of trials n_i, the sample of transition counts $(n_{i1}, n_{i2} \cdots, n_{im})$ can be considered as a sample of size n_i from a multinomial distribution* with probabilities

---

* Multinomial Distribution: Let a random experiment be repeated n times (n_i for Markov process considered in text). On each repetition the experiment terminates in but one of m mutually exclusive and exhaustive ways, say state 1, state 2...state m. Let P_j (P_{ij} in text above) be the constant throughout the n independent repetitions (j=1...m). Define the random variable X_j to be equal to the number of outcomes (transitions) which are in state j(j=1,2...m-1). Furthermore, let $X_1, X_2, \ldots X_{m-1}$ (n_{i1}, ...$n_{i,m-1}$ in text above) be non-negative integers so that $X_1+X_2+\ldots+X_{m-1} \leq n$, $\sum_{j=1}^{m-1} n_{ij} \leq n_i$. Then the probability that exactly X_i terminations of the experiment are in state i,...., exactly X_{m-1} transactions of the experiment are in state m-1, is given by:

$$\frac{n!}{X_1! X_2! \ldots X_m!} P_1^{X_1} P_2^{X_2} \ldots P_m^{X_m} \text{ where } X_m = n-(X_1+X_2+\ldots+X_{m-1})$$
$$P_m = 1-(P_1+P_2+\ldots+P_{m-1})$$

This is the multinomial probability density function of m-1 independent random variables, and is obviously an extension of the binomial density for which m=2.

$(P_{i1}, P_{i2}, \ldots P_{im})$ such that

$$\sum_{j=1}^{m} P_{ij} = 1$$

The probability of this outcome can therefore be given as:

$$\frac{n_i!}{n_{i1} n_{i2} \cdots n_{im}!} P_{i1}^{n_{i1}} P_{i2}^{n_{i2}} P_{i3}^{n_{i3}} \cdots P_{im}^{n_{im}}$$

such that

$$\sum_{j=1}^{m} n_{ij} = n_i \quad \text{and} \quad \sum_{j=1}^{n} P_{ij} = 1$$

Extending this argument for the m initial states (i=1...m) when the breakdown of the total number of trials n into $(n_1, n_2, \ldots n_m)$ is given, the probability of the realization of transition counts as given in the table above is given by

$$\prod_{i=1}^{m} \frac{n_i!}{n_{i1}! n_{i2}! \cdots n_{im}!} P_{i1}^{n_{i1}} P_{i2}^{n_{i2}} \cdots P_{im}^{n_{im}}$$

It should be noted that the row sums of the table $(n_i, n_2 \ldots n_m)$ are also random variables, and therefore the unconditional likelihood function $f(P_{ij})$ of the sample observation consists of another factor giving the joint distribution of these random variables. It can be shown that this distsribution is independent of the probability elements P_{ij} and therefore can be denoted implicitly by $A(n_{ij})$. Therefore, the likelihood function $f(P_{ij})*$ and its natural logarithm $L(P_{ij})$ for the observed transition count table can be expressed as:

$$F(P_{ij}) = A(n_{ij}) \prod_{i=1}^{m} \frac{n_i!}{n_{i1}! n_{i2}! \cdots n_{im}!} P_{i1}^{n_{i1}} P_{i2}^{n_{i2}} \cdots P_{im}^{n_{im}} \quad (C.1)$$

---

* The likelihood function is the joint density distribution considered as a function of the parameters. Thus the right hand side of C.1 may be considered as the joint density function $f(X_{ij}, P_{ij})$ or the likelihood function $f(P_{ij})$.

$$L(P_{ij}) = \ell\eta B(n_{ij}) + \sum_{i=1}^{m} \sum_{j=1}^{m} n_{ij} \ell\eta P_{ij} \qquad (C.2)$$

where $\ell\eta B(n_{ij})$ contains all terms independent of the P_{ij}'s. From (C.2) it may be also noted that the n_{ij}'s are a set of sufficient statistics for $f(P_{ij})$ and $L(P_{ij})$.

To derive the maximum likelihood estimates, maximize (C.2) under the condition

$$\sum_{j=1}^{m} P_{ij} = 1 \quad (i=1, 2...m)$$

Incorporating this condition in (C.2) yields the following:

$$L(P_{ij}) = \ell\eta B(n_{ij}) + \sum_{i=1}^{m} \sum_{j=1}^{m-1} n_{ij} \ell\eta P_{ij}$$

$$+ \sum_{i=1}^{m} n_{in} \ell\eta(1-P_{i1}-P_{i2}\cdots P_{i,m-1}) \qquad (C.3)$$

From the structure of the log likelihood function $L(P_{ij})$ it is clear that the estimate for P_{ij} can be obtained separately for the m values of i=1, 2, ... m. For a specific value of i, the following expression may be extracted from (C.3).

$$L_i(P_{ij}) = \ell\eta B_i(n_{ij}) + \sum_{j=1}^{m-1} n_{ij} \ell\eta P_{ij} + n_{im} \ell\eta (1-P_{i1} - P_{i2} \cdots -P_{i,m-1}) \qquad (C.4)$$

Differentiating (C.4) with respect to P_{ij} (j=1,2...m-1) and setting it equal to zero:

$$\frac{\partial L_i(P_{ij})}{\partial P_{i1}} = 0 = 0 + \frac{n_{ij}}{P_{ij}} - \frac{n_{im}}{(1-P_{i1} - P_{i2} - P_{i,m-1})}$$

$$\vdots$$

$$\frac{\partial L_i(P_{ij})}{\partial P_{i,m-1}} = 0 = 0 + \frac{n_{i,m-1}}{P_{i,m-1}} - \frac{n_{im}}{(1-P_{i1} - P_{i2}\cdots -P_{i,m-1})}$$

These expressions may be consolidated into:

$$\frac{n_{i1}}{P_{i1}} = \frac{n_{i2}}{P_{i2}} = \cdots\cdots = \frac{n_{i,m-1}}{P_{i,m-1}} = \frac{n_{im}}{(1-P_{i1}-P_{i2}\cdots-P_{i,m-1})}$$

which may be expressed as the following set of m equations:

$$\frac{n_{i1}}{n_{i1}} P_{i1} = P_{i1} \qquad\qquad \frac{n_{i3}}{n_{i1}} P_{i1} = P_{i3}$$

$$\frac{n_{i2}}{n_{i1}} P_{i1} = P_{i2} \qquad\qquad \frac{n_{im}}{n_{i1}} P_{i1} = 1-P_{i1}-P_{i2}\cdots-P_{i,m-1}$$

Adding both sides of the above equations together yields

$$\left[\frac{n_{i1}}{n_{i1}} + \frac{n_{i2}}{n_{i1}} + \frac{n_{i3}}{n_{i1}} + \cdots + \frac{n_{im}}{n_{i1}}\right] P_{i1} = P_{i1} + P_{i2} + P_{i3}$$

$$+ \cdots + 1 - P_{i1} = P_{i2} - P_{i,m-1}$$

$$\frac{\hat{P}_{i1}}{n_{i1}} \sum_{j=1}^{m} n_{ij} = 1$$

$$\hat{P}_{ij} = \frac{n_{ij}}{n_i} \qquad\qquad\qquad\qquad\qquad (C.5)$$

In a similar manner, the estimates of all the other elements of P may be derived. Thus,

$$\hat{P}_{ij} = \frac{n_{ij}}{n_i} \quad i, j = 1, 2, \ldots m$$

Now suppose that the transition probabilities are not stationary. In this case there will be T transition matrices of the form:

$$P(t) = \begin{bmatrix} P_{11}(t) & P_{12}(t) & P_{13}(t) & \cdots & P_{1m}(t) \\ P_{21}(t) & P_{22}(t) & P_{23}(t) & \cdots & P_{2m}(t) \\ \vdots & \vdots & \vdots & & \vdots \\ P_{m1}(t) & P_{m2}(t) & P_{m3}(t) & & P_{mm}(t) \end{bmatrix} \quad t=1,2,\ldots T$$

If $n(t)$ transitions are observed from time $t-1$ to time t and if $n_{ij}(t)$ represents the number of transitions from state i at time $t-1$ to state j at time t and if

$$n_i(t-1) = \sum_{j=1}^{m} n_{ij}(t)$$

then the transition frequencies from i at time $t-1$ to state j at time t may be represented by the following table:

states	1	2	...	m	Row Sum
1	$n_{11}(t)$	$n_{12}(t)$	......	$n_{1m}(t)$	$n_1(t-1)$
2	$n_{21}(t)$	$n_{22}(t)$	......	$n_{2m}(t)$	$n_2(t-1)$
m	$n_{m1}(t)$	$n_{m2}(t)$	......	$n_{mm}(t)$	$n_m(t-1)$

By an argument identical to that for the stationary transition probabilities, estimates for the nonstationary transition probabilities, $P_{ij}(t)$ can be determined by the maximum likelihood technique. In this case the transition probability estimates for $P_{ij}(t)$ will be:

$$P_{ij}(t) = \frac{n_{ij}(t)}{n_i(t-1)} \tag{C.6}$$

C.2 Chi Square Approximation to Multinomial Density

In order to test various hypotheses concerning the properties of the transition probabilities, i.e. Markov stationarity, etc., the joint probability density of the transition frequencies (C.4) will be required. However, the multinomial distribution is very difficult to use in making such tests. Therefore, the Chi Square approximation to the multinomial distribution is very difficult to use in making such tests. It should be noted that these tests are only good if there are several (usually more than 5) observations of transitions for each transition. The approximation is derived using the following identities:

$$\ln n! = n\ln n - n + (1/2)\ln n + (1/2)\ln 2n + O(\tfrac{1}{n}) \tag{C.7}$$

using Sterling's approximation for factorials of large numbers

$$\ln(1+X) = X - (1/2)X^2 + O(X^3) \tag{C.8}$$
using expansion of $\ln$ for small values of $\overline{X}$

First take the logarithm of the general form of the multinomial distribution:

$$\ln f(X_1 X_2 \ldots X_m; P_1, P_2 \ldots P_m) = \ln \frac{n!}{X_1! X_2! \ldots X_m!} P_1^{X_1} P_2^{X_2} \ldots$$

$$\ell\, f(x_1, X_2 \ldots X_m, P_1 \ldots P_m) = \ln n! - \sum_{i=1}^{m} X_i! + \sum_{i=1}^{m} X_i \ell\, {}^\backprime P_i$$

Applying Sterling's formula (C.7): \hfill (C.9)

$$\ln f(X_1 \ldots X_m, P_1 \ldots P_m) = n\ln n - n + (1/2)\ln n + (1/2)\ln 2\pi$$

$$- \sum_{i=1}^{m} (X_1 \ln X_i - X_i + (1/2)\ln X_1 + (1/2)\ln 2\pi + \sum_{i=1}^{m} X_i \ln P_i$$

$$= \text{constant} - \sum_{i=1}^{m} X_i \ln X_i + \sum_{i=1}^{m} X_i - (1/2) \sum_{i=1}^{m} \ln X_i + \sum_{i=1}^{m} X_i \ln P_i$$

$$= \text{constant} - \sum_{j=1}^{m} X_i \ln X_i - 1/2 \sum_{i=1}^{m} \ln X_i + \sum_{i=1}^{m} X_i \ln P_i$$

where all terms not depending on the i's are lumped in the constant. Now introduce the transformation

$$Z_i = \frac{X_i - nP_i}{\sqrt{nP_i}}$$

so that

$$X_i = nP_i \left(1 + \frac{Z_i}{\sqrt{nP_i}}\right) \tag{C.10}$$

Taking the logarithm of X_i yields:

$$\ln X_i = \ln nP_i + \ln\left(1 + \frac{Z_i}{\sqrt{nP_i}}\right)$$

$$\simeq \ln(nP_i) + \frac{Z_i}{\sqrt{nP_i}} - \frac{Z_i^2}{2nP_i} \quad \text{using (C.8).}$$

Substituting this expression for (B.10) and (B.9) yields:

$$\ln f(X_1 \ldots X_m; P_1 \ldots P_m) = \text{constant} - \sum_{i=1}^{m}\left\{nP_1\left(1 + \frac{Z_i}{\sqrt{nP_i}}\right)\right.$$

$$\left.\left[\ln nP_i + \frac{Z_i}{\sqrt{nP_i}} - \frac{Z_i^2}{2nP_i}\right]\right\} - 1/2 \sum_{i=1}^{m}\left\{\ln nP_i + \frac{Z_i}{\sqrt{nP_i}} - \frac{Z_i^2}{2nP_i}\right\}$$

$$+ \sum_{i=1}^{m} nP_i \left(1 + \frac{Z_i}{\sqrt{nP_i}}\right)\ln = \text{constant} - \sum_{i=1}^{m}\{nP_i \ln nP_i\} - \sum_{i=1}^{m} nP_i \cdot$$

$$\cdot \frac{Z_i}{\sqrt{nP_i}} + \sum_{i=1}^{m} nP_i \frac{Z_i^2}{2nP_i} \frac{Z_i^3}{2(nP_i)^{3/2}} - 1/2 \sum_{i=1}^{m}\{\ln nP_i\}$$

$$- 1/2 \sum_{i=1}^{m}\left\{\frac{Z_i}{\sqrt{nP_i}}\right\} + (1/4)\sum_{i=1}^{m}\left\{\frac{Z_i^2}{nP_i}\right\} + \sum_{i=1}^{m}\{nP_i \ln P_i\}$$

$$+ \sum_{i=1}^{m}\left\{\frac{nP_i Z_i}{\sqrt{nP_i}}\ln P_i\right\} = \text{constant} - \sum_{i=1}^{m}\sqrt{nP_i}\, Z_i + \sum_{i=1}^{m} \frac{Z_i^2}{2}\sum_{i=1}^{m}\sqrt{nP_i}$$

$$Z_i \ln nP_i - \sum_{i=1}^{m} Z_i^2 - 1/2 \sum_{i=1}^{m} \frac{Z_i}{\sqrt{nP_i}} + 1/4 \sum_{i=1}^{m} \frac{Z_i^2}{nP_i}$$

$$+ \sum_{i=1}^{m} \sqrt{nP_i}\, Z_i \ln P_i$$

where only terms involving Z_i have been kept out of the constant and Z_i^3 and higher terms have been dropped. This expression may be further simplified to:

$$\ln\{f(X_1,\ldots,X_m;P_1,\ldots,P_m)\} = \text{constant} - \sum_{i=1}^{m} \sqrt{nP_i}\ Z_i$$

$$- 1/2 \sum_{i=1}^{m} Z_i^2 - \sum_{i=1}^{m} \sqrt{nP_i}\ Z_i\ \ln n\} - 1/2 \sum_{i=1}^{m} \frac{Z_i}{\sqrt{nP_i}}$$

$$+ 1/4 \sum_{i=1}^{m} \frac{Z_i^2}{nP_i}$$

This expression may be further simplified by noting that:

$$\sum_{i=1}^{m} \sqrt{nP_i}\ Z_i = \sum_{i=1}^{m} \{\sqrt{nP_i}\ (\frac{X_i - nP_i}{\sqrt{nP_i}})\} = \sum_{i=1}^{m} X_i - n \sum_{i=1}^{m} P_i = n - n = 0 \quad (C.11)$$

so that

$$\ln f(X_1,\ldots X_m; P_1 \ldots P_m) = \text{constant} - \sum_{i=1}^{m} \frac{Z_i^2}{2}$$

$$- 1/2 \sum_{i=1}^{m} \frac{Z_i}{\sqrt{nP_i}} + 1/4 \sum_{i=1}^{m} \frac{Z_i^2}{nP_i}$$

Since the third and fourth terms are much smaller than the second, they may also be dropped leaving only:

$$\ln f(x_1,\ldots X_m : P_1,\ldots P_m) = \text{constant} - \sum_{i=1}^{m} \frac{Z_i^2}{2}$$

or equivalently

$$f(X_1,\ldots X_m; P_1 \ldots P_m) = e^{\text{constant}} \exp\left[-\sum_{i=1}^{m} \frac{Z_i}{2}\right]^2 \quad (C.12)$$

If $K \equiv e^{\text{constant}}$ and $\chi^2 \equiv \sum_{i=1}^{m} Z_i^2$, then (C.11) may be rewritten as:

$$f(X_1 \ldots X_m; P_1 \ldots P_m) = K\, e^{-\chi^2/2} \qquad (C.13)$$

The term

$$\chi^2 = \sum_{i=1}^{m} Z_i^2 = \sum_{i=1}^{m} \frac{(X_i - nP_i)^2}{nP_i} = \text{sum of (observed values} - \text{expected values})^2 / \text{expected values}$$

since nP_i is the expected value of n_i.

The multinomial density has now been approximated by a multiple normal distribution of the Z_i's, which are not, however, independent as can be seen from (C.11). Therefore, in this case, the number of independent Z_i's (and X_i's since the Z_i's and X_i's are linear transforms of each other) is m-1 and this number is the number of degrees of freedom of χ^2.

The degrees of freedom will be reduced by one for each additional independent linear relation among the X_i's that is specified. Also, it is possible to calculate the value of K in any given application by summing (C.13) over all the allowed independent sets of values of X_i (the values of P_i being given) and making use of the fact that the sum of these possibilities must be unity.

C.3 Testing Hypothesis Using Chi-Square Density

At this point it is seen that the multinomial density

$$f(X_1 \ldots X_m; P_1 \ldots P_m) = \frac{N!}{X_1! X_2! \ldots X_m!} P_1^{X_1} P_2^{X_2} \ldots P_m^{X_m}$$

may be approximated by the multinomial density

$$f(X_1 \ldots X_m; P_1 \ldots P_m) = K \cdot \left[\exp\left(-(1/2) \sum_{i=1}^{m} \frac{(X_i - nP_i)^2}{nP_i}\right)\right] = K e^{-\chi^2/2}$$

for the case when the X_i's are large. The number of degrees of freedom for χ^2 is equal to the number of independent variables among the X_i's. It is proven in most statistics texts that a χ^2 with J degrees of freedom will have the density:

$$f(\chi^2) = \frac{1}{(\frac{J}{2}) 2^{J/2}} (\chi^2)^{(J-2)/2} \exp(-\chi^2/2)$$

The null hypothesis for the Chi-Square test is that the observed frequency of events, $X_1 \ldots X_m$ is from a process which has probabilities for the events of $P_1, P_2 \ldots P_m$. If α is defined as the percent of the time when the null hypothesis is wrongly rejected, then the values of χ^2 for various values of α or the confidence interval (α) if χ^2 is known, can be found in various tables. Figure C.1 shows a typical χ^2_α distribution. The null hypothesis will be rejected if the value of χ^2

$$= \sum_{i=1}^{m} \frac{(\text{no. of observed values of } i^{th} \text{ event} - \text{expected no. of values for } i^{th} \text{ event})^2}{\text{expected no. of values for } i^{th} \text{ event}}$$

is greater than χ^2_α.

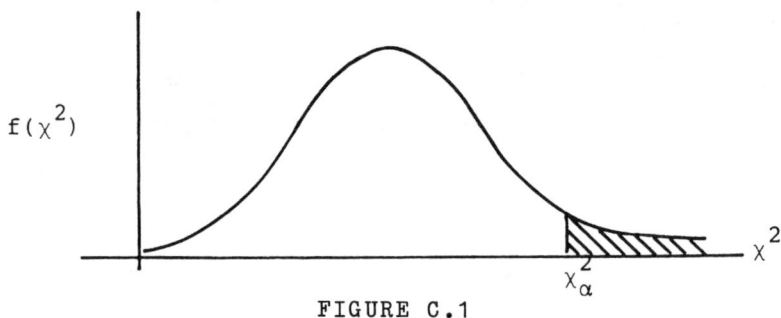

FIGURE C.1

In the following section there will be several examples of the application of the Chi-Square test.

C.4 <u>Test to Determine if the Observed n_{ij}'s are from a Markov Chain with Transition Probabilities</u>

To test the null hypothesis that for a given i, the observed transition frequencies N_{ij} (multinomially distributed) are from an m-state stationary Markov Chain with transition probabilities P_{ij}, compute the χ^2 statistic derived from the multinomial approximation to the multinomial density.

$$\chi^2 = \sum_j \frac{(\text{observed values} - \text{expected values})^2}{\text{expected values}}$$

$$= \sum_{j=1}^{m} \frac{(n_{ij} - n_i P_{ij})^2}{n_i P_{ij}}$$

The expected number of values under the null hypothesis is simply the number initially in state i times the probability of a transition from state i to state j, $n_i P_{ij}$. The χ^2 statistic will have $(m-1)$ degrees of freedom, m from the n_{ij} (j=1...m) variables minus one for the linear relation

$$\sum_{j=1}^{m} n_{ij} = n_i$$

The null hypothesis will not be rejected unless $\chi^2 \geq \chi^2_{\alpha, m-1}$.

The test for the null hypothesis for a given i will be asymptotically independent from that of a different i so that the χ^2 statistic for all i

$$\chi^2 = \sum_{i=1}^{m} \sum_{j=1}^{m} \frac{(n_{ij} - n_i P_{ij})^2}{n_i P_{ij}} = \sum_{i=1}^{m} \sum_{j=1}^{m} \frac{n_i (P_{ij} - P_{ij})^2}{P_{ij}}$$

will simply be a χ^2 statistic with m (for i=1...m) times (m-1) (for each i), that is m(m-1), degrees of freedom. It should be noted that if any of the P_{ij} are zero, then the degrees of freedom should be reduced by one for each such P_{ij}.

C.5 Test to Determine Stationarity of Transition Probabilities

To test the null hypothesis that for a given i the observed transition frequencies $n_{ij}(t)$ (t=1,2...T; j=1...m) are from an m-state stationary Markov Chain with transition probabilities P_{ij} (estimates of P_{ij} which are assumed unknown), calculate the χ^2 statistic

$$\chi^2 = \sum_{t=1}^{T} \sum_{j=1}^{m} \frac{(\text{observed value of } n_{ij}(t) - \text{expected value under null hypothesis})}{\text{expected value under null hypothesis}}$$

$$= \sum_{t=1}^{T} \sum_{j=1}^{m} \frac{(n_{ij}(t) - \hat{P}_{ij} n_i(t-1))^2}{\hat{P}_{ij} n_i(t-1)}$$

$$= \sum_{t=1}^{T} \sum_{j=1}^{m} = \frac{(n_{ij}(t) - \frac{n_{ij}}{n_i} n_i(t-1))^2}{\frac{n_{ij}}{n_i} n_i(t-1)}$$

where

P_{ij} = estimate of stationary probability P_{ij}

$$= \frac{\text{no. of transitions from state } i \text{ to } j}{\text{no. of transitions from state } i} = \frac{\sum_{t=1}^{T} n_{ij}(t)}{\sum_{j=1}^{m} \sum_{t=1}^{T} n_{ij}(t)} = \frac{n_{ij}}{n_i}$$

$n_i(t-1)$ = no. of transitions from state i at time $t-1$

$$= \sum_{j=1}^{m} n_{ij}(t)$$

The expected number of values under the hull hypothesis is simply the average number of transitions from state i at time t-1 to state j at t given that the probability of transition from state i to state j is the stationary $\hat{P}_{ij}$ which is the estimate of P_{ij}. In other words, expected number of transitions under null hypothesis will be $n_i(t-1)\hat{P}_{ij}$. As was seen previously, the maximum likelihood estimate of $P_{ij}(t)$, $P_{ij}(t)$ is found by the formula

$$\hat{P}_{ij}(t) = \frac{n_{ij}(t)}{n_i(t-1)}$$

Substituting this formula into the expression for χ^2 yields:

$$\chi^2 = \sum_{t=1}^{\tau} \sum_{j=1}^{m} \frac{n_{i(t-1)} \left(\frac{n_{ij}(t)}{n_j(t-1)} - P_{ij} \right)^2}{P_{ij}} = \sum_{t=1}^{\tau} \sum_{j=1}^{m} n_i(t-1) \frac{(P_{ij}(t) - P_{ij})^2}{P_{ij}}$$

This χ^2 statistic will have its number of degrees of freedom computed to be as follows:

No. of $n_{ij}(t)$ Parameters)	Degrees of Freedom
$(t=1...\tau;\ j=1...m)$	$+m\tau$
No. of linear relations required for P_{ij} estimation $(j-1...m)$	$-m$
No. of linear relations required for continuity between transition periods	$-(\tau-1)$

(the number of observations from period to period remains the same since all samples must either transit or remain in same state). Relations are of form:

$$\sum_{j=1}^{m} n_{ij}(1) = \sum_{j=1}^{m} n_{ij}(s) = ... = \sum_{j=1}^{m} n_{ij}(\tau)$$

Total degrees of freedom $\quad m\tau - m - (\tau-1) = (m-1)(\tau-1)$

The null hypothesis will not be rejected unless

$$\chi^2 > \chi^2_{\alpha[(m-1)(t-1)]}$$

Since the test for a given i will be asymptotically independent from that of a different i the χ^2 statistic for all i

$$\chi^2 = \sum_{i=1}^{m} \sum_{t=1}^{\tau} \sum_{j=1}^{m} \frac{n_{i(t-1)} (\hat{P}_{ij}(t) - P_{ij})^2}{\hat{P}_{ij}}$$

will simply be a χ^2 statistic with m (for i=1...m) times (m-1)(-1)(for each i) or m(m-1)(-1) degrees of freedom.

C.6 Test to Determine Independence of Observed Transition Frequencies n

To test the null hypothesis that the observed stationary transition frequencies N_{ij} (i,j=1...m) are independent against the alternative hypothesis that the process is Markov, calculate the χ^2 statistic

$$\chi = \frac{\sum_{i=1}^{m}\sum_{j=1}^{m} (\text{observed no. of transitions from i to j} - \text{expected number of transitions under null hypothesis})^2}{\text{expected number of transitions under null hypothesis}}$$

$$= \sum_{i=1}^{m}\sum_{j=1}^{m} \frac{(N_{ij}-N_i \hat{P}_i)^2}{N_i \hat{P}_i} = \sum_{i=1}^{m}\sum_{j=1}^{m} \frac{N_i(\frac{N_{ij}}{N_i} - \hat{P}_j)^2}{\hat{P}_j} =$$

$$\sum_{i=1}^{m}\sum_{j=1}^{m} \frac{N_i(\hat{P}_{ij}-\hat{P}_i)^2}{\hat{P}_j}$$

where

$\hat{P}_j$ = estimate of $\hat{P}_j$, the independent probability of transition to state j =

$$\frac{\text{number of transitions to state j}}{\text{total number of transitions}} = \sum_{i=1}^{m} N_{ij} \Big/ \sum_{i=1}^{m}\sum_{j=1}^{m} N_{ij}$$

The expected number of transitions from state i to state j under the null hypothesis that the N_{ij} are independent will be simply the product of the number initially in state i, N_i, and the probability of transiting to state j from any state,

$$\sum_{i=1}^{m} N_{ij} \Big/ \sum_{i=1}^{m}\sum_{j=1}^{m} N_{ij},$$

that is equal to

$$N_i \sum_{i=1}^{m} N_{ij} \Big/ \sum_{i=1}^{m}\sum_{j=1}^{m} N_{ij}$$

The χ^2 statistic will have its number of degrees of freedom computed as follows:

No. of N_{ij} parameter $+m^2$

$(i=1...m, j=1...m)$

No. of linear relations required for $-m$

$\hat{P}_j$ estimation $(j=1...m)$

No. of linear relations relating N_{ij} and N_i $-(1-1)$

$= \sum_{j=1}^{m} N_{ij} + N_i$ $(i=1...m-1)$ Note: the relation $\sum_{j=1}^{m} N_{mj}$

$= N_m$ is dependent upon the other relations.

Total number of degrees of freedom $m^2 - m - (m-1) = \underline{(m-1)^2}$

The null hypothesis will not be rejected unless

$$\chi^2 \geq \chi^2_{\alpha, ((m-1)^2)}$$

If the null hypothesis is rejected then it is only assured that the N do not come from an independent process. To test for the Markovity of a process, it is necessary to perform the following tests.

C.7 Test to Determine if Process is First or Second Order Markov Chain

To test the null hypothesis that the observed stationary transition frequencies for a given j, N_{ijk} $(i,k=1...m)$, which are the number of transitions from state i at time t-2, state j at time t-1 and state k at time t, are from a first order Markov chain against the alternative hypothesis that the process is second order Markov, calculate the χ^2 statistic:

$$\chi^2 = \frac{\sum_{i=1}^{n} \sum_{k=1}^{m} (\text{no. of transitions from i to j to k} - \text{no. of transitions expected under null hypothesis})^2}{\text{no. of transitions from i to j to k under null hypothesis}}$$

$$\chi^2 = \sum_{i=1}^{m} \sum_{k=1}^{m} \frac{(N_{ijk} - N_{ij}\hat{P}_{jk})^2}{N_{ij}\hat{P}_{jk}}$$

$$= \sum_{i=1}^{m} \sum_{k=1}^{m} \frac{N_{ij}(\frac{N_{ijk}}{N_{ij}} - \hat{P}_{jk})^2}{\hat{P}_{jk}} = \sum_{i=1}^{m} \sum_{k=1}^{m} \frac{N_{ij}(\hat{P}_{ijk}-\hat{P}_{jk})}{\hat{P}_{jk}}$$

where: $N_{ij} = \sum_{k=1}^{m} N_{ijk}$ = number of observations that are in state i at time t-2 and j at time t-1

$\hat{P}_{jk}$ = estimate of probability of transition from state j at time t-1 to state k at time t

$$= \sum_{i=1}^{m} N_{ijk} / \sum_{i=1}^{m} \sum_{k=1}^{m} N_{ijk} = N_{jk}/N_j$$

$N_j = \sum_{i=1}^{m} N_{ij}$ = number of observations in state j at time t-1

$\hat{P}_{ijk} = \frac{N_{ijk}}{N_{ij}}$ = maximum likelihood estimate of the probability of transition $\hat{P}_{ijk}$ from state i at time t-2, state j at time t-1 to state k at time t

The expected number of transitions from state i to state j to state k under the null hypothesis that the process is first order Markov will simply be the product of the number of initial transitions from state i to state j, N_{ij}, times the estimated probability of transition from state i to state k, $\hat{P}_{jk}$.

The χ^2 statistic will have its number of degrees of freedom computed as follows:

No. of N_{ijk} parameters (i,k=1...m) $+m^2$

No. of linear relations required to estimate $\hat{P}_{jk}$ (k=1...m) $-m$

No. of independent linear relations among N_{ij} and N_{ijk}'s, i.e. $-(m-1)$

$$\sum_{k=1}^{m} N_{ijk} = N_{ij} \quad (i=1...m-1)$$

(remembering $\sum_{k=1}^{m} N_{mjk} = N_{mj}$ is dependent on previous relations)

Total No. of degrees of freedom $\quad m^2 - m - (m-1) = (m-1)^2$

The null hypothesis that the process is first order Markov will not be rejected unless

$$\chi^2 \geq \chi^2_{\alpha, ((m-1)^2)}$$

Since the test for a given j will be asymptotically independent from that of a different j, the χ^2 statistic for all j

$$\chi^2 = \sum_{i=1}^{m} \sum_{j=1}^{m} \sum_{k=1}^{m} \frac{N_{ij}(\hat{P}_{ijk} - \hat{P}_{jk})^2}{\hat{P}_{jk}}$$

will simply be a χ^2 statistic with m (for j=1...m) times (m-1)^2 (for each j) or $m(m-1)^2$ degrees of freedom.

C.8 Test to Determine if Two Markov Chains are Identical

At times it is of interest to investigate if two different processes are identical. To test the null hypothesis that for a given i, two processes are identical Markov chains, compute the χ^2 statistic:

$$\chi^2 = \sum_{h=1}^{2} \sum_{j=1}^{m} \frac{(\text{no. of transitions from i to j} - \text{expected no. of transitions from i to j under null hypothesis})^2}{\text{expected number of transitions from i to j under null hypothesis}}$$

$$= \sum_{h=1}^{2} \sum_{j=1}^{m} \frac{\left[N_{ij}(h) - \frac{N_{ij}(1) + N_{ij}(2)}{\sum_{j=1}^{m}(N_{ij}(1) + N_{ij}(2))} \sum_{j=1}^{m} N_{ij}(h) \right]^2}{\frac{[N_{ij}(1) + N_{ij}(2)] \sum_{j=1}^{m} N_{ij}(h)}{\sum_{j=1}^{m}[N_{ij}(1) + N_{ij}(2)]}}$$

$$= \sum_{n=1}^{2} \sum_{j=1}^{m} \frac{\left[N_{ij}(h) - \frac{\overline{N}_{ij}}{\overline{N}_i} N_i(h) \right]}{\frac{\overline{N}_{ij}}{\overline{N}_i}}$$

$$= \sum_{n=1}^{2} \sum_{j=1}^{m} \frac{\left[N_i(h) \frac{N_{ij}(h)}{N_i(h)} - \frac{\overline{N}_{ij}}{\overline{N}_i} \right]^2}{\frac{\overline{N}_{ij}}{\overline{N}_i}}$$

where

$N_{ij}(h)$ = transition frequency for the hth process

$N_i(h) = \sum_{j=1}^{m} N_{ij}(h)$

$\overline{N}_{ij} = \sum_{h=1}^{2} N_{ij}(h)$ = total observations of transitions from state i to state j

$\overline{N}_i = \sum_{j=1}^{m} \overline{N}_{ij}$ = total observations of transitions initiated at state i

The expected number of transitions from state i to state j under the null hypothesis that the processes are identical will simply be the number initially in state i, $N_i(h)$ times

the probability of transit from state i to state j which is estimated to be the total number of transitions from state i to state j divided by total transitions initially in state i,

$$\frac{\overline{N}_{ij}}{\overline{N}_i}$$

If the estimated transition probability $\frac{\overline{N}_{ij}}{\overline{N}_i}$

is denoted as $\hat{P}_{ij}$ and if the maximum likelihood estimate of the transition probability of process h is denoted by

$$\hat{P}_{ij}(h) = \frac{N_{ij}(h)}{N_i(h)}$$

then the χ^2 statistic may be expressed as:

$$\chi^2 = \sum_{h=1}^{2} \sum_{j=1}^{m} \frac{N_i(h)(\hat{P}_{ij}(h) - \hat{P}_{ij})^2}{\hat{P}_{ij}}$$

This χ^2 statistic will have its number of degrees of freedom computed as follows:

	Degrees of freedom
No. of $N_{ij}(h)$ parameters $(j=1...m;\ h=1,2)$	$2m$
No. of linear relations required to determine $(j=1...m)\ \hat{P}_{ij}$	$-m$
Linear relation $\sum_{j=1}^{n} N_{ij}(h) = N_i(h)$	-1
Total degrees of freedom	$2m-m-1 = m-1$

The null hypothesis that the processes are identical will not be rejected unless

$$\chi^2 > \chi^2_{\alpha,\ (m-1)}$$

Since the test for a given i will be asymptotically independent from that of a different i, the χ^2 statistic for all i

$$\chi^2 = \sum_{i=1}^{m} \sum_{h=1}^{2} \sum_{j=1}^{m} \frac{N_i(h)\ P_{ij}(h) - \hat{P}_{ij}^{\ 2}}{\hat{P}_{ij}}$$

will simply be χ^2 statistic with m (for i=1...m) times m-1 (for each i) or m(m-1) degrees of freedom.

In a similar manner a χ^2 test can be developed to test whether k different processes are identical. The formulae will be the same as those developed above except h=1,2...k and the number of degrees of freedom will be $(k-1)(m)(m-1)$.

C.9 Example of Chi-Square Tests

Suppose that a process is observed at four intervals in time, time 0, 1, 2, and 3 and the number of transitions observed between states 1, 2, and 3 are as follows:

Transitions from time = 0 to time = 1

States	1	2	3	Row Total
1	125	5	16	146
2	7	106	15	128
3	11	18	142	171
				445

Transitions from time = 1 to time = 2

States	1	2	3	Row Total
1	124	3	16	143
2	6	109	14	129
3	22	9	142	173
				445

Transitions from time = 2 to time = 3

States	1	2	3	Row Total
1	146	2	4	152
2	6	111	4	121
3	40	36	96	172
				445

using the maximum likelihood estimator for the P_{ij}

$$\hat{P}_{ij}(t) = \frac{N_{ij}(t)}{N_i(t-1)} = \frac{\text{row entry}}{\text{row total}}$$

The following tables are generated from the ones above.

	State	1	2	3
Time = 1	1	0.856	0.034	0.011
	2	0.055	0.828	0.117
	3	0.064	0.105	0.831
Time = 2	1	0.867	0.021	0.112
	2	0.047	0.845	0.108
	3	0.127	0.052	0.821
Time = 3	1	0.961	0.013	0.026
	2	0.050	0.917	0.033
	3	0.233	0.209	0.558

To test whether the process is stationary (as per section C.5) it is necessary to obtain the pooled transition probability estimators P_{ij} which are found from the pooled data of the three tables.

Pooled Transition Data

State	1	2	3	Row Total
1	395	10	36	441
2	19	326	33	378
3	73	63	300	516
Total				1335

The pooled transition probabilities

$$\hat{P}_{ij} = \frac{\text{pooled table entry}}{\text{pooled table row sum}}$$

will be as follows:

State	1	2	3
1	0.896	0.023	0.081
2	0.050	0.862	0.088
3	0.141	0.122	0.737

The approximate χ^2 statistic for the null hypothesis of stationary is (from Section B.5):

$$\chi^2 = \sum_{i=1}^{3} \sum_{t=1}^{3} \sum_{j=1} N_i(t-1) \frac{(\hat{P}_{ij}(t) - \hat{P}_{ij})}{\hat{P}_{ij}}$$

with $m(m-1)(t-1)$ or $3(2)(2) = 12$ degrees of freedom.

The value of χ^2 in this example is found to be:

$$\chi^2 = 65.1$$

The value of

$$\chi^2_{\alpha, 12}$$

with $\alpha = 0.01$ can be found in Chi-Square tables to be 26.2. Since

$$\chi^2 > \chi^2_{0.01, 12}$$

the null hypothesis must be rejected and it should be assumed that the process is not stationary.

C.10 The Likelihood Ratio Tests

An alternative approach to testing various null hypothesis for Markov Chain is the likelihood ratio technique. The method is basically as follows (detailed proof of the method may be found in most statistics texts):

1. Determine the likelihood function of the process in the case of stationary Markov Chains; this function is of the form of equation (C.1).

2. Maximize the likelihood function with respect to the parameters assuming any value allowed by the null hypothesis. Call it $f(P_{ij})$.

3. Maximize the likelihood function with respect to the parameters assuming any value allowed by the alternative hypothesis. Call it $g(P_{ij})$.

4. Take the ratio of $f(P_{ij})$ and $g(P_{ij})$, call it

$$\lambda = \frac{f(P_{ij})}{g(P_{ij})}$$

5. Take $-2 \ln \lambda$. According to the Neyman-Pearson-Cramer theorems, this function will have a χ^2 distribution with the number of degrees of freedom the same as found by the equivalent Chi-Square test of the previous sections.

For an example of this technique, consider the test of section C.4 for determining whether the observed N_{ij}'s are from a Markov Chain with transition probabilities P_{ij}. In this case the likelihood function

$$f(P_{ij}) = A(N_{ij}) \prod_{i=1}^{m} \frac{N_{i}!}{N_{i1}! N_{i2}! \ldots N_{im}!} (P_{i1})^{N_{i1}} (P_{i2})^{N_{i2}} (P_{im})^{N_{im}}$$

will be maximized under the null hypothesis by setting

$$P_{ij} = P_{ij}$$

in the above expression. The likelihood function will be maximized with respect to the alternative hypothesis (P_{ij} any set of stationary values) by substituting the maximum likelihood estimates of P_{ij}, P_{ij} into the above expression. That is,

$$\lambda = \frac{f(P_{ij})}{g(\hat{P}_{ij})} = \frac{A(N_{ij}) \prod_{i=1}^{m} \frac{N_i!}{N_{i1}! N_{i2}! \ldots N_{im}!} (P_{i1})^{N_{i1}} (P_{i2})^{N_{i2}} \ldots (P_{im})^{N_{im}}}{A(N_{ij}) \prod_{i=1}^{m} \frac{N_i!}{N_{i1}! N_{i2}! \ldots N_{im}!} (\hat{P}_{i1})^{N_{i1}} (\hat{P}_{i2})^{N_{i2}} \ldots (\hat{P}_{im})^{N_{im}}}$$

$$\lambda = \prod_{i=1}^{m} \prod_{j=1}^{m} \left(\frac{P_{ij}}{\hat{P}_{ij}}\right)^{N_{ij}}$$

Remembering that $\hat{P}_{ij} = \frac{N_{ij}}{N_i}$; λ becomes

$$\lambda = \prod_{i=1}^{m} \prod_{j=1}^{m} \left[\frac{P_{ij} N_i}{N_{ij}}\right]^{N_{ij}} \text{ so that}$$

$$-2 \ln \lambda = \left[\prod_{i=1}^{m} \prod_{j=1}^{m}\right]^{N_{ij}} (\ln \frac{N_{ij}}{N_i P_{ij}})$$

and this expression will be distributed as a χ^2 density with $m(m-1)$ degrees of freedom.

As another example of this technique, consider the test of section C.5 for determining whether the chain is stationary. In this case of a non-stationary Markov Chain, the likelihood function will be of the form:

$$f(P_{ij}(t)) = A(N_{ij}(t)) \prod_{i=1}^{m} \prod_{t=1}^{\tau} \frac{N_i(t)!(P_{ij}(t)^{N_{i1}(t)} (P_{im}(t))^{N_{im}(t)}}{N_{i1}(t)! N_{i2}(t)! \ldots N_{im}(t)!}$$

This function will be maximized under the null hypothesis when the $P_{ij}(t) = \hat{P}_{ij}$ = the maximum likelihood estimates of stationary transition probabilities. It will be maximized under the alternative hypothesis (non-stationary) when the $P_{ij}(t) = P_{ij}(t)$. Thus, λ will be in this case:

$$\lambda = \frac{A(N_{ij}(t)) \prod_{i=1}^{m} \prod_{t=1}^{\tau} \frac{N_i(t)!}{N_{i1}(t)! \ldots N_{im}(t)!} (\hat{P}_{i1})^{N_{i1}(t)} \ldots (\hat{P}_{1m})^{N_{i1}(t)}}{A(N_{ij}(t)) \prod_{i=1}^{m} \prod_{t=1}^{\tau} \frac{N_i(t)}{N_{i1}(t)! \ldots N_{im}(t)!} (\hat{P}_{i1}(t))^{N_{i1}(t)} \ldots \hat{P}_{im}(t)^{N_{im}(t)}}$$

$$= \prod_{j=1}^{m} \prod_{i=1}^{m} \prod_{t=1}^{\tau} \left(\frac{\hat{P}_{ik}}{\hat{P}_{ij}(t)}\right)^{N_{ij}(t)}$$

Thus

$$-2 \ln \lambda = \sum_{j=1}^{m} \sum_{i=1}^{m} \sum_{t=1}^{\tau} N_{ij}(t) \ln \left(\frac{\hat{P}_{ij}(t)}{\hat{P}_{ij}}\right)$$

and this expression should be distributed as a χ^2 density with the same number of degrees of freedom found in section C.5, that is $m(m-1)(\tau-1)$. Evaluating this expression for the example given in the previous section it is found that:

$$-2 \ln \lambda = 97.644$$

Therefore, under this test technique the null hypothesis will again be rejected. It should be noted that there is considerable discrepancy between the χ^2 values determined for the two different techniques. This is due to the fact that both methods are just crude approximations for the actual multinomial densities for N_{ij} or $N_{ij}(t)$ not large. However, they both do converge to the same value as the N_{ij} or $N_{ij}(t)$ becomes large, so the methods are consistent. The choice of method depends on the null hypothesis being investigated. Either will be acceptable if numerous data is available. If numerous data is not available then either test should be used with care.

C.11 Continuous Time Parameter Markov Processes

Since processes are often sampled at discrete points in time, the methods of Markov Chains are often appropriate. However, some processes are continuously under observation, i.e. the operation of a critical machine within a power system, and the times of each transition are noted. For these continuous time parameter processes, methods of statistical inference similar to those discussed for Markov Chains have been developed. The development of these tests are beyond the scope of this book but they generally are similar to the likelihood ratio technique discussed above. The method is essentially as follows:

1. Develop expression for likelihood of process with respect to age dependent failure and repair rate.

2. Observe the process for a time T.

3. Note the time the process is in each state.

4. Note the number of transitions from one state to another for each type of transition.

5. Derive estimates for the age dependent failure rates and/or age dependent repair rates from the data obtained during the time T in steps (2) and (3).

6. Maximize the likelihood function under the null hypothesis.

7. Maximize the likelihood function under the alternative hypothesis (usually using the estimates derived instep (5).

8. Calculate minus two times the logarithm of the ratio of likelihood functions of step (6) and step (7).

9. Compare the value calculated in step (8) with the value with the appropriate number of degrees of freedom.

Details concerning the Markov processes may be found in Billingsley, P. <u>Statistical Inference for Markov Processes</u>, 1961, Chicago: University of Chicago Press.

C.12 References

1. Mood, A.M. and Graybill, F.A. "Introduction to the Theory of Statistics", 2nd Edition, McGraw-Hill, New York, 1963.

2. Weatherburn, C.E. "Mathematical Statistics", Cambridge, University Press, 1961.

3. Crow, E.L., Davis, F.A., and Maxfield, M.W. "Statistics Manual", Dover Publications, Inc., New York, 1960.

4. Dynkin, E.B. "Theory of Markov Processes", Prentice-Hall, Inc., Englewood Cliffs, New Jersey, Pergamon Press, Oxford, 1961.

5. Hogg, R.V. and Craig, A.T. "Introduction to Mathematical Statistics", 2nd Edition, MacMillan, New York, 1965.

APPENDIX D - NON-MARKOVIAN SYSTEMS

Ernst G. Frankel

If a system or component hazard rate is $f(t) = \lambda t$, then the system is non-Markovian because its condition depends on its past. Using the birth-death equation for a two component, off-line redundant system:

$$\underline{P}_0(t+dt) = \underline{P}_0(t)[1-\lambda t dt] + 0 dt$$

$$\underline{P}_1(t+dt) = \underline{P}_0(t)\lambda t \, dt + \underline{P}_1(t)(1-\lambda t dt)$$

and
$$\underline{P}_2(t+dt) = \underline{P}_1(t)\lambda t dt + \underline{P}_2(t)$$

$$\underline{P}_0'(t) = -\lambda t \underline{P}_0(t)$$

$$\underline{P}_1'(t) = \lambda t \underline{P}_0(t) - \lambda t \underline{P}_1(t)$$

and
$$\underline{P}_2'(t) = \lambda t \underline{P}_1(t)$$

$$R(t) = \underline{P}_0(t) + \underline{P}_1(t) = e^{-\lambda t^2/2}(1+\lambda t^2/2)$$

and

$$\underline{P}_i(t) = \frac{(\lambda t^2/2)^i \, e^{-\lambda t^2/2}}{i!} \quad i = 0, 1, \ldots, n$$

if we have n off-line stand-by components. This non-Markovian process is therefore still a Poisson process. The reliability of an n component off-line stand-by system is then

$$R(t) = \sum_{i=0}^{n-1} \frac{(\lambda t^2/2)^i \, e^{-\lambda t^2/2}}{i!}$$

in general if $f(t)$ is a function of time the reliability of an n component off-line system is

$$R(t) = \sum_{i=0}^{n-1} \frac{[H(t)]^i e^{-H(t)}}{i!}$$

where

$$H(t) = \int_0^t f(\tau) d\tau$$

Similarly if we have a system with on-line redundancy of, say, two components

$$\underline{P}_0'(t) = -2\lambda t \underline{P}_0(t)$$

$$\underline{P}_1'(t) = 2\lambda t \underline{P}_0(t) - \lambda t \underline{P}_1(t)$$

$$\underline{P}_2'(t) = \lambda t \underline{P}_1(t)$$

and

$$\underline{R}(t) = \underline{P}_0(t) + \underline{P}_1(t) = 2e^{-\lambda t^2/2} - e^{-\lambda t^2}$$

and for an n component on-line redundant system

$$\underline{R}(t) = 1 - [1-e^{-\lambda t^2/2}]^n$$

for the general case of $f(t)$ a function of time

$$R(t) = [1-(1-e^{-H(t)})^n]$$

In other words, the reliability of such an on-line redundant system is a binomial series.

In the above, we assumed a failure rate $h(t)$ or the probability of failure in the increment $t \leq \tau < t+dt$, is dependent on how long the equipment has been operating. In a two-component system, if one component system has been operating for t hours and the second component had failed and been replaced at $t+\tau$, then the respective probabilities of failure are $h(t)dt$ and $h(t-\tau)dt$. It is often convenient to treat some simple Non-Markovian systems in the Markov sense. This can be done by increasing the states, so that each state is described by a constant transition probability. If, for example, a single component system has a probability of failure from 0 to t equal to a gamma distribution with a Failure Distribution function of

$$F(t) = 1-e^{-\lambda t} - \lambda t\, e^{-\lambda t} = 1-e^{-\lambda t}(1+\lambda t)$$

we could assume that this system undergoes two phases, each of length $1/\lambda$ as

$$\int_0^\infty e^{-\lambda t}dt = \int_0^\infty \lambda t\, e^{-\lambda t}dy = \frac{1}{\lambda}$$

If we then define state zero as the first operating state and state one the second operating state, with state two the failed state, then the transition probability matrix becomes:

$$P = \begin{bmatrix} 1-\lambda & \lambda & 0 \\ 0 & 1-\lambda & \lambda \\ 0 & 0 & 1 \end{bmatrix}$$

$$R(t) = e^{-\lambda t}(1+\lambda t)$$

Employing this concept to a redundant system with a stand-by component, then

$$P = \begin{bmatrix} 1-\lambda & \lambda & & & \\ & 1-\lambda & \lambda & & \\ & & 1-\lambda & \lambda & \\ & & & & 1 \end{bmatrix}$$

and

$$R(t) = e^{-\lambda t}[1+\lambda t+(\lambda t)^2/2 + (\lambda t)^3/6]$$

Considering next a Non-Markovian maintained one-component system with a failure distribution function

$$F(t) = 1-e^{-\lambda t}-\lambda t e^{-\lambda t}$$

and a repair distribution function

$$G(t) = 1-e^{-\mu t}$$

then the transition probability matrix

$$P = \begin{bmatrix} 1-\lambda & \lambda & 0 \\ 0 & 1-\lambda & \lambda \\ \mu & 0 & 1-\mu \end{bmatrix}$$

where states zero and one are operating and

$$A(\infty) = P_0(\infty) + \underline{P}_1(\infty) = \frac{2\mu}{2\mu+\lambda}$$

if the repair distribution function is also a gamma function

$$G(t) = 1-e^{-\mu t} - \mu t\, e^{-\mu t} \text{ and}$$

$$P = \begin{bmatrix} 1-\lambda & 0 & \lambda \\ \mu & 1-\mu & 0 \\ 0 & \mu & 1-\mu \end{bmatrix}$$

$$A(\infty) = P_0(\infty) = \frac{\mu}{\mu+2\lambda}$$

APPENDIX E - INTRODUCTION TO FLOW GRAPHS AND GERT

Ernst G. Frankel

In analysis we often desire to investigate the sensitivity of a system to changes in system parameters themselves as well as to changes in input or output. Normally, a system in reliability theory is represented by a set of linear algebraic equations relating state and transition probabilities. A simple system consists of a single branch which consists of two nodes connected by a link, with transmittance a.

If x_1 is the input at node 1, then $a \cdot x_1 = x_2$ is the output at node 2. In general, the signal transmitted over a branch is equal to the product of the branch transmittance and the input node. If more than one signal arrives at a node, the value of the node is equal to the sum of all incoming signals or

$$x_n = \Sigma \text{ Incoming signals}$$

Nodes which have incoming branches are called dependent nodes. If a node has only incoming branches, it is also called a sink. Similarly, a node which only radiates branches is called a source. A succession of branches in series is called a path. The term path normally denotes possible connections between any two nodes of a system. In an open path no node appears more than once, while a closed path returns to its originating node. A closed path is also called a loop. If the closed path returns to the origin node without touching another node, it is called a selfloop. Loops normally indicate feedback.

Cascade Path

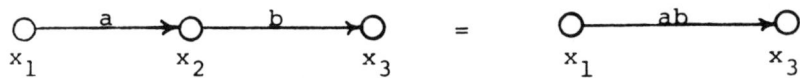

Parallel Path

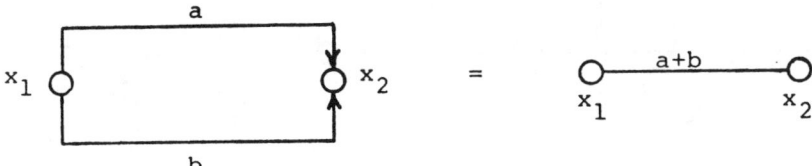

Contraction

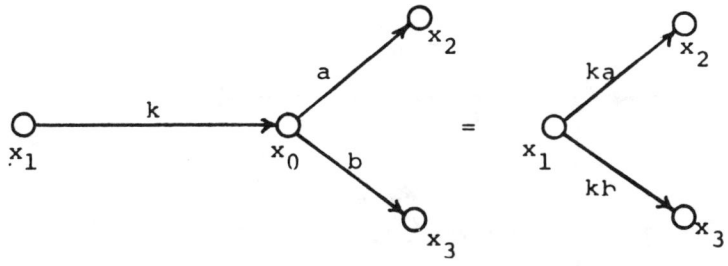

To change the sign of any node, the signs of entering as well as leaving transmittance must be changed.

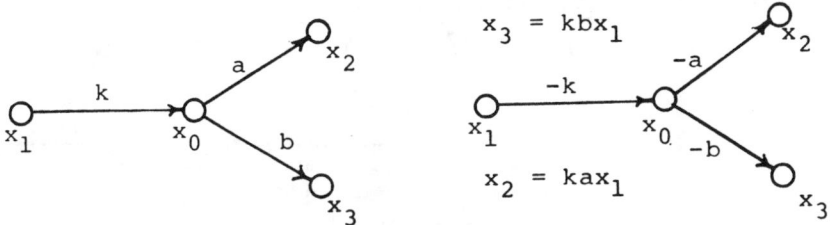

Now, using these fundamental concepts to simplify a flow graph representing the set of equations, we find

$$x_1 = a\, x_0 + c\, x_2$$
$$x_2 = b\, x_0$$
$$x_3 = f\, x_2 + d\, x_1 + a\, x_0$$

This flow graph can be simplified into a single branch as shown:

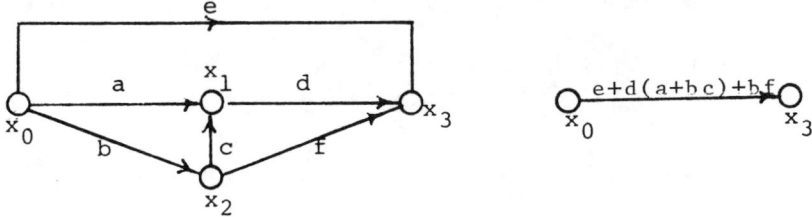

Next, considering a selfloop,

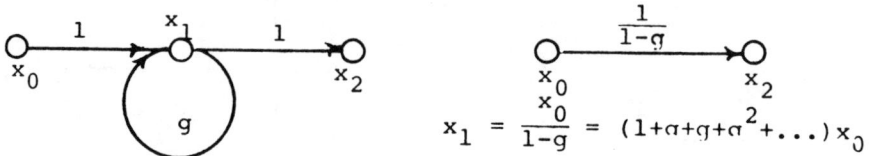

Single branch representation of selfloop. Using this expression, we can reduce the following graph:

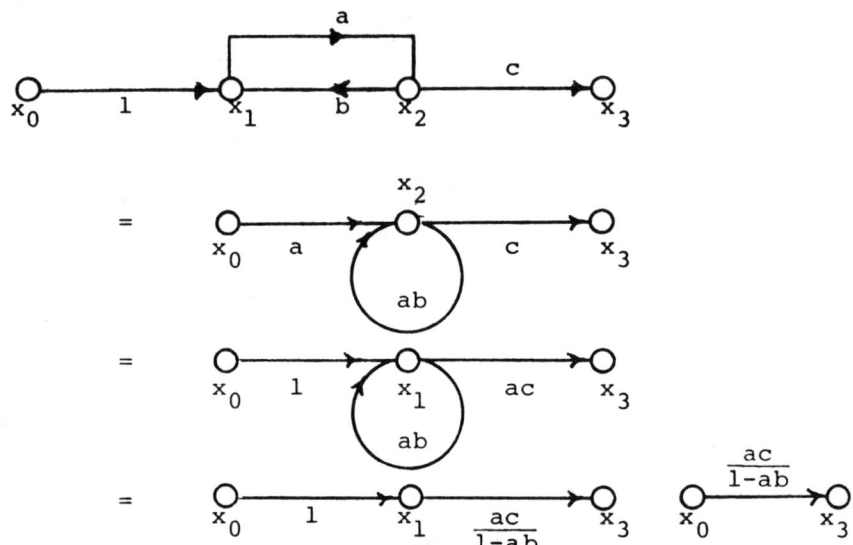

Mason's Reduction

The preceding rules were collected and reduced to rule for the solution of complex system analysis, called 'Mason's Reduction'.

If L_i are the transmittance of the i loops (closed paths) in the system and G_j are the transmittance of the j open paths connecting the two nodes whose relationship is to be found, then we can define the graph determinant.

$$\Delta = [1 - \sum_i L_i + \sum_{i,j} L_i L_j - \sum_{i,j,k} L_i L_j L_k + \ldots]$$

where

$\sum_i L_i$ = sum of all loop transmittances

$\Sigma L_i L_j$ = sum of products of transmittances of all pairs of NON-TOUCHING loops

$\Sigma L_i L_j L_k$ = sum of products of all triplets or NON-TOUCHING loops, etc.

We also define as path factor Δ_j (cofactor), the graph determinant in which the transmittance of any loop touching path G_j is equal to zero.

$$\frac{\text{Output}}{\text{Input}} = \frac{\Sigma G_j \Delta_j}{\Delta} = \frac{\Sigma \text{path} \cdot \text{path factor}}{\text{Graph Determinant}}$$

The above is the statement of 'Mason's Reduction'. As an example, let us consider the following set of algebraic equations; and find $\dfrac{x_5}{x_1}$ or output to input ratio:

$$x_2 = 8x_1 - 3x_3 - 6x_4$$

$$x_3 = 4x_2 - 2x_3 + 9x_4$$

$$x_4 = 5x_2 - 10x_3$$

$$x_4 = 5x_2 - 10x_3$$

or in the flow graph form:

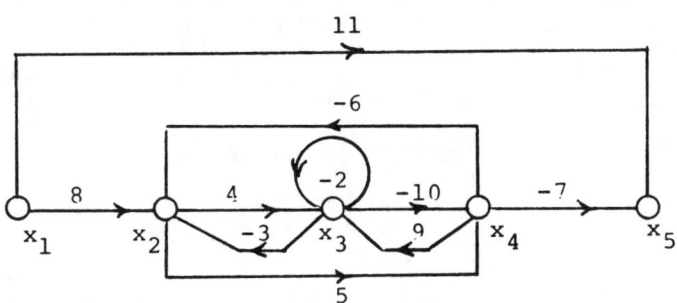

Loops $L_1 = 2 - 3 - 2 = -12$

$L_2 = 3 - 4 - 3 = -90$

$L_3 = \text{Selfloop} = -2$

$$L_4 = 2 - 4 - 2 = -30$$

$$L_5 = 2 - 4 - 3 - 2 = +135$$

$$L_6 = 2 - 3 - 4 - 2 = 240$$

$$\Sigma L_i = L_1 + L_2 + \ldots + L_6 = -29$$

$$\Sigma L_i L_j = L_3 L_4 = 60$$

$$\Sigma L_i L_j L_k = 0$$

$$\Delta = 1 - (-29) + 60 = 90$$

$$G_1 = 1 - 2 - 4 - 5 = -280 \qquad \Delta_1 = 1 - (-2) = 3$$

$$G_2 = 1 - 2 - 3 - 4 - 5 = 2240 \qquad \Delta_2 = 1$$

$$G_3 = 1 - 5 = 11 \qquad \Delta_3 = \Delta = 90$$

$$\frac{x_5}{x_1} = \text{Graph Gain} = \frac{G_1 \Delta_1 + G_2 \Delta_2 + G_3 \Delta_3}{\Delta} = \frac{239}{9}$$

Another useful tool in flow graph methods is path inversion. This permits the elimination of feedback loops facilitating evaluation of sensitivity of system to changes in parameters. There are two well-known path inversion methods in use today, and we will explain both by the use of the same simple example. Consider sink x_3 and two sources x_1 and x_2, and let it be required to change x_1 into a sink and x_3 into a source without altering the transmittance of the system.

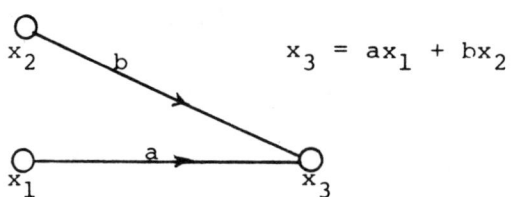

$$x_3 = ax_1 + bx_2$$

Loren's Method (Non-Linear and Linear Transmittances)

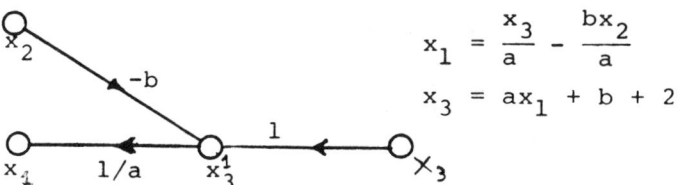

$$x_1 = \frac{x_3}{a} - \frac{bx_2}{a}$$

$$x_3 = ax_1 + b + 2$$

First, invert direction of transmittance into NEW sink x_1 and replace by reciprocal of old transmittance $1/a$. Call source of this branch x_3^1 and lead other original source x_2 into this dummy node with the negative of its old transmittance. Finally, connect the new source node x_3 with unit transmittance to the dummy node x_3^1.

A simpler inversion technique (only for linear systems) is due to Mason.

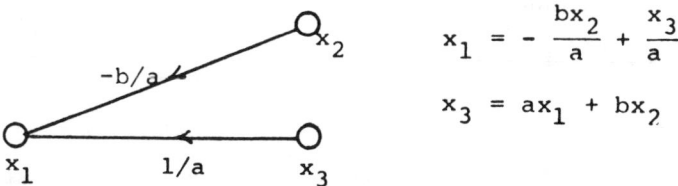

$$x_1 = -\frac{bx_2}{a} + \frac{x_3}{a}$$

$$x_3 = ax_1 + bx_2$$

In this method the branch from the new source x_3 to the new sink x_1 is inverted by making the transmittance equal to the reciprocal of the original value. The other source x_2 is now let into the new sink x_1 with a transmittance equal to the negative value of its old transmittance divided by the old transmittance of the inverted branch. In general, we use inversion to reduce feedback loops, minimize instability difficulties, and facilitate handling of non-linear transmittance. Considering the example,

$$L_1 = ab \qquad G_1 = a^2 \qquad \frac{x_4}{x_1} = \frac{a^2}{1-2ab} \quad \text{(Mason's Reduction)}$$

$$L_2 = ab$$

By Loren's method:

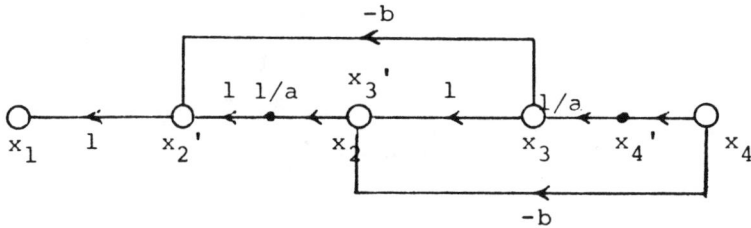

By Mason's method:

$$\frac{x_1}{x_4} = \frac{1-2ab}{a^2}$$

Sensitivity of Systems to Small Transmittance Changes

Sensitivity is defined as the ratio of the rate of change of graph gain or system gain to the rate of change of branch transmittance gain, or

$$S_b^G = \frac{\partial G/G}{\partial b/b} = \frac{d \ln G}{d \ln b} = \text{ratio of percentage change of}$$

$G(\frac{\text{Output}}{\text{Input}}) = \text{Graph Gain}$) to percentage change in a branch transmittance b.

The next logical extension of the network flow graph technique is to develop the ability to study conditional probabilistic networks that are stochastic in nature. A stochastic network is one in which (a) there is a probability associated with each branch, thus indicating the relative likelihood of selection of that branch, and (b) a probability distribution of time to complete that activity, given it is selected. GERT, an acronym for Graphical Evaluation and Review Technique, is the technique that has been developed to analyze such networks. GERT is a powerful tool for reliability and risk analysis since it has inherent in it all the advantages associated with other techniques and provides an exact evaluation of systems in which transitions between state i, j are expressed by conditional probabilities $p_{ij}(t)$ as a function of time and a time function $f_{ij}(t)$, which is usually represented by its statistical density function.

Prior to the discussion of GERT the reader is reminded that the Moment Generating Function (M.G.F.) used in the following paragraphs is no more than a particular expected value, namely,

$$M.G.F. = E[e^{st}] = E[g(s)]$$

where here g(t) is used in both continuous and discrete cases. (See Appendix A.)

$$M.G.F. = E[e_{st}]$$

$$= \int_t e^{st} f(t) \, dt \quad \text{for t continuous}$$

$$= \sum_t e^{st} f(t) \, dt \quad \text{for t discrete}$$

It should also be noted that while GERT is somewhat restricted in the types of networks it can evaluate, the full potential of this technique will be realized upon resolution of the conceptual and computational problems which are associated in the handling of certain classes of problems. In general, GERT can be applied to all problems where times of transition can be described by a well-behaved statistical distribution, for which a moment generating function or Laplace transform can readily be obtained. The method is particularly suited for semi-Markov problems.

GERTS or GERT Simulation and their special purpose derivatives use a number of imaginative node definitions which provide a large amount of flexibility in analyzing and evaluating system performance, resource use, schedules, and more. GERTS nodes can have deterministic or stochastic outputs and inputs. Nodes can have a deterministic (semi-circular) or probabilistic (triangular) input or output, in which case not all incoming activities are required to realize the node nor do all outgoing activities have to be performed.

In addition to the nodes shown there are regular or standard node, which only perform the function of receiving and routing jobs. Statistical Distributions are designated by codes such as:

    BE   Beta Distribution
    BP   Beta Distributions fitted to three parameters
    CO   Constant Distribution
    ER   Erlang Distribution
    EX   Exponential Distribution
    GA   Gamma Distribution
    LO   Log Normal Distribution
    NO   Normal Distribution
    PO   Poisson Distribution
    TR   Triangular Distribution

UN   Uniform Distribution

It is noted that GERT permits use of most practical distributions in simulating system performance. The main characteristics of GERT can be summarized as follows:

GERT, GERT-S, and VERTS Network Models

- Use imaginative node definitions which permit analysis and evaluation of complex systems reliability and availability subject to many diverse:

      1. design or structural alternatives
      2. operating procedures
      3. maintenance strategies

- Solution is by simulation for all but the smallest models.

- Analysis automatically provides statistical information about system under study.

- Nodes are structural components of the system or events.

- Branches represent transactions, actions, or activities.

- Nodes and branches of a network can be developed to model any array of sequential and parallel components, servers, etc.

- Nodes and branches describe the structural aspects of the system and govern the flow of operations, transactions, or actions including component interactions and externally or randomly imposed actions.

- Model allows designer/user to easily modify or extend system structure, operating procedures, or maintenance strategies.

- Alpha-numeric representation of network model serves as input to a program which in turn allows output performance measures to be readily obtained.

GERT networks are flow graphs which can be solved using flow graph solution techniques such as Mason's reduction. The most basic of GERT network method is the exclusive-or type network. Here one and only one of the incoming branches is realized to reach a node. Similarly at each node the output is probabilistic and only one output branch needs to be realized. To simplify the computations the product of the transition probability p_{ij} and the Moment Generation Function $MGF_{ij}(s)$ is usually represented by an equivalent function

$w_{ij}(s)$. A simple problem might look as follows:

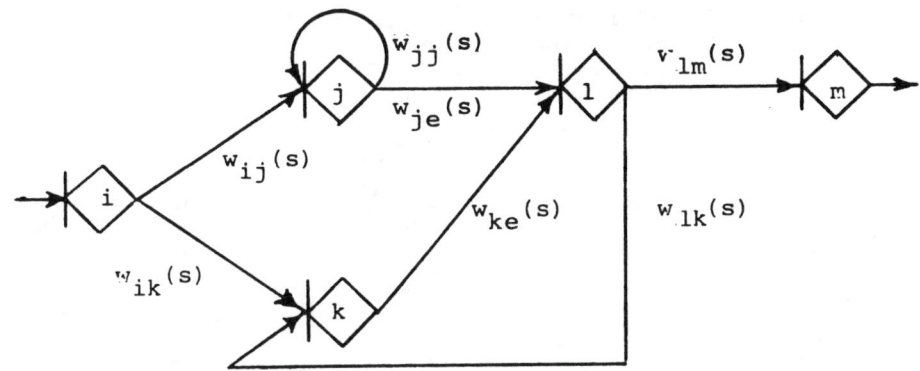

$w_{ij}(s) = p_{ij} \, MGF_{ij}(s)$ = conditional or transition probability ij multiplied by the Moment Generating Function of the transaction ij = equivalent function

States i, j, k, l, m,...

p_{ij} can be made time variant

$MGF_{ij}(s)$ can be a function of one variable such as time or several decision variables such as time, cost, etc.

$$\frac{\partial w_{ij}(s)}{\partial s} \Big/ w_{ij}(0) \Big|_{s=0} = \text{expected resource use}$$

Basic GERT Rationale

$$\frac{\partial^n m_{kz}(s)}{\partial s^n} \Big/ w_{kz}(0) \Big|_{s=0} = n^{th} \text{ moment of expected resource use in transactions or transitions from state k to state z}$$

To obtain the answer to a complex stochastic conditional network problem, we usually employ the Topological Equation of flow graph theory, also called the graph determinant

$$\Delta = H(s) = 1 + \sum_m \sum_i (-1)^m L_i(m) = 0$$

where

$L_i(m)$ = Loop Product as defined before, that is, non-touching loops product

Considering a GERT type network

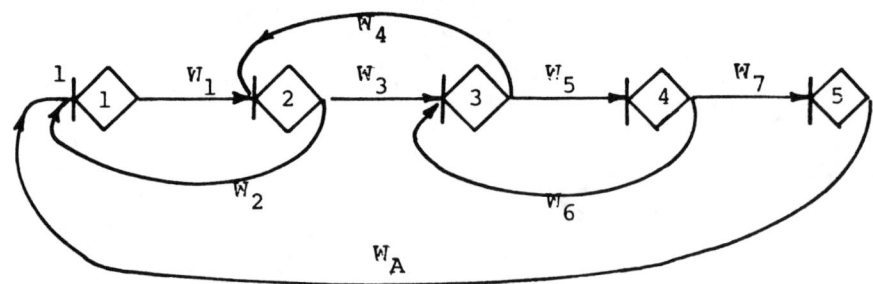

we can obtain the input to output equivalent transmittance of the network by making $H(s) = 0$ when

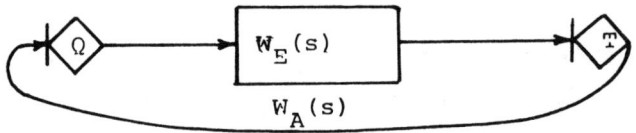

and

$$H(s) = 1 - W_E W_A = 0 \text{ and } W_A = \frac{1}{W_E}$$

or

$$H(s) = 1 - W_1 W_2 - W_3 W_4 - W_5 W_6 + W_1 W_2 W_5 W_6 - W_1 W_3 W_5 W_7 W_A = 0$$

$$W_E = \frac{W_1 W_3 W_5 W_7}{1 - W_1 W_2 - W_3 W_4 - W_5 W_6 + W_1 W_2 W_5 W_6}$$

As $H(s)$ is linear we can write it in terms not including W_A and including W_A as follows:

$$H(s) = H(s)\big|_{W_A=0} + W_A \frac{\partial H(s)}{\partial W_A} = 0$$

and

$$W_E = \frac{1}{W_A} = \frac{-\partial H(s)/\partial W_A}{H(s)\big|_{W_A=0}}$$

when

$$H(s)|_{W_A=0} = 1 - W_1W_2 - W_3W_4 - W_5W_6 + W_1W_2W_5W_6$$

and

$$\frac{\partial H(s)}{\partial W_A} = -W_1W_3W_5W_7$$

Considering next a simple feedback network:

$$W_E(s) = \frac{p_a e^{st_a}}{1-p_b e^{st_b}} = \left(\frac{W_a}{1-W_b}\right)$$

when $p_E = W_E(0) = \dfrac{W_E(s)}{1-p_b}$ and the

$$\text{MGF of } t_E = M_E(s) = \frac{W_E(s)}{W_E(0)} = \frac{(1-p_b)e^{st_a}}{[1-p_b e^{st_b}]}$$

$$\mu_{1E} = (1-p_b)[t_a e^{st_a}(1-p_b e^{st_b})^{-1} + e^{st_a}(1-p_b e^{st_a})^{-2}$$

$$p_b t_b e^{st_b}] = [1-p_b][t_a(1-p_b)^{-1} + (1-p_b)^{-2} p_b t_b]$$

$$= t_a + t_b(p_b/1-p_b)$$

$$\mu_{2E} = (1-p_b)[t_a^2(1-p_b)^{-1} + t_a(1-p_b)^{-2} p_b t_b$$

$$+ t_a(1-p_b)^{-2} p_b t_b + 2(1-p_b)^{-3}(p_b t_b)^2$$

$$+ (1-p_b)^{-2} p_p t_b^2]$$

$$\text{Var} = \mu_{2E} - (\mu_{1E})^2 = p_b t_b^2 (1-p_b)-2$$

The GERT networks discussed so far are called Exclusive OR networks and are of use in representing semi-Markov processes. In such networks we can obtain for

	$W_E(s)$	$M_E(s)$
Series System	$p_a p_b M_a(s) M_b(s)$	$M_a(s) M_b(s)$
Parallel System	$p_a M_a(s) + p_b M_b(s)$	$\frac{1}{p_a+p_b}[p_a M_a(s) + p_b M_b(s)]$
Selfloop System	$p_a M_a(s)[1-p_b M_b(s)]^{-1}$	$(1-p_b)M_a(s)[1-p_b M_b(s)]^{-1}$

or in terms of link equivalent functions:

Series System	$W_a W_b$	$e^{s(t_a+t_b)}$
Parallel System	$W_a + W_b$	$\frac{1}{p_a+p_b}[p_a e^{st_a} + p_b e^{st_b}]$
Selfloop System	$\frac{W_a}{1-W_b}$	$(1-p_b)e^{st_a}[1 - p_b e^{st_b}]^{-1}$

A node in an Exclusive-OR network requires but one of the arriving links to be realized. An Exclusive-OR node in such a network similarly emanates one or more stochastic links, one and only one may be realized. There are other nodes as for example an AND node which requires all incoming links to be realized for the node to be accomplished. It similarly is usually associated with a deterministic output which requires all emanating links to be performed. An example of an AND node is the node used in a critical path network. There are also other nodes such as the Inclusive-OR nodes or mixed nodes where the nodal inputs obey one modal rule, say the Exclusive-OR characteristics while the nodal outputs obey another, say the deterministic characteristics.

GERT network representations are extremely useful in the solution of reliability problems, particularly problems with a semi-Markovian behavior. The method can also be used to solve non-Markovian problems, both ergodic and non-ergodic.

Time-varying transition probabilities and statistical distributions of time of transitions between states can be included. Even where the problem is very large and appears unwieldy, use of GERT provides effective insights into the structure of the problem.

APPENDIX F - STATISTICAL TABLES

Ernst G. Frankel

Tables of Standard Normal Curve and Table of Exponential Functions

Coordinates of "Standard Normal Curve"

±0.00 : 0.3989	±1.50 : 0.1295	±3.00 : 0.0044
±0.25 : 0.3867	±1.75 : 0.0863	±3.25 : 0.0020
±0.50 : 0.3521	±2.00 : 0.0540	±3.50 : 0.0009
±0.75 : 0.3011	±2.25 : 0.0317	±3.75 : 0.0004
±1.00 : 0.2420	±2.50 : 0.0175	±4.00 : 0.0001
±1.25 : 0.1826	±2.75 : 0.0091	

Area Under "Standard Normal Curve"

0.0 : 0.0000	2.0 : 0.4772
0.1 : 0.0398	2.1 : 0.4821
0.2 : 0.0793	2.2 : 0.4861
0.3 : 0.1179	2.3 : 0.4893
0.4 : 0.1554	2.4 : 0.4918
0.5 : 0.1915	2.5 : 0.4938
0.6 : 0.2257	2.6 : 0.4953
0.7 : 0.2580	2.7 : 0.4965
0.8 : 0.2881	2.8 : 0.4974
0.9 : 0.3159	2.9 : 0.4981
1.0 : 0.3413	3.0 : 0.4987
1.1 : 0.3643	3.1 : 0.4990
1.2 : 0.3649	3.2 : 0.4993
1.3 : 0.4032	3.3 : 0.4995
1.4 : 0.4192	3.4 : 0.4997
1.5 : 0.4332	3.5 : 0.4998
1.6 : 0.4452	3.6 : 0.4998
1.7 : 0.4554	3.7 : 0.4999
1.8 : 0.4641	3.8 : 0.4999
1.9 : 0.4713	3.9 : 0.5000
	4.0 : 0.5000

$$A(x) = \int_0^\infty \frac{1}{\sqrt{2\pi}} e^{-x^2/2} \, dx$$

If data is not standard normal, i.e. $\mu = 0$, $\sigma = 1$, then

$$A(x) = \int_0^\infty \frac{1}{\sqrt{2\pi}} e^{-(x-\mu)^2/2\sigma^2} \, dx$$

where

μ = population or sample mean

σ = standard deviation of sample

TABLE OF EXPONENTIAL FUNCTIONS

λ_t	R	Q
x	e^{-x}	$1-e^{-x}$
0.00000	1.00000	0.00000
.00000	.99999	.00000
.00000	.99999	.00000
.00000	.99999	.00000
.00000	.99999	.00000
.00000	.99999	.00000
.00000	.99999	.00000
.00000	.99999	.00000
.00000	.99999	.00000
.00000	.99999	.00000
.00001	.99999	.00001
.00001	.99998	.00001
.00001	.99998	.00001
.00001	.99998	.00001
.00001	.99998	.00001
.00001	.99998	.00001
.00001	.99998	.00001
.00001	.99998	.00001
.00001	.99998	.00001
.00001	.99998	.00001
.00002	.99998	.00002
.00002	.99997	.00002
.00002	.99997	.00002
.00002	.99997	.00002
.00002	.99997	.00002
.00002	.99997	.00002
.00002	.99997	.00002
.00002	.99997	.00002
.00002	.99997	.00002
.00002	.99997	.00002
.00003	.99997	.00003
.00003	.99996	.00003
.00003	.99996	.00003
.00003	.99996	.00003
.00003	.99996	.00003
.00003	.99996	.00003
.00003	.99996	.00003
.00003	.99996	.00003
.00003	.99996	.00003
.00003	.99996	.00003

λ_t	R	Q
x	e^{-x}	$1-e^{-x}$
.00004	.99996	.00004
.00004	.99995	.00004
.00004	.99995	.00004
.00004	.99995	.00004
.00004	.99995	.00004
.00004	.99995	.00004
.00004	.99995	.00004
.00004	.99995	.00004
.00004	.99995	.00004
.00004	.99995	.00004
.00005	.99995	.00005
.00005	.99994	.00005
.00005	.99994	.00005
.00005	.99994	.00005
.00005	.99994	.00005
.00005	.99994	.00005
.00005	.99994	.00005
.00005	.99994	.00005
.00005	.99994	.00005
.00005	.99994	.00005
.00006	.99994	.00006
.00006	.99993	.00006
.00006	.99993	.00006
.00006	.99993	.00006
.00006	.99993	.00006
.00006	.99993	.00006
.00006	.99993	.00006
.00006	.99993	.00006
.00006	.99993	.00006
.00006	.99993	.00006
.00007	.99993	.00007
.00007	.99992	.00007
.00007	.99992	.00007
.00007	.99992	.00007
.00007	.99992	.00007
.00007	.99992	.00007
.00007	.99992	.00007
.00007	.99992	.00007
.00007	.99992	.00007
.00007	.99992	.00007

λ_t	R	Q
x	e^{-x}	$1-e^{-x}$
.00008	.99992	.00008
.00008	.99991	.00008
.00008	.99991	.00008
.00008	.99991	.00008
.00008	.99991	.00008
.00008	.99991	.00008
.00008	.99991	.00008
.00008	.99991	.00008
.00008	.99991	.00008
.00008	.99991	.00008
.00009	.99991	.00009
.00009	.99990	.00009
.00009	.99990	.00009
.00009	.99990	.00009
.00009	.99990	.00009
.00009	.99990	.00009
.00009	.99990	.00009
.00009	.99990	.00009
.00009	.99990	.00009
.00009	.99990	.00009
.0001	.99990	.00010
.0002	.99980	.00020
.0003	.99970	.00030
.0004	.99960	.00040
.0005	.99950	.00049
.0006	.99940	.00059
.0007	.99930	.00069
.0008	.99910	.00079
.0009	.99910	.00089
.0010	.99900	.00099
.0011	.99890	.00109
.0012	.99880	.00119
.0013	.99870	.00129
.0014	.99860	.00139
.0015	.99850	.00149
.0016	.99840	.00159
.0017	.99830	.00169
.0018	.99820	.00179
.0019	.99810	.00189

λ_t	R	Q
x	e^{-x}	$1-e^{-x}$
.0020	.99800	.00199
.0021	.99790	.00209
.0022	.99780	.00219
.0023	.99770	.00229
.0024	.99760	.00239
.0025	.99750	.00249
.0026	.99740	.00259
.0027	.99730	.00269
.0028	.99720	.00279
.0029	.99710	.00289
.0030	.99700	.00299
.0031	.99690	.00309
.0032	.99680	.00319
.0033	.99670	.00319
.0034	.99660	.00339
.0035	.99650	.00299
.0036	.99640	.00359
.0037	.99630	.00369
.0038	.99620	.00379
.0039	.99610	.00389
.0040	.99600	.00399
.0041	.99590	.00409
.0042	.99580	.00419
.0043	.99570	.00429
.0044	.99560	.00439
.0045	.99551	.00448
.0046	.99541	.00458
.0047	.99531	.00468
.0048	.99521	.00478
.0049	.99511	.00488
.0050	.99501	.00498
.0051	.99491	.00508
.0052	.99481	.00518
.0053	.99471	.00528
.0054	.99461	.00538
.0055	.99451	.00548
.0056	.99441	.00558
.0057	.99431	.00568
.0058	.99421	.00578
.0059	.99411	.00588

λ_t	R	Q
x	e^{-x}	$1-e^{-x}$

x	e^{-x}	$1-e^{-x}$
.0060	.99401	.00598
.0061	.99391	.00608
.0062	.99381	.00618
.0063	.99371	.00628
.0064	.99362	.00637
.0065	.99352	.00647
.0066	.99342	.00657
.0067	.99332	.00667
.0068	.99322	.00677
.0069	.99312	.00687
.0070	.99302	.00697
.0071	.99292	.00707
.0072	.99282	.00717
.0073	.99272	.00737
.0074	.99262	.00737
.0075	.99252	.00747
.0076	.99242	.00757
.0077	.99232	.00767
.0078	.99223	.00776
.0079	.99213	.00786
.0080	.99203	.00796
.0081	.99193	.00806
.0082	.99183	.00816
.0083	.99173	.00826
.0084	.99163	.00836
.0085	.99203	.00846
.0086	.99143	.00856
.0087	.99133	.00866
.0088	.99123	.00876
.0089	.99113	.00886
.0090	.99104	.00895
.0091	.99094	.00905
.0092	.99084	.00915
.0093	.99074	.00925
.0094	.99064	.00935
.0095	.99054	.00945
.0096	.99044	.00955
.0097	.99034	.00965
.0098	.99024	.00975
.0099	.99014	.00985

λ_t	R	Q
x	e^{-x}	$1-e^{-x}$
.0100	.99004	.00995
.0101	.98995	.01004
.0102	.98985	.01014
.0103	.98975	.01024
.0104	.98965	.01034
.0105	.98955	.01044
.0106	.98945	.01054
.0107	.98935	.01064
.0108	.98925	.01074
.0109	.98915	.01084
.0110	.98906	.01093
.0111	.98896	.01103
.0112	.98886	.01113
.0113	.98876	.01123
.0114	.98866	.01133
.0115	.98856	.00143
.0116	.98846	.01153
.0117	.98836	.01163
.0118	.98826	.01173
.0119	.98817	.01182
.0120	.98807	.01192
.0121	.98797	.01202
.0122	.98787	.01212
.0123	.98777	.01222
.0124	.98767	.01232
.0125	.98757	.01242
.0126	.98747	.01252
.0127	.98738	.01261
.0128	.98728	.01271
.0129	.98718	.01281
.0130	.98708	.01291
.0131	.98698	.01301
.0132	.98688	.01311
.0133	.98678	.01321
.0134	.98668	.01331
.0135	.98659	.01340
.0136	.98649	.01350
.0137	.98639	.01350
.0138	.98629	.01370
.0139	.98619	.01380

λ_t	R	Q
x	e^{-x}	$1-e^{-x}$
.0140	.98609	.01390
.0141	.98599	.01400
.0142	.98590	.01409
.0143	.98580	.01419
.0144	.98570	.01429
.0145	.98560	.01439
.0146	.98550	.01449
.0147	.98540	.01459
.0148	.98530	.01469
.0149	.98521	.01478
.0150	.98511	.01488
.0151	.98501	.01498
.0152	.98491	.01508
.0153	.98481	.01518
.0154	.98471	.01528
.0155	.98461	.01538
.0156	.98452	.01547
.0157	.98442	.01557
.0158	.98342	.01567
.0159	.98422	.01577
.0160	.98412	.01587
.0161	.98402	.01597
.0162	.98393	.01606
.0163	.98383	.01616
.0164	.98373	.01626
.0165	.98363	.01636
.0166	.99353	.01646
.0167	.98343	.01656
.0168	.98334	.01665
.0169	.98324	.01675
.0170	.98314	.01685
.0171	.98304	.01695
.0172	.98294	.01705
.0173	.98284	.01715
.0174	.98275	.01724
.0175	.98265	.01734
.0176	.98255	.01744
.0177	.98245	.01754
.0178	.98235	.01764
.0179	.98225	.01774

λ_t	R	Q
x	e^{-x}	$1-e^{-x}$
.0180	.98216	.01783
.0181	.98206	.01793
.0182	.98196	.01803
.0183	.98186	.01813
.0184	.98176	.01823
.0185	.98167	.01832
.0186	.98157	.01842
.0187	.98147	.01852
.0188	.98137	.01862
.0189	.98127	.01872
.0190	.98117	.01882
.0191	.98108	.01891
.0192	.98098	.01901
.0193	.98088	.01911
.0194	.98078	.01921
.0195	.98068	.01931
.0196	.98059	.01940
.0197	.98049	.01950
.0198	.98039	.01960
.0199	.98029	.01970
.0200	.98019	.01980
.0201	.98010	.01989
.0202	.98000	.01999
.0203	.97990	.02009
.0204	.97980	.02019
.0205	.97970	.02029
.0206	.97961	.02038
.0207	.97951	.02048
.0208	.97941	.02058
.0209	.97931	.02068
.0210	.97921	.02078
.0211	.97912	.02087
.0212	.97902	.02097
.0213	.97892	.02107
.0214	.97882	.02117
.0215	.97872	.02127
.0216	.97863	.02136
.0217	.97853	.02146
.0218	.97843	.02156
.0219	.97833	.02166

λ_t	R	Q
x	e^{-x}	$1-e^{-x}$
.0220	.97824	.02175
.0221	.97814	.02185
.0222	.97804	.02195
.0223	.97794	.02205
.0224	.97784	.02215
.0225	.97775	.02224
.0226	.97765	.02234
.0227	.97755	.02244
.0228	.97745	.02254
.0229	.97736	.02263
.0230	.97726	.02273
.0231	.97716	.02283
.0232	.97706	.02293
.0233	.97696	.02303
.0234	.97687	.02312
.0235	.97677	.02322
.0236	.97667	.02332
.0237	.97657	.02342
.0238	.97648	.02351
.0239	.97638	.02361
.0240	.97628	.02371
.0241	.97618	.02381
.0242	.97609	.02390
.0243	.97599	.02400
.0244	.97589	.02410
.0245	.97579	.02420
.0246	.97570	.02420
.0247	.97560	.02439
.0248	.97550	.02249
.0249	.97540	.02459
.0250	.97530	.02469
.0251	.97521	.02478
.0252	.97511	.02488
.0253	.97501	.02498
.0254	.97491	.02508
.0255	.97482	.02517
.0256	.97472	.02527
.0257	.97462	.02537
.0258	.97453	.02547
.0259	.97443	.02556

λ_t	R	Q
x	e^{-x}	$1-e^{-x}$
.0260	.97433	.02566
.0261	.97423	.02576
.0262	.97414	.02585
.0263	.97404	.02595
.0264	.97394	.02605
.0265	.97384	.02615
.0266	.97375	.02624
.0267	.97365	.02634
.0268	.97355	.02644
.0269	.97245	.02654
.0270	.97336	.02663
.0271	.97326	.02673
.0272	.97316	.02683
.0273	.97306	.02693
.0274	.97297	.02702
.0275	.97287	.02712
.0276	.97277	.02722
.0277	.97268	.02731
.0278	.97258	.02741
.0279	.97248	.02751
.0280	.97238	.02761
.0281	.97229	.02770
.0282	.97219	.02780
.0283	.97209	.02790
.0284	.97199	.02800
.0285	.97190	.02809
.0286	.97180	.02819
.0287	.97170	.02829
.0288	.97161	.02838
.0289	.97151	.02848
.0290	.97141	.02858
.0291	.97131	.02868
.0292	.97122	.02877
.0293	.97112	.01887
.0294	.97102	.01897
.0295	.97093	.02906
.0296	.97083	.02916
.0297	.97073	.02926
.0298	.97063	.02936
.0299	.97054	.02945

λ_t	R	Q
x	e^{-x}	$1-e^{-x}$
.0300	.97044	.02955
.0305	.96996	.03003
.0310	.96947	.03052
.0315	.96899	.03100
.0320	.96850	.03149
.0325	.96802	.03197
.0330	.96802	.03197
.0335	.96705	.03294
.0340	.96657	.03342
.0345	.96608	.03991
.0350	.96560	.03439
.0355	.96512	.03487
.0360	.96464	.03535
.0365	.96415	.03584
.0370	.96367	.03632
.0375	.96319	.03680
.0380	.96271	.03728
.0385	.96223	.03776
.0390	.96175	.03824
.0395	.96126	.03973
.0400	.96078	.03921
.0405	.96030	.03969
.0410	.95982	.04017
.0415	.95934	.04065
.0420	.95886	.04113
.0425	.95839	.04160
.0430	.95791	.04208
.0435	.95743	.04256
.0440	.95695	.04304
.0445	.95647	.04352
.0450	.95599	.04400
.0455	.95551	.04448
.0460	.95504	.04495
.0465	.95456	.04543
.0470	.95408	.04591
.0475	.95361	.04638
.0480	.95313	.04686
.0485	.95265	.04734
.0490	.96218	.04781
.0495	.95170	.04829

λ_t	R	Q
x	e^{-x}	$1-e^{-x}$
.0500	.95122	.04877
.0505	.95075	.04924
.0510	.95027	.04972
.0515	.94980	.05019
.0520	.94932	.05067
.0525	.94885	.05114
.0530	.94838	.05162
.0535	.94790	.05290
.0540	.94743	.05256
.0545	.94695	.05304
.0550	.94648	.05351
.0555	.94601	.05398
.0560	.94553	.05446
.0565	.94506	.05493
.0570	.94459	.05540
.0575	.94412	.05587
.0580	.94364	.05635
.0585	.94317	.05682
.0590	.94270	.05729
.0595	.94223	.05776
.0600	.94176	.05823
.0605	.94129	.05870
.0610	.94082	.05917
.0615	.94035	.05964
.0620	.93988	.06011
.0625	.93941	.06058
.0630	.93894	.06105
.0635	.93847	.06152
.0640	.93800	.06199
.0645	.93753	.06246
.0650	.93706	.06293
.0655	.93659	.06340
.0660	.96613	.06386
.0665	.93566	.06433
.0670	.93519	.06480
.0675	.93472	.06527
.0680	.93426	.06573
.0685	.93379	.06620
.0690	.93332	.06667
.0695	.93286	.06713

λ_t	R	Q
x	e^{-x}	$1-e^{-x}$

.0700	.93239	.06760
.0705	.93192	.06807
.0710	.93146	.06853
.0715	.93099	.06911
.0720	.93053	.06947
.0725	.93006	.06993
.0730	.92960	.07039
.0735	.92913	.07086
.0740	.92867	.07132
.0745	.92820	.07179
.0750	.92774	.07225
.0755	.92727	.07272
.0760	.92681	.07318
.0765	.92635	.07364
.0770	.92588	.07411
.0775	.92542	.07457
.0780	.92496	.07503
.0785	.92450	.07549
.0790	.92403	.07596
.0795	.92357	.07642
.0800	.92311	.07688
.0805	.92265	.07734
.0810	.92219	.07780
.0815	.92173	.07826
.0820	.92127	.07892
.0825	.92081	.07918
.0830	.92035	.07964
.0835	.91989	.08010
.0840	.91943	.08056
.0845	.91897	.08102
.0850	.91851	.08148
.0855	.91805	.08194
.0860	.91759	.08240
.0865	.91713	.08286
.0870	.91667	.08332
.0875	.91621	.08378
.0880	.91576	.08423
.0885	.91530	.08469
.0890	.91484	.08515
.0895	.91438	.08561

λ_t	R	Q
x	e^{-x}	$1-e^{-x}$
.0900	.91393	.08608
.0905	.91347	.08652
.0910	.91301	.08698
.0915	.91256	.08743
.0920	.91210	.08789
.0900	.91393	.08606
.0905	.91347	.08652
.0910	.91301	.08698
.0915	.91256	.08743
.0920	.91210	.08789
.0925	.91164	.08835
.0930	.91119	.08880
.0935	.91073	.08926
.0940	.91028	.08971
.0945	.90982	.09017
.0950	.90937	.09062
.0955	.90891	.09108
.0960	.90846	.90153
.0965	.90800	.09199
.0970	.90755	.09244
.0975	.90710	.09289
.0980	.90644	.09335
.0985	.90619	.09380
.0990	.90574	.09425
.0995	.90528	.09471
.1000	.90483	.09516
.1005	.90438	.09561
.1010	.90393	.09606
.1015	.90348	.09651
.1020	.90302	.09697
.1025	.90257	.09742
.1030	.90212	.09787
.1035	.90167	.09832
.1040	.90122	.09877
.1045	.90077	.09922
.1050	.90032	.09967
.1055	.89987	.10012
.1060	.89942	.10057
.1065	.89387	.10102
.1070	.89852	.10147

λ_t	R	Q
x	e^{-x}	$1-e^{-x}$
.1075	.89807	.10192
.1080	.89762	.10237
.1085	.89717	.10282
.1090	.89673	.10326
.1095	.89628	.10371
.1100	.89583	.10416
.1105	.89538	.10461
.1110	.89493	.10506
.1115	.89448	.10550
.1120	.89404	.10595
.1125	.89359	.10640
.1130	.89315	.10684
.1135	.89270	.10729
.1140	.89225	.10774
.1145	.89181	.10818
.1150	.89136	.10863
.1155	.89092	.10907
.1160	.89047	.10952
.1165	.89003	.10996
.1170	.88958	.11041
.1175	.88914	.11085
.1180	.88869	.11130
.1185	.88825	.11174
.1190	.88780	.11219
.1195	.88736	.11263
.1200	.88692	.11307
.1205	.88647	.11352
.1210	.88603	.11396
.1215	.88559	.11440
.1220	.88514	.11485
.1225	.88470	.11529
.1230	.88426	.11574
.1235	.88382	.11617
.1240	.88337	.11662
.1245	.88293	.11706
.1250	.88249	.11750
.1255	.88205	.11794
.1260	.88161	.11838
.1265	.88117	.11882
.1270	.88073	.11926

λ_t	R	Q
x	e^{-x}	$1-e^{-x}$
.1275	.88029	.11970
.1280	.87985	.12014
.1285	.87941	.12058
.1290	.87897	.12102
.1295	.87853	.12146
.1300	.87809	.12190
.1305	.87765	.12234
.1310	.87721	.12278
.1315	.87677	.12322
.1320	.87634	.12365
.1325	.87590	.12400
.1330	.87546	.12453
.1335	.87502	.12497
.1340	.87459	.12640
.1345	.87415	.12584
.1350	.87381	.12628
.1355	.87327	.12672
.1360	.87284	.12715
.1365	.87240	.12759
.1370	.87197	.12802
.1375	.87153	.12846
.1380	.87109	.12890
.1385	.87066	.12933
.1390	.87022	.12977
.1395	.86979	.13020
.1400	.86935	.13064
.1405	.86892	.13107
.1410	.86848	.13151
.1415	.86805	.13194
.1420	.86762	.13237
.1425	.86718	.13281
.1430	.86675	.13324
.1435	.86632	.13367
.1440	.86588	.13411
.1445	.86545	.13454
.1450	.86502	.13497
.1455	.86502	.13541
.1460	.86415	.13584
.1465	.86372	.13627
.1470	.86329	.13670

λ_t	R	Q
x	e^{-x}	$1-e^{-x}$
.1475	.86286	.13713
.1480	.86243	.13756
.1485	.86200	.13799
.1490	.86156	.13843
.1495	.86113	.13886
.1500	.86070	.13929
.1505	.86027	.13972
.1510	.85984	.14015
.1515	.85941	.14058
.1520	.85898	.14101
.1525	.85855	.14144
.1530	.85812	.14187
.1535	.85770	.14220
.1540	.85727	.14272
.1545	.85684	.14315
.1550	.85641	.14358
.1555	.85598	.14401
.1560	.85555	.14444
.1565	.85513	.14486
.1570	.85470	.14529
.1575	.85427	.14572
.1580	.85384	.14615
.1585	.85342	.14657
.1590	.85299	.14700
.1595	.85257	.14742
.1600	.85214	.14785
.1605	.85171	.14828
.1610	.85129	.14870
.1615	.85086	.14913
.1620	.85044	.14955
.1625	.85001	.14998
.1630	.84959	.15040
.1635	.84916	.15083
.1640	.84874	.15125
.1645	.84831	.15168
.1650	.84789	.15210
.1655	.84746	.15253
.1660	.84704	.15295
.1665	.84662	.15337
.1670	.84619	.15380

λ_t	R	Q
x	e^{-x}	$1-e^{-x}$
.1675	.84577	.15422
.1680	.84535	.15464
.1685	.84535	.15464
.1690	.84450	.15549
.1695	.84408	.15591
.1700	.84366	.15633
.1705	.84324	.15675
.1710	.84282	.15717
.1715	.84240	.15759
.1720	.84197	.15802
.1725	.84155	.15844
.1730	.84113	.15886
.1735	.84071	.15928
.1740	.84029	.15970
.1745	.83987	.16012
.1750	.83945	.16054
.1755	.83903	.16096
.1760	.83861	.16138
.1765	.83819	.16180
.1770	.83777	.16222
.1775	.83736	.16263
.1780	.83694	.16305
.1785	.83652	.16347
.1790	.83610	.16389
.1795	.83568	.13431
.1800	.83527	.16472
.1805	.83485	.16514
.1810	.83443	.16556
.1815	.83401	.16598
.1820	.83360	.16639
.1825	.83318	.16681
.1830	.83276	.16723
.1835	.83235	.16764
.1840	.83193	.16806
.1845	.83151	.16848
.1850	.83110	.16889
.1855	.83068	.16931
.1860	.84027	.16972
.1865	.82985	.17014
.1870	.82944	.17055

λ_t	R	Q
x	e^{-x}	$1-e^{-x}$
.1875	.82902	.17097
.1880	.82861	.17138
.1885	.82820	.17179
.1890	.82788	.17221
.1895	.82737	.17262
.1900	.82695	.17304
.1905	.82654	.17345
.1910	.82613	.17386
.1915	.82571	.17428
.1920	.82530	.17469
.1925	.82489	.17510
.1930	.82448	.17551
.1935	.82406	.17593
.1940	.82365	.17634
.1945	.82324	.17675
.1950	.82283	.17716
.1955	.82242	.17757
.1960	.82201	.17798
.1965	.82160	.17839
.1970	.82119	.17880
.1975	.82078	.17921
.1980	.82036	.17963
.1985	.81995	.18004
.1990	.81954	.18045
.1995	.81914	.18085
.2000	.81873	.18126
.2005	.81832	.18167
.2010	.81791	.18208
.2015	.81750	.18249
.2020	.81709	.18290
.2025	.81668	.18331
.2030	.81627	.18372
.2035	.81587	.18412
.2040	.81546	.18453
.2045	.81505	.18494
.2050	.81464	.18535
.2055	.81424	.18575
.2060	.81383	.18616
.2065	.81342	.18657
.2070	.81301	.18698

λ_t	R	Q
x	e^{-x}	$1-e^{-x}$
.2075	.81261	.18738
.2080	.81220	.18779
.2085	.81180	.18819
.2090	.81139	.18860
.2095	.81098	.18901
.2100	.81058	.18941
.2105	.81017	.18982
.2110	.80977	.19022
.2115	.80936	.19063
.2120	.80896	.19103
.2125	.80856	.19143
.2130	.80815	.19184
.2135	.80775	.19224
.2140	.80734	.19265
.2145	.80694	.19305
.2150	.80654	.19345
.2155	.80613	.19386
.2160	.80573	.19426
.2165	.80533	.19466
.2170	.80493	.19506
.2175	.80452	.19547
.2180	.80412	.19587
.2185	.80372	.19527
.2190	.80332	.19667
.2195	.80292	.19707
.2200	.80251	.19748
.2205	.80211	.19788
.2210	.80171	.19828
.2215	.80131	.19868
.2220	.80091	.19908
.2225	.80051	.19948
.2230	.80011	.19988
.2235	.79971	.20028
.2240	.79931	.20068
.2245	.79891	.20108
.2250	.79851	.20148
.2255	.79811	.20188
.2260	.79771	.20228
.2265	.79731	.20268
.2270	.79691	.20307

λ_t	R	Q
x	e^{-x}	$1-e^{-x}$

.2275	.79652	.20347
.2280	.79612	.20387
.2285	.79572	.20427
.2290	.79532	.20467
.2295	.79493	.20506
.2300	.79453	.20546
.2305	.79413	.20586
.2310	.79373	.20626
.2315	.79334	.20665
.2320	.79294	.20705
.2325	.79254	.20754
.2330	.79215	.20784
.2335	.79175	.20824
.2340	.79136	.20863
.2345	.79096	.20903
.2350	.79057	.20942
.2355	.79017	.20982
.2360	.78978	.21021
.2365	.78938	.21061
.2370	.78899	.21100
.2375	.78859	.21140
.2380	.78820	.21179
.2385	.78780	.21219
.2390	.78741	.21258
.2395	.78702	.21297
.2400	.78662	.21337
.2405	.78623	.21376
.2410	.78584	.21415
.2415	.78544	.21455
.2420	.78505	.21494
.2425	.78466	.21533
.2430	.78427	.21572
.2435	.78387	.21612
.2440	.78348	.21651
.2445	.78309	.21690
.2450	.78270	.21729
.2455	.78231	.21768
.2460	.78192	.21807
.2465	.78153	.21846
.2470	.78114	.21885

λ_t	R	Q
x	e^{-x}	$1-e^{-x}$
.2475	.78075	.21924
.2480	.78035	.21964
.2485	.77996	.22003
.2490	.77958	.22041
.2495	.77919	.22080
.2500	.77880	.22119
.2505	.77841	.22158
.2510	.77802	.22197
.2515	.77763	.22236
.2520	.77724	.22275
.2525	.77685	.22314
.2530	.77646	.22353
.2535	.77607	.22392
.2540	.77569	.22430
.2545	.77530	.22469
.2550	.77491	.22508
.2555	.77452	.22547
.2560	.77414	.22585
.2565	.77375	.22624
.2570	.77336	.22663
.2575	.77298	.22701
.2580	.77259	.22740
.2585	.77220	.22779
.2590	.77182	.22817
.2595	.77143	.22856
.260	.77105	.22894
.261	.77028	.22971
.262	.76951	.23048
.263	.76874	.23125
.264	.76797	.23202
.265	.76720	.23279
.266	.76643	.23356
.267	.76567	.23432
.268	.76490	.23509
.269	.76414	.23585
.270	.76337	.23662
.271	.76261	.23728
.272	.76185	.23814
.273	.76109	.23890
.274	.76032	.23966

λ_t	R	Q
x	e^{-x}	$1-e^{-x}$

.275	.75957	.24042
.276	.75881	.24118
.277	.75805	.24194
.278	.75729	.24270
.279	.75653	.24346
.280	.75578	.24421
.281	.75502	.24497
.282	.75427	.24572
.283	.75351	.14648
.284	.75276	.24723
.285	.75201	.24798
.286	.75126	.24873
.287	.75051	.24928
.288	.74976	.25023
.289	.74901	.25098
.290	.74826	.25173
.291	.74751	.25248
.292	.74676	.25323
.293	.74602	.25397
.294	.74527	.25472
.295	.74453	.25546
.296	.74378	.25621
.297	.74304	.25695
.298	.74230	.25769
.299	.74155	.25844
.300	.74081	.25918
.301	.74007	.25992
.302	.73933	.26066
.303	.73859	.26240
.304	.73786	.26213
.305	.73712	.26387
.306	.73638	.26361
.307	.73565	.26434
.308	.73491	.26508
.309	.73418	.26581
.310	.73344	.26655
.311	.73271	.26728
.312	.73198	.26801
.313	.73124	.26875
.314	.73051	.26948

λ_t	R	Q
x	e^{-x}	$1-e^{-x}$
.315	.72878	.27021
.316	.72905	.27094
.317	.72833	.27166
.318	.72760	.27239
.319	.72687	.27312
.320	.72614	.27385
.321	.72542	.27457
.322	.72469	.27530
.323	.72397	.27602
.324	.72325	.27674
.325	.72252	.27747
.326	.72180	.27819
.327	.72108	.27891
.328	.72036	.27963
.329	.71964	.28035
.330	.71892	.28107
.331	.71820	.28179
.332	.71748	.28251
.333	.71677	.28322
.334	.71605	.28394
.335	.71533	.28466
.336	.71462	.28537
.337	.71390	.28609
.338	.71319	.28680
.339	.71248	.28751
.340	.71177	.28822
.341	.71105	.28894
.342	.71034	.28965
.343	.70963	.29036
.344	.70892	.29107
.345	.70822	.29177
.346	.70751	.29248
.347	.70680	.29319
.348	.70609	.29390
.349	.70539	.29460
.350	.70468	.29531
.351	.70398	.29601
.352	.70328	.29671
.353	.70257	.29742
.354	.70187	.29812

λ_t	R	Q
x	e^{-x}	$1-e^{-x}$
.355	.70117	.29882
.356	.70047	.29952
.357	.69977	.30022
.358	.69907	.30092
.359	.69837	.30162
.360	.69767	.30202
.361	.69697	.30302
.362	.69628	.30371
.363	.69558	.30441
.364	.69489	.30510
.365	.69419	.30580
.366	.69350	.30649
.367	.69280	.30719
.368	.69211	.30788
.369	.69142	.30857
.370	.69073	.30926
.371	.69004	.30995
.372	.68935	.31064
.373	.68866	.31133
.374	.68797	.31202
.375	.68728	.31271
.376	.68660	.31339
.377	.68591	.31408
.378	.68523	.31476
.379	.68454	.35145
.380	.68386	.31613
.381	.68317	.31682
.382	.68249	.31750
.383	.68181	.31818
.384	.68113	.31886
.385	.68045	.31954
.386	.67966	.32022
.387	.67909	.32090
.388	.67841	.32158
.389	.67773	.32226
.390	.67705	.32294
.391	.67638	.32362
.392	.67570	.32429
.393	.67502	.32497
.394	.67435	.32564

λ_t	R	Q
x	e^{-x}	$1-e^{-x}$
.395	.67378	.32631
.396	.67300	.32699
.397	.67233	.32766
.398	.67166	.32833
.399	.67099	.32900

DISTRIBUTION OF X^2

DF	Probability, p						
	0.99	0.975	0.95	0.90	0.80	0.75	0.50
1	0.000157	0.000982	0.00393	0.0158	0.0642	0.10153	0.455
2	0.0201	0.0506	0.103	0.211	0.446	0.5753	1.386
3	0.115	0.216	0.352	0.584	1.005	1.2125	2.366
4	0.297	0.484	0.711	1.064	1.649	1.9225	3.357
5	0.554	0.831	1.145	1.610	2.343	2.674	4.351
6	0.872	1.237	1.635	2.204	3.070	3.454	5.348
7	1.239	1.689	2.167	2.833	3.822	4.254	6.346
8	1.646	2.179	2.733	3.490	4.594	5.070	7.344
9	2.088	2.700	3.325	4.168	5.380	5.898	8.343
10	2.558	3.247	3.940	4.865	6.179	6.737	9.342
11	3.053	3.816	4.575	5.578	6.989	7.584	10.341
12	3.571	4.404	5.226	6.304	7.807	8.438	11.340
13	4.107	5.008	5.892	7.042	8.634	9.299	12.340
14	4.660	5.628	6.571	7.790	9.467	10.165	13.339
15	5.229	6.262	7.261	8.547	10.307	11.036	14.339
16	5.812	6.907	7.962	9.312	11.152	11.912	15.338
17	6.408	7.564	8.672	10.085	12.002	12.791	16.338
18	7.015	8.231	9.390	10.865	12.857	13.675	17.338
19	7.633	8.906	10.117	11.651	13.716	14.562	18.338
20	8.260	9.591	10.851	12.443	14.578	15.452	19.337
21	8.897	10.283	11.591	13.240	15.445	16.344	20.337
22	9.542	10.982	12.338	14.041	16.314	17.239	21.337
23	10.196	11.688	13.091	14.848	17.187	18.137	22.337
24	10.856	12.400	13.848	15.659	18.062	19.037	23.337
25	11.524	13.119	14.611	16.473	18.940	19.939	24.337
26	12.198	13.844	15.379	17.292	19.820	20.843	15.336
27	12.879	14.573	16.151	18.114	20.703	21.749	26.336
28	13.565	15.308	16.928	18.933	21.588	22.657	27.336
29	14.256	16.047	17.708	19.768	22.475	23.566	28.336
30	14.953	16.791	18.492	20.599	23.364	24.476	29.236

DISTRIBUTION OF χ^2 (Note: For degrees of freedom greater than 30, the quantity $\sqrt{2\chi^2}$ is approximately normally distributed with mean $\sqrt{2(DF)-1}$ and variance 1.

DF	0.50	0.25	0.20	0.10	Probability, p 0.05	0.025	0.01	0.001
1	0.455	1.322	1.642	2.706	3.841	5.024	6.635	10.827
2	1.386	2.772	3.219	4.605	5.991	7.377	9.210	13.815
3	2.366	4.108	4.642	6.251	7.815	9.348	11.345	16.268
4	3.357	5.385	5.989	7.779	9.488	11.143	13.277	18.465
5	4.351	6.625	7.289	9.236	11.070	12.832	15.086	20.517
6	5.348	7.840	8.558	10.645	12.592	14.449	16.812	22.457
7	6.346	9.037	9.803	12.017	14.067	16.013	18.475	24.322
8	7.344	10.218	11.030	13.362	15.507	17.534	20.090	26.125
9	8.343	11.388	12.242	14.684	16.919	19.023	21.666	27.877
10	9.342	12.548	13.442	15.987	18.307	20.483	23.209	29.588
11	10.341	13.701	14.631	17.275	19.675	21.920	24.725	31.264
12	11.340	14.845	15.812	18.549	21.026	23.336	26.217	32.909
13	12.340	15.984	16.985	19.812	22.362	24.735	27.688	34.528
14	13.339	17.117	18.151	21.064	23.685	26.119	29.141	36.123
15	14.339	18.245	19.311	22.307	24.996	27.488	30.578	37.697
16	15.338	19.368	20.465	23.542	26.296	28.845	32.000	39.252
17	16.338	20.488	21.615	24.769	27.587	30.191	33.409	30.790
18	17.338	21.605	22.760	25.989	28.869	31.526	34.805	42.312
19	18.338	22.717	23.900	27.204	30.144	32.852	36.191	43.820
20	19.337	23.817	25.038	28.412	31.410	34.169	37.566	45.315
21	20.337	14.935	26.171	29.615	32.671	35.479	38.932	46.797
22	21.337	26.039	27.301	30.813	33.924	36.780	30.289	48.268
23	22.337	27.141	28.429	32.007	35.172	38.075	41.638	49.728
24	23.337	28.241	29.553	33.196	36.415	39.364	42.980	51.179
25	24.337	29.339	30.675	34.382	37.652	40.646	44.314	52.620
26	25.336	30.434	31.795	35.563	38.885	41.923	45.642	54.052
27	26.336	31.528	32.912	36.741	40.113	43.194	46.963	55.476
28	27.336	32.620	34.027	37.916	41.337	44.460	48.278	56.893
29	28.336	33.711	35.139	39.087	42.557	45.722	49.588	58.302
30	29.236	34.799	36.250	40.256	43.773	46.980	50.892	59.703

REQUIRED NUMBER OF FAILURES FOR VARIOUS VALUES OF CONFIDENCE AND PRECISION (EXPONENTIAL DISTRIBUTION)

Precision –	Confidence			
	85%	90%	95%	99%
5%	830	1082	1537	2655
10%	207	271	384	664
15%	92	120	171	295
20%	52	67	96	166
25%	33	43	61	106
30%	23	30	43	74
35%	17	22	31	54

Example: 43 failures are required to be 90% confident that the estimated MTBF is within 25% of the true value.

TABLE OF t

Degrees of Freedom	Probability				
	0.50	0.10	0.05	0.02	0.01
1	1.000	6.31	12.71	31.82	63.66
2	0.816	2.92	4.30	6.96	9.92
3	0.765	2.35	3.18	4.54	5.84
4	0.741	2.13	2.78	3.75	4.60
5	0.727	2.02	2.57	3.36	4.03
6	0.718	1.94	2.45	3.14	3.71
7	0.711	1.90	2.36	3.00	3.50
8	0.706	1.86	2.31	2.90	3.36
9	0.703	1.83	2.26	2.82	3.25
10	0.700	1.81	2.23	2.76	3.17
11	0.697	1.80	2.20	2.72	3.11
12	0.695	1.78	2.18	2.68	3.06
13	0.694	1.77	2.16	2.65	3.01
14	0.692	1.76	2.14	2.62	2.98
15	0.691	1.75	2.13	2.60	2.95
16	0.690	1.75	2.12	2.58	2.92
17	0.689	1.74	2.11	2.57	2.90
18	0.688	1.73	2.10	2.55	2.88
19	0.688	1.73	2.09	2.54	2.86
20	0.687	1.72	2.09	2.53	2.84
21	0.686	1.72	2.08	2.52	2.83
22	0.686	1.72	2.07	2.51	2.82
23	0.685	1.71	2.07	2.50	2.81
24	0.685	1.71	2.06	2.49	2.80
25	0.684	1.71	2.06	2.48	2.79
26	0.684	1.71	2.06	2.48	2.78
27	0.684	1.70	2.05	2.47	2.77
28	0.683	1.70	2.05	2.47	2.76
29	0.683	1.70	2.04	2.46	2.76
30	0.683	1.70	2.04	2.46	2.75
35	0.682	1.69	2.03	2.44	2.72
40	0.681	1.68	2.02	2.42	2.71
45	0.680	1.68	2.02	2.41	2.69
50	0.679	1.68	2.01	2.40	2.68
60	0.678	1.67	2.00	2.39	2.66
70	0.678	1.67	2.00	2.38	2.65
80	0.677	1.66	1.99	2.38	2.64
90	0.677	1.66	1.99	2.37	2.63
100	0.677	1.66	1.98	2.36	2.63
125	0.676	1.66	1.98	2.36	2.62
150	0.676	1.66	1.98	2.35	2.61
200	0.675	1.65	1.97	2.35	2.60
300	0.675	1.65	1.97	2.34	2.59
400	0.675	1.65	1.97	2.34	2.59
500	0.674	1.65	1.96	2.33	2.59
1000	0.674	1.65	1.96	2.33	2.58
	0.674	1.64	1.96	2.33	2.58

GAMMA FUNCTION

Values of $\Gamma(n) = \int_0^\infty e^{-x} x^{n-1} dx$; $\Gamma(m+1) = n\Gamma(n)$

n	$\Gamma(n)$	n	$\Gamma(n)$	n	$\Gamma(n)$
1.00	1.00000	1.35	0.89115	1.69	0.90678
1.01	0.99433	1.36	0.89018	1.70	0.90864
1.02	0.98884	1.37	0.88031	1.71	0.91057
1.03	0.98355	1.38	0.88854	1.72	0.91258
1.04	0.97844	1.39	0.88785	1.73	0.91466
1.05	0.97350	1.40	0.88726	1.74	0.91683
1.06	0.96874	1.41	0.88676	1.75	0.91906
1.07	0.96516	1.42	0.88636	1.76	0.92137
1.08	0.95973	1.43	0.88604	1.77	0.92376
1.09	0.95546	1.44	0.88580	1.78	0.92623
1.10	0.95135	1.45	0.88565	1.79	0.92877
1.11	0.94739	1.46	0.88560	1.80	0.93138
1.12	0.94359	1.47	0.88563	1.81	0.93408
1.13	0.93993	1.48	0.88575	1.82	0.93685
1.14	0.93642	1.49	0.88595	1.83	0.93969
1.15	0.93304	1.50	0.88623	1.84	0.84261
1.16	0.92980	1.51	0.88659	1.85	0.94561
1.17	0.92670	1.52	0.88704	1.86	0.94869
1.18	0.92373	1.53	0.88757	1.87	0.95184
1.19	0.92088	1.54	0.88818	1.88	0.95507
1.20	0.91817	1.55	0.88887	1.89	0.95838
1.21	0.91558	1.56	0.88964	1.90	0.96177
1.22	0.91311	1.57	0.89049	1.91	0.96523
1.23	0.91075	1.58	0.89142	1.92	0.96878
1.24	0.90852	1.59	0.89243	1.93	0.97140
1.25	0.90640	1.60	0.89352	1.94	0.97610
1.26	0.90440	1.61	0.89468	1.95	0.97988
1.27	0.90250	1.62	0.89592	1.96	0.98374
1.28	0.90072	1.63	0.89724	1.97	0.98768
1.29	0.89904	1.64	0.89864	1.98	0.99171
1.30	0.89747	1.65	0.90012	1.99	0.99581
1.31	0.80600	1.66	0.90167	2.00	1.00000
1.32	0.89464	1.67	0.90330		
1.33	0.89338	1.68	0.90500		
1.34	0.89222	1.69	0.90678		

INDEX

Absorbing state, 166, 171, 173-174, 244
Action strategies, 206
Active redundancy, 145, 149
Adjoint, 333
Age dependent failure rate, 12, 13, 17, 53, 114, 118, 121, 254
Aggregate operating efficiency, 162
Aging, 17
Allocation constraint, 210, 218
AND gate, 275, 276, 282
AND node, 276, 280
Associativity, 329, 331
Availability, 11, 142, 219

Bathtub curve, 17
Bayes theorem, 98, 303-305
Beta factor, 65
Binomial distribution, 315, 319
Binomial probability function, 295
Binomial theorem, 292
Birth and death equations, 148, 167, 171
Black and Proschan Model, 209, 236, 240
Bounding, 69
Breipohl, 224

Casualty repairs, 141
Catastrophic component failure, 223
Catastrophic failure event, 275
Catastrophic failure rate, 126, 157, 223
Catastrophic risk, 285, 286
Chapman-Komolgorov equation, 100, 101, 102
Chebycheff's theorem, 314
Chi Square, 346
Chi Square approximation, 351-355
Chi Square density, 355-356
Chi Square tests, 365, 369
Cofactor, 169, 331-332
Collectively exhaustive events, 298
Column vector, 327
Command event, 271, 275
Command faults, 275
Common Cause failure, 65, 70, 274, 275
Commutativity, 331
Complex Series Redundant System, 48
Component allocation problem, 54-55
Component equivalent function, 280
Component failure rate, 215
Component interaction, 253
Component maintenance, 331
Component standby, 218
Compound probability function, 307
Conditional multivariable distributions, 85
Conditional probabilistic networks, 269

Conditional probability, 301, 302
Continuous random variables, 309
Continuous time parameter Markov processes, 370
Contraction, 377
Cost equation, 249
Critical components, 143
Critical failure, 79
Cumulative function, 310, 311

Decision theory, 8
Decision matrix, 247, 248
Decision tree, 191
Dependent nodes, 376
Derating factor, 254, 255, 264
Determinant, 331
Distribution function, 313
Dynamic allocation, 210
Dynamic programming, 210, 214, 216, 234

Eigenvalues, 333-334
Eigenvectors, 334
Environmental stress factor, 254
Equivalent function, 118, 281, 388
Ergodic Markov chain, 105, 108
Ergodic process, 93, 94, 179
Erlang distribution, 133, 273
Essentiality factor, 240, 262
Event space, 299
Exclusive-OR networks, 275, 276, 280, 388
Expected availability, 143
Expected fatal failure rate, 240
Expected lifetime, 15
Expected reliability, 143
Expected time to failure, 15, 16, 17
Expected value, 308, 312
Expected value of stochastic functions, 92
Exponential distributions, 319-320
Exponential failure density, 25, 35, 46
Exponential redundant system, 38

Failure, 4, 11, 70, 71
Failure class, 79
Failure density, 12, 13, 17
Failure mode and effects analysis, 64, 69, 78-80
Failure mode frequency, 78
Failure probability, 273, 273
Fault tree, 64, 72, 74
Fault tree analysis, 69, 271-277
Feedback, 3
Flow graphs, 266, 275, 376
Four state system, 246
Frequency distribution, 313
Functional equation, 211, 223, 229, 230, 233
Fundamental matrix, 168, 171

Gamma distribution, 321
Gates, 275-276
Gauss Jordan method, 222
Gaussian wear-rate failure distribution, 265
Geisler and Kurr Model, 208
GERT, 269, 277, 281, 382-383
GERT networks, 276, 279, 281, 384
GERTS, 383, 384
Gouray, 208

Hazard rate, 12, 13
Homogeneous Markov chains, 96, 101, 105
Human error, 9, 29, 287
Hypergeometric distribution, 316, 318

Identity matrix, 330
Impacts, 126
Inclusive-OR node, 280, 389
Independence of events, 302
Independent components, 33
Inherent reliability, 3
Instantaneous availability, 142, 145, 149, 156
Instantaneous failure rate, 264
Instantaneous repair rate, 148
Integrated systems, 6
Interaction factor, 254
Interaction functions, 253, 265
Interactive systems reliability, 256
Interval estimate, 27, 29
Inverse matrix, 332

Joint probability function, 307, 311

Kurtosis, 314

Lagrange multiplier method, 227, 228, 229
Laplace transformation, 120, 123, 138, 342-345, 382-383
Left eigenvectors, 334
Likelihood function, 25, 348
Likelihood ratio test, 368
Linear stress model, 132
Log likelihood function, 348, 349, 369
Log normal distribution, 273, 278, 322
Logical gates, 70
Loop, 376, 378
Loop product, 385-386
Loren's method, 381-382

Maintainability, 5, 6, 7, 8
Maintainability engineering analysis, 81
Maintained parallel systems, 153
Maintained series systems, 151
Maintained systems, 141, 281
Maintenance, 141, 227
Major failure, 80

Make-shift repair, 150
Marginal analysis, 57
Marginal probability function, 307, 311
Markov chain, 95-97, 105, 170, 243, 356, 361, 363, 368-369
Markov process, 94, 96, 266, 360
Mason's reduction, 280, 378, 381-384
Matrix, 329
Matrix algebra, 327-335
Maximum likelihood estimate, 25, 28, 348
Maximum likelihood techniques, 346
Mean time before failure, 15, 167, 170
Mean time to repair, 198
Mean value, 92
Mellin integral transform, 281, 282
Min. cut sets, 74
Minimal cut, 67-70
Minimal path, 67
Minimum cut sets, 274, 275, 276
Minor failure, 80
Mixed system, 52
Moment, 309, 312, 313
Moment generating function, 281, 282, 315, 382, 383
Moments of a stochastic process, 92
MTBF, 15, 146, 162, 169, 171, 179
MTR, 198
MTTR, 148
Multinomial approximation, 355, 356
Multinomial distribution, 87, 347, 351-355
Multistage decision process, 216-218
Multivariable probability distributions, 83
Mutually exclusive events, 297

Negative binomial distributions, 310
Network looping, 281
Network models, 383
Neyman-Pearson-Cramer Theorem, 369
Node, 376
Non-homogeneous Markov chains, 100, 102
Non-Markovian systems, 372-375
Normal distribution, 321
Normal wear-rate distribution, 265, 266

Off-line redundant system, 58-60, 122, 154, 172, 175, 178, 182, 372
On-line, 145
On-line parallel systems, 129, 136, 137, 145, 153, 203
On-line redundant systems, 44, 271
On-line standby system, 155-156, 220
Operability, 5, 6, 7, 8
Operational wear failure rate, 126
Optimal achievable reliability, 210
Optimum maintenance policies, 233-236
Optimum overhaul and inspection interval, 205
Optimum spare part provisioning, 207, 243
OR gate, 275, 276, 282

Orthogonal matrix, 334
Overhauls, 141

Parallel systems, 36-42, 279
Parallel systems in series, 49
Partial failure, 150
Path, 376
Path factor, 378
Poisson density function, 121, 215
Poisson distribution, 316, 319
Poisson failure process, 114, 117, 122, 255
Poisson process, 65, 166, 321, 372
Policy vector, 246
Preventive maintenance scheduling, 234, 236
Primary events, 275
Principle of Optimality, 211, 213, 215
Probabilistic importance, 277
Probability, 292
Probability density function, 294, 309
Probability function, 305
Probability measure, 300
Probability of failure, 272
Probability state vector, 116, 121
Probability theory, 11
Probability transients, 336
Public risk, 284, 286

Random failure, 18, 19
Random variables, 305
Real vector, 327
Recurrence equation, 211, 218, 220, 226-227, 229
Redundancy, 296
Redundant parallel systems, 129, 138, 163
Redundant on-line systems, 170, 171, 176, 204
Redundant standby on-line system, 44
Redundant system with identical components, 43
Reliability, 1, 11, 12, 144, 173-174
Reliability analysis, 273, 290
Reliability index, 4
Reliability theory, 11
Repair cost, 207
Repair strategy, 204
Response times, 3
Right eigenvectors, 334
Risk, 9, 11, 285, 286
Risk analysis, 11, 273, 286, 290
Risk attitudes, 285
Row vector, 327

Safety margins, 2
Sample space, 299
Sample variance, 323
Scalar vector, 327
Scheduled downtime, 143, 205, 281
Secondary event, 271

Selfloop, 376, 377
Semi-Markov technique, 276-277
Sensitivity, 382
Sequential failures, 276, 281
Sequential sample space, 299
Sequential testing, 27
Series system, 33-36, 41, 54, 127, 137, 170, 264
Sink, 376
Source, 376
Spare part inventory provisioning, 207
Spare part provisioning models, 236
s-t graph, 275, 276
Staggered redundant system, 42-44
Standby redundant system, 122
State probability, 376
Stationarity test, 357-358, 366
Stationary Markov chain, 99
Stationary stochastic process, 92
Stationary transition frequency, 360, 361
Statistical sampling, 323-326
Steady-state availability, 142, 143, 156-157, 162, 165
Steady state probability vector, 110
Stochastic matrix, 110, 115, 126, 127, 147, 168, 195, 243
Stochastic network, 206
Stochastic process, 88, 89
Stochastic process theory, 83
Stochastic transition probability matrix, 123
Switching, 134, 135, 218
Symmetric matrix, 331
System availability, 142
System failure probability, 278
System performance, 243
System reliability, 5, 6

Test data, 4
Three-state Markov chain, 109, 339
Time average of a stochastic process, 92
Time dependence, 276, 282
Tolerance limits, 4
Top events, 271, 272, 277, 375
Topological equation, 385
Transition frequency, 360
Transition probability, 244, 281, 346, 356, 357, 376
Transition probability matrix, 107, 108, 118, 244
Transmittance, 376, 378
Transpose, 169, 195, 327, 331
Two-state Markov process, 103, 127, 338

Uncertainty importance, 278
Uncertainty in reliability analysis, 277
Unscheduled downtime, 143
Uptime availability, 142, 150
User risk, 284

Variance of time to failure, 170, 176

Venn diagram, 297

Wear, 256-263, 267
Wear-in failure, 19
Wear-out failure, 18, 19
Wear-out failure density distributions, 260, 267
Weibull distribution, 215, 233, 262, 273, 321
Weibull wear-rate failure distribution, 264

Yule-Fary distribution, 132

Z-transform 335-341